# Computational Bioinorganics

# Computational Bioinorganics

From Description to Prediction

*Edited by Jean-Didier Maréchal*

*Universitat Autònoma de Barcelona*
*Bellaterra, SP*

WILEY

This edition first published 2026


*Registered Office(s)*
John Wiley & Sons, Inc., 111 River Street, Hoboken, NJ 07030, USA
John Wiley & Sons Ltd. New Era House, 8 Oldlands Way, Bognor Regis, West Sussex, PO22 0NO, UK
Wiley-VCH GmbH, Boschstr. 12, 69469 Weinheim, Germany
John Wiley & Sons Singapore Pte Ltd, 134 Jurong Gateway Road, #04-307H, Singapore 600134
For details of our global editorial offices, customer services, and more information about Wiley products visit us at www.wiley.com.

The manufacturer's authorized representative according to the EU General Product Safety Regulation is Wiley-VCH GmbH, Boschstr. 12, 69469 Weinheim, Germany, e-mail: Product_Safety@wiley.com.

Wiley also publishes its books in a variety of electronic formats and by print-on-demand. Some content that appears in standard print versions of this book may not be available in other formats.



*Library of Congress Cataloging-in-Publication Data has been applied for:*

**Hardback:** 9781119415138
**ePDF:** 9781119415091
**epub:** 9781119415244
**oBook:** 9781119415152

**Cover Design:** Wiley
**Cover Image:** Courtesy of the Author

Set in 9.5/12.5pt STIXTwoText by Lumina Datamatics

# Contents

**Preface** *ix*

**1 What Could Bring Theory to Modern Bioinorganics: A Conversation?** *1*
*Gerard Roelfes and Jean-Didier Maréchal*

**2 Computational Prediction and Modeling of Metal-binding Sites in Proteins** *9*
*José-Emilio Sánchez-Aparicio, Giuseppe Sciortino, and Jean-Didier Maréchal*
2.1 Introduction *9*
2.2 Experimental Tools *10*
2.3 Computational Tools *10*
2.3.1 Pattern Recognition Algorithms *11*
2.3.2 The Backbone Preorganization Hypothesis *14*
2.3.3 Protein–Ligand Dockings *17*
2.3.4 GaudiMM Platform *19*
2.3.5 Quantum Mechanical and Hybrid Approaches *21*
2.3.6 Molecular Dynamics, Integrative or Multiscale Approaches *22*
2.4 Conclusions *25*
References *25*

**3 Quantum-mechanics Approaches in Bioinorganic Chemistry: Targeting the Oxidation State** *33*
*Marcel Swart*
3.1 Introduction *33*
3.2 Experimental Observations *35*
3.3 Quantum-chemistry Approaches *37*
3.3.1 Chemical Bonding Analyses *38*
3.3.2 Oxidation States from Wavefunctions or Densities *41*
3.4 Classical Examples *43*
3.5 Intriguing Examples *46*
3.6 Summary and Outlook *48*
Acknowledgments *49*
References *49*

4 **Computation and Spectroscopy** *57*
*Eugenio Garribba*
4.1 Introduction *57*
4.2 Software, Computational Techniques, and Statistical Indicators *58*
4.3 Electron Paramagnetic Resonance *61*
4.4 Electron Spin Echo Envelope Modulation and Electron-nuclear Double Resonance *67*
4.5 Nuclear Magnetic Resonance *69*
4.6 Electron Absorption Spectroscopy *77*
4.7 Circular Dichroism and Magnetic Circular Dichroism *81*
4.8 Vibrational Spectroscopy *83*
4.9 Mossbauer Spectroscopy *84*
4.10 Conclusions *88*
References *89*

5 **Multiscale Computational Modeling for the Discovery and Characterization of New Metalloenzyme-catalyzed Reactions** *99*
*Ferran Feixas and Marc Garcia-Borràs*
5.1 Importance of Metalloenzymes *99*
5.2 Multi-scale Computational Microscope for Modeling Metalloenzymatic Reaction Mechanisms *100*
5.2.1 Classical Molecular Dynamics Simulations *102*
5.2.2 Truncated Active Site Models at the Quantum Mechanics Level *103*
5.2.3 Hybrid Quantum Mechanics/Molecular Mechanics *103*
5.3 Practical Considerations for Computational Modeling of Metalloenzymes *104*
5.3.1 Selecting the Starting Structure *104*
5.3.2 Initial Exploration of Conformational Dynamics *105*
5.3.3 Placement of Substrate or Reaction Intermediate *106*
5.4 The Mechanistic Versatility of P450s Through the Lens of a Computational Microscope: Practical Examples *107*
5.4.1 Practical Example: Anti-Markovnikov Alkene Oxidations Catalyzed by P450s *108*
5.4.2 Molecular Dynamics Simulations for Modeling the Protein Environment *112*
5.4.3 Insightful Revelations from QM/MM Calculations *114*
5.4.4 The Role of Local Electric Fields in Enzymatic Reactions *116*
5.5 Conclusions *117*
Acknowledgments *118*
References *119*

6 **Force Fields and Metals** *123*
*José-Emilio Sánchez-Aparicio, Lorena Roldán-Martín, Jean-Didier Maréchal, and Giuseppe Sciortino*
6.1 Introduction *123*

6.2 Overview on Molecular Mechanics and How to Deal with Metals *124*
6.3 Bonded Model *126*
6.3.1 Cu(II) Binding to Polyhistidine Peptides *126*
6.3.2 Interaction of Oxaliplatin with Insulin *127*
6.3.3 Artificial Metallohydratase *128*
6.4 Nonbonded Models *131*
6.4.1 The Dummy Atom *131*
6.5 The 12-6(-4) Lennard–Jones Potentials *132*
6.6 Conclusions *134*
References *135*

**7 Modeling Thermally Activated Processes** *139*
*Pietro Vidossich and Alessandra Magistrato*
7.1 Introduction *139*
7.2 Modeling Transitions *140*
7.2.1 Theoretical Framework *141*
7.2.2 Collective Variables *142*
7.2.3 Biased Methods *144*
7.3 Illustrative Studies *147*
7.3.1 Metal Coordination *147*
7.3.2 Enzymatic Reactions *150*
7.4 Conclusion *153*
References *154*

**8 Metal-induced Folding of Oligopeptides** *159*
*L. Roldán-Martín, L. Rodríguez-Santiago, J. Alí-Torres, J. D. Maréchal, and M. Sodupe*
8.1 Introduction *159*
8.2 Metal–Ligand Interactions with Short Peptides ($n = 2$–8) *160*
8.2.1 $Gly_n$ and $Ala_n$ Metal Complexes *160*
8.2.2 Metal–Peptide Complexes Involving Residues Other than Gly and Ala *163*
8.3 Metal-oligopeptides ($n > 8$) *165*
8.3.1 Construction of Three-dimensional Models. Application to $Cu^{2+}$-$A\beta_{1-16}$ *166*
8.3.2 Impact of $Cu^{2+}$ and $Al^{3+}$ Binding on the $A\beta_{1-42}$ Conformational Landscape *169*
8.4 Conclusions *172*
References *172*

**9 Computational Studies of Metallodrugs** *179*
*James A. Platts and Matthew Turner*
9.1 Introduction *179*
9.2 QM Studies of Metallodrugs and Their Solvation *179*
9.3 QM Studies of Metallodrug–Biomolecule Interactions *181*

9.4 QM/MM Modeling of Drug–Biomolecule Interactions *184*
9.5 Classical MM Description of Drugs and Their Interactions *186*
9.6 Docking and QSAR Studies *190*
9.7 Conclusion *191*
References *192*

**10 Computational Design of Artificial Metalloenzymes** *195*
*Laura Tiessler-Sala, Maria Fatima Lucas, and Emanuele Monza*
10.1 Introduction *195*
10.2 Typical Computational Pipeline *196*
10.3 Critical Aspects and Frontiers of ArM Design *198*
10.3.1 Challenges in Modeling Metallic Centers *198*
10.3.2 Unnatural Amino Acids Can Unlock New Reactivity *200*
10.3.3 Beyond TS Stabilization *202*
10.3.4 The New Era: *De Novo* Protein Design with DL *204*
10.4 Concluding Remarks *205*
References *206*

**11 How Computational Chemistry Can Contribute to the Understanding of the Effects of Metal Ions in Biological Systems: Aluminum as a Case Study** *213*
*J.I. Mujika and X. Lopez*
11.1 Introduction *213*
11.2 Aluminum *215*
11.2.1 Aluminum Speciation in Water *216*
11.2.2 Mapping Aluminum Speciation with Biological Building Molecules *217*
11.3 The Pro-oxidant Activity of Aluminum *217*
11.4 Unveiling the Al(III) – Aβ Complex *223*
11.4.1 Step 1: Most Favorable First Coordination Spheres of Aluminum *225*
11.4.2 Step 2: The Building of Preliminary Al.Aβ Complexes *227*
11.4.3 Step 3: Refinement of the Al.Aβ Complexes *228*
11.5 *In Silico* Design of Novel Aluminum Chelators *231*
11.6 Concluding Remarks *237*
References *237*

**Index** *239*

# Preface

Dear Reader,

If you hold this book in your hands, one of these questions – if not all – may have crossed your mind at some point: What molecular foundations enable life to orchestrate the synergy between its organic toolkits and metallic moieties? How can molecular modeling help unveil their underlying processes? How can computation assist in metalloenzyme design? Is it possible to assess the mechanism of metallodrugs as efficiently as organic ones? This book, "Computational *Bioinorganics: From Description to Prediction*," aims to provide a good picture of the current state-of-the-art in computational bioinorganics and hopefully provides suitable answers that could serve you well.

When the editorial team at Wiley asked me if I was interested in editing a book on advances in computational bioinorganics, I saw the opportunity to give the community a comprehensive book that would help to push the boundaries of our understanding of the role of metals in biology through computation. It would also crystallize my two scientific passions. One is bioinorganic chemistry, which I discovered as an undergraduate student and continues to surprise me daily. The other is computational chemistry, which allows us to "see" how molecules work at the atomistic level, unveiling Nature's fantastic molecular machinery.

This book aims to overview a broad spectrum of modern computational bioinorganic chemistry areas, clarifying which methodological approaches and which kinds of physical and dimensional models can be applied. For this, you will be aided by leading scientists, to whom I am tremendously grateful for sharing their expertise. All agreed to create chapters accessible to a broad audience, including undergraduate students, experimental scientists, and computational researchers from other fields. To do so, each chapter aimed to present the underlying methods, focusing on their physical chemistry aspects and highlighting their strengths, challenges, and limitations while minimizing mathematical content. Relevant illustrative examples complement the chapters. This format should provide valuable insights into computational bioinorganics for all readers.

Computational bioinorganics is a vivid field of research that has rapidly evolved over the last two decades. Traditionally, the emphasis has been on models that accurately represent the first coordination sphere of the metal, which is primarily treated using quantum mechanical approaches (QM or QM/MM). These models are essential for studying processes governed by electronic properties, such as redox properties, spectroscopy, and catalysis. Nowadays, however, the entire field of bioinorganic chemistry facilitates the study of other types of processes. For instance, improvements in parameterization strategies for molecular mechanics force fields have eased the simulation of dynamic processes in metalloenzymes, folding patterns in metallopeptides, and binding metallic cofactors and/or ligands to their hosts. Furthermore, new structural bioinformatics tools and AI-based approaches are also enhancing the field, and there is no doubt that their contributions will continue to grow in the years ahead. Modern computational bioinorganic chemistry is also characterized by a Pleiades of multiscale applications. In this type of protocols, molecular processes are studied using various physical and dimensional models to grasp a complete picture of a process that could involve substrate binding, electronic shift of the metal and catalysis.

This book accounts for a representative set of chapters that cover a wide range of computational bioinorganics topics. It includes, among others, the use of computation for spectroscopic analysis, the study of redox processes, the interaction of natural and contaminant metals with neuropeptides, the identification of metal binding sites in proteins and their transit from site to site, the study of the interaction of metallodrugs with their target, and finally metalloenzyme catalysis, optimization or design. I want to stress that the subtitle *From Description to Prediction* is not innocent. It emphasizes that molecular modeling in bioinorganics is now more than a descriptive tool that aims to explain what happens in the pot; it is showing enough maturity to become a predictive tool and ultimately drive the synthesis of new biometallic systems.

*Computational Bioinorganics: From Description to Prediction* aims to provide a comprehensive collection of chapters that should help anyone interested in applying computation in this field. We wish you a fruitful reading and that you'll find inspiration on how computation can be a perfect companion for your studies.

Manresa, April 3rd 2025 *Jean-Didier Maréchal*

1

# What Could Bring Theory to Modern Bioinorganics: A Conversation?

*Gerard Roelfes*[1] *and Jean-Didier Maréchal*[2]

[1] *Biomolecular Chemistry and Catalysis, Stratingh Institute for Chemistry, University of Groningen, Groningen, The Netherlands*
[2] *Insilichem, Departament de Química, Universitat Autònoma de Barcelona, Bellaterra, Spain*

In this exercise, Gerard Roelfes plays the part of the experimental bioinorganic chemist and Jean-Didier Maréchal, the part of the computational one.

**Jean-Didier Maréchal (question [Q]):** My entire scientific life has been dedicated to metals in biological systems, and I feel like an expert bioinorganic chemist. But what has struck me the most during the more than 20 years I spent on the subject is the increase in the number of fields in which bioinorganics is involved! From your perspective, what characterizes nowadays the fields of application of bioinorganics?

**Gerard Roelfes (answer [A]):** Indeed, the field of bioinorganic chemistry has evolved a lot over the past decades. When I started in this field (many years ago.......), a significant focus was on structural and functional models for metalloproteins and enzymes. The models were small molecule compounds that were used to study the spectroscopic properties of these coordination complexes, to discover the spectroscopic signatures of reactive intermediates (for example, metal oxo, superoxide, and peroxo complexes), and to study their reactivity, and in this way, learn about mechanism. While this is still a significant focus of attention, the field's center of gravity has switched more to the biological systems themselves. This is, among others, due to the developments in spectroscopy, which make it possible to study biological systems directly without using model complexes. Also, the developments in biochemistry and molecular biology over the past decades have made it possible, even for nonexperts, to create proteins with any desired mutation, handle cell cultures, etc. So, bioinorganic chemistry is now being studied in more complex environments.

**Gerard Roelfes (Q):** I observe that molecular modeling is becoming increasingly relevant in molecular sciences. Let's consider bioinorganic chemistry. However, I feel that

*Computational Bioinorganics: From Description to Prediction*, First Edition. Edited by Jean-Didier Maréchal.
 Published 2026 by John Wiley & Sons Ltd.

the focus still often is on simplified complexes (to reduce computation effort?) and idealized conditions (e.g. gas phase), which may not translate very well to the (biological) system of interest. Why do we still focus on systems of small dimensions?

**Jean-Didier Maréchal (A):** You are right; little could be argued regarding the increase in the application of molecular modeling in molecular sciences. From small systems to solids, nanoparticles, or polymers, molecular modeling is now widely used in fields ranging from interstellar chemistry to atmospheric phenomena, homogeneous and heterogeneous catalysis, and biochemical processes. The Nobel Prizes in Chemistry awarded in 2013 to Levitt, Warshell, and Karplus and, more recently, in 2024 to Baker, Hassabis, and Jumper are clear examples of the rise of molecular modeling in chemistry and biology. However, molecular modeling is a generic term that covers many methods and answering that question obliges to discuss essential aspects of molecular modeling, especially in calculating energy and exploring conformational space.

Regarding energy calculation, methods can be classified depending on their physical models. Two prominent families exist: quantum mechanics (QM) and molecular mechanics (MM). On one side, QM methods consider the quantum nature of the system, and their primary interest in many chemistry fields lays on their ability to evaluate the electronic properties of the systems explicitly. On the other side, MM methods are based on classical physics and do not explicitly consider the electrons as separated entities but rather incorporate them, together with the nuclei, in atomic potentials. In this kind of method, the cohesion of the molecular structure and its energetic behavior are provided by a network of parametric functions that reproduce bonded (e.g. covalent) and nonbonded (e.g. electrostatic) interactions between atoms. The ensemble of functions that defines the system is called the force field.

There is a wide range of methods in the QM and MM realms. But your question regarding small systems mostly refers to QM models used for biomimetic models or truncated parts of a metalloprotein. The use of this strategy lays on two fundamental aspects QM methods offer: the accuracy in representing the metals' electronic properties on one side and a high computational cost on the other. QM approaches, like post-Hartree-Fock or functional density theories (DFT), are the most indicated to study any phenomenon that implies electronic processes like catalysis, red-ox, changes in the coordination sphere, and many kinds of spectroscopies. Those processes are the focus of interest in many studies in bioinorganics and are fundamentally driven by the first coordination sphere. Unfortunately, QM calculations are computationally too expensive to be performed on a complete protein, and the dimension of the system must be reduced. The most critical aspect of these modeling exercises is to correctly ascertain the dimensional model so that the protein's principal features are well reproduced despite reducing the number of atoms involved. For a metalloenzyme, for example, we tend to model its active site by including the first coordination sphere and incorporating as many residues of the second coordination sphere as possible.

Another reason that often conditions the use of QM over MM is that dealing with the electronic properties of the metal are challenging for MM methods. Parametrization of the force field functions, one of the critical steps in the way, is generally more complex in those cases than for organic molecules, and only described one unique coordination state at a time. Yet, molecular modeling studies are increasingly escaping the scope of small-dimension truncated models. For example, QM/MM calculations are an option when we look at reactivity. In those approaches, you can deal with the whole protein system, including the protein region that needs accurate electronic representation to be handled in the QM part and the rest of the system to be incorporated in the MM one. However, these types of calculations remain limited in terms of geometric exploration.

**Gerard Roelfes (Q):** This would explain why QM and small models have been so prominent in bioinorganic chemistry. However, I would like to get a better sense of what modeling can do and the challenges of the field, especially regarding the "real" dimensions. What are the limits regarding size and conditions?

**Jean-Didier Maréchal (A):** Well, when it comes to dealing with systems of large dimensions, modeling requires MM approaches. This drastically reduces the cost of calculating the energy by several orders of magnitude regarding QM; therefore, MM methods allow for the work of systems of large dimensions like proteins and DNA, and they can even be solvated in solvents or bound to membranes. The difference in computational cost between QM and MM methods is not limited to the size of the systems (number of atoms), but also the number of conformations could be explored (sampling). QM methods allow sampling that rarely explores phenomena higher than the scale fs or ps. In a chemical context, a typical application would characterize a reaction's transition state and the systems' local rearrangement. On the other side, MM methods enable calculations on a comprehensive set of conformations far wider than QM affords. Allied with the power of current computational architectures, this means that MM techniques can simulate events occurring at the microsecond timescale and in some cases milliseconds. This enables the study of stability and dynamical properties like folding, conformational changes, or even binding between partners.

As a last pointer, there are many ways to explore the conformational landscape, and their application depends on the focus of attention and the cost of calculating the energy. Calculations can focus on identifying global minima (e.g., minimization schemes or genetic algorithms), providing sampling that offers a vast statistical ensemble with thermodynamical meaning (e.g. Monte Carlo), or also generating the trajectory of the system's molecular motions, even at a long-time scale (e.g. molecular dynamics).

As you can see, molecular modeling encompasses numerous methodologies applicable to small models and accurate electronic descriptions, large systems and massive conformational sampling, and everything in the middle. In this landscape, modelers can overcome the limitations of each separate method by combining them in integrative pipelines. For example, pure QM methods on small models of the active site of a metalloenzyme can be used to investigate the core of the catalytic properties of an enzyme, QM/MM the impact of residues in the binding site in this process, and large-scale molecular dynamics the stability of the enzyme-substrate complex. This multiscale philosophy enables pushing the frontiers of molecular modeling in bioinorganics.

**Gerard Roelfes (Q):** Still, I feel that certain metals are more frequently studied than others. Which metals are "easy" and which are challenging, and why?

**Jean-Didier Maréchal (A):** It is a question difficult to answer. That depends. In a way, when we talk about the "easiness" of the modeling regarding the nature of the metal, what we need to consider is how electronically complex are the properties to assess (e.g., red-ox potential, solvation, catalysis, structural function, etc.). Some tendencies exist. For example, whatever the QM method, metals with open shells (unpaired number of electrons) are more challenging to model than those with closed shells. In the MM paradigm, the modeling of processes without changes in the coordination and electronic structure of the metal can be dealt with relatively straight forward formalisms now while processes resulting from a dynamic behavior of the first coordination sphere of the metal (e.g., ligand exchanges) are difficult to address. It is a case-by-case scenario.

**Jean-Didier Maréchal (Q):** Let's talk about timescales, and with timescale, I refer to the time it takes experimentalists to get a new set of experiments. Frequently, computational chemists have the feeling that their experimental colleagues expect simulations – "because it is all computer-based" – to be fast and easy to set up. That is very rarely the case, especially with bioinorganic systems. But I also have to say that I have the feeling that in chemical biology, experimental scientists can produce works a lot faster than they did years ago. That is particularly true when dealing with mutational works. So, how long would it take to experimentally generate a new mutant of an enzyme? What do you think modelers need to consider regarding schedules?

**Gerard Roelfes (A):** I consider this from my experience in enzyme design. If we have a working enzyme and the catalysis and subsequent analysis are set up, generating, and testing a new mutant is very easy. The techniques and methods for this are well-established and easy to set up. Typically, this can be performed in 1–2 weeks.

If you are trying to design something new, then obviously, it will take more time because we have to create the enzyme, test the conditions, analyze for catalysis, etc. There are many more variables in this. If you are lucky and everything works as planned, it can be performed within a month. But, of course, everything never happens as planned in an exploratory project, and you have to invest time to troubleshoot.

**Gerard Roelfes (Q):** In my view, molecular modeling is very valuable in helping to rationalize the outcome of experiments. I often say that the power of molecular modeling is to help "pull me out of the tunnel" I am in. That means that based on experimental results only, often you formulate a hypothesis of what is going on and then focus on trying to (dis)prove this experimentally. On several occasions, the molecular modeler then later came to me to say: "I have computed this, and I don't think it can work like that. Instead, you should consider this," and offered an explanation that I had never considered. This is very refreshing. My question is, how accurate are these computations nowadays?

**Jean-Didier Maréchal (A):** We have already mentioned accuracy partially. In summary, it depends on (i) the quality of the physical model you use, (ii) the adequacy of the structural models' dimensions regarding the experimental conditions, and (iii) the possibility of assessing the correct timescale (sampling). Accuracy increases as experimental and calculated systems converge in terms of size and physical models. For example, comparing results from molecular dynamics calculation of a protein simulated in a box of water may be more representative of experimental data obtained *in vitro* but less from those obtained *in vivo*.

The better the three conditions are satisfied, the smaller the discrepancy should be. However, you would rarely expect experimental and theoretical values to agree at 1 kcal $mol^{-1}$ range (something that can already be a substantial deviation in specific questions like enantioselective profiles in different mutants of a metalloenzyme for example). There is little doubt that, with time, the convergence will improve, but I understand that experimentalists struggle to figure out what you can expect from current state-of-the-art modeling. Drawing an uncertainty bar on modeling results is sometime tricky especially in condensed matter.

By today's standards, I believe that current state-of-the-art modeling is trustworthy enough to be informative for numerous questions in bioinorganics, and behaviors observed in the simulations are likely to be significant in the real system. One example that comes to my mind is the identification of breathing motions of proteins with large-scale molecular dynamics or normal mode analysis.

But it is important to highlight that the accuracy between experimental and calculated values should be mainly focused on tendencies rather than exact numbers. For example, in classical molecular dynamics, the study of the difference in flexibility between two mutants can show the correct tendency, and the same would apply for enantioselective paths studied by a QM or QM/MM approach.

**Gerard Roelfes (Q):** But how independent are simulations from the initial inputs from the modelers, then? How much is the outcome of molecular modeling biased by what the modeler thinks is going on?

**Jean-Didier Maréchal (A):** Your last questions concern the biases of modeling studies and the accuracy of the results. Well, both aspects are related.

Regarding biases, modelers tend to minimize them but frequently, imposing some "constraints" on the starting models or during the calculations is necessary because of the lack of structural data. Manual handling of the simulations to fit, at whatever cost, the experimental values – hence leading to non-realistic biases – can still be found, but this kind of strong biasing is less pronounced today than in the past. The increased savoir-faire and computational power have a lot to do with this, as they give more explorative capacities to the simulations and one would expect that the starting point of the simulation will not limit the calculations to reach the energetically relevant geometries.

However, in many cases, human guidance is impossible to avoid. A prototypical example that comes to my mind is protein–ligand docking campaigns. Protein–ligand dockings are a family of MM tools that look for the best orientations of small molecules into cavities of biomolecules. These techniques are widely applied in numerous fields, such as drug design. Generally, calculations are targeted to the primary binding site, with a minimal degree of protein flexibility allowed. Such methods are very good at pointing out good and bad binders. Still, the absence of information on how ligand channeling could impact the binding process is a form of bias that can create deviation between theoretical and experimental values.

However, biasing is not necessarily a negative thing. I sometimes introduce modeling to my students as a hypothesis-driven methodology. For example, biasing simulations by imposing some structural features is frequently the best way to ascertain the correct solution. In your field, Gerard, when designing new metalloenzymes, it may be challenging to assess the difference in enantioselectivity between two mutants. In modeling, imposing geometric criteria to isolate the transition states of the two stereoisomers generally allows for the identification of which pathway is energetically favorable and the impact of the mutations.

Just let me make a last remark on this aspect. Everybody would like to use molecular modeling as a "magic wand." In such a hypothetical situation, it would be enough to provide an initial sequence of a protein to obtain a correct and indisputable answer regarding structure, reactivity, substrate recognition, etc. Advances toward such situations are constantly growing. The irruption of AI-based approaches to complement standard physical-based models and the constant increase of computational power are the game changers of the last decade as it has been demonstrated by the 2024 Nobel Prize in Chemistry. However, there is still a lot of work to do, especially when dealing with biometallic systems where the study of the electronic properties of the metallic center can be related to large-scale conformational changes of the protein. So, "human" intervention and biasing will still be needed.

**Jean-Didier Maréchal (Q):** My last question is: In what field will bioinorganics grow the most in the following years?

**Gerard Roelfes (A):** I think the trend we have been observing for the past years, that is, that bioinorganic chemistry increasingly moves into "real" biological systems, will continue. This means that molecular modeling will have to develop methods to deal with this complexity further. That means integrating further methods spanning the different length and time scales.

**Gerard Roelfes (Q):** What we, of course, would like is for molecular modeling to provide an accurate design of bioinorganic systems, which we could then validate experimentally. There is certainly development in this direction, but we are not there yet. How can experimentalists help? What would molecular modelers need from experimentalists to help them transition from "rationalization" to "design"?

**Jean-Didier Maréchal (A):** With the Nobel prizes in chemistry awarded to researchers who contributed to modeling biochemical systems for description (2013) or design (2024), I am confident that the predictive power of modeling will continue to improve rapidly. Designing bioinorganic systems still presents challenges, but new tools and protocols are emerging. Exposing them has been one of the objectives of this book and hopefully more experimentalists will be convinced of their usefulness.

Based on multiscale approaches, AI, or a mixture of both, protocols for the efficient design of biometallic species are within reach. The prediction of mutations for tailored metalloenzyme activities, the creation of scaffolds that bear *de novo* metal-binding sites and provide new-to-nature biocatalytic profiles, the development of biomimetic models that copy metalloenzyme mechanisms, or the design of metal-induced folding processes is not futuristic; it is already happening. Indeed, standardized approaches have not yet been defined in many cases, and the accuracy of the prediction can only be assessed through experimental validation.

Experimentalists could help in this effort not only as collaborators eager to understand and optimize their systems but also as supporters to test the modeller's hypothesis. For that, they should embrace computational predictions despite their uncertainties. With such an alliance, trust, and throughout iterative works, the precision of the computational prediction will increase, and positive outcomes will rise. Resources are limited, and computer-aided design of bioinorganic systems is still in its infancy. But the results from the experimental-theory partnership will be outstanding.

# 2

# Computational Prediction and Modeling of Metal-binding Sites in Proteins

*José-Emilio Sánchez-Aparicio, Giuseppe Sciortino, and Jean-Didier Maréchal*

*Departament de Química, Universitat Autònoma de Barcelona, 08193 Cerdanyola del Vallés, Barcelona, Spain*

## 2.1 Introduction

More than 30% of the genome encodes biomolecules that integrate metal ions in their structure, bound, or unbound with a prosthetic group [1]. Metal ions show a wide range of functions in biology, including the stabilization of structural features (i.e. Ca, Zn), the control of signaling cascades (Ca), as well as mechanisms that involve the dynamical processes of their electronic structures like electron transfer (Fe, Cu, Mo), acid-base (Zn, Fe, Ni, Mn), or redox catalysis (Mn, V, Fe, Co, Ni, Cu, W). Metals can be firmly bound to their receptor to lead to metalloproteins (e.g. Fe, Cu, or Zn in cytochrome, blue proteins, and Zn fingers), but can also interact weakly and so generally provide an activator role on metal-dependent enzymes (e.g. Ca, Mg, and Mn in kinases or adenylate cyclase). Additionally, metal ions are also involved in diseases, such as neurodegenerative pathologies involving amyloid plaque formation in Alzheimer's disease, transmissible spongiform encephalopathies, the Lewy bodies of Parkinson's disease, and many others [2]. The study of the interaction between metal ions and proteins is fundamental in biology, pharmacy, and medicine [1, 3]. For years now, numerous fields of research focused on the function of metals in biological systems, including therapy (e.g. metallodrugs), diagnosis (e.g. biosensors), and industrial processes (e.g. metal-containing biocatalysts) [3, 4]. Until recently, very little could be performed in those fields if the location of the metal in the protein is unknown. This chapter aims at this problem and summarizes the progress performed by computational chemistry and structural bioinformatics tools in this area in recent years.

*Computational Bioinorganics: From Description to Prediction*, First Edition. Edited by Jean-Didier Maréchal.

## 2.2 Experimental Tools

Despite the progress over the last years, the exact mechanism of metal binding to biological receptors and the prediction of where a bare metal ion can, in principle, bind on a biological scaffold is still an open question. Experimental techniques like electrospray ionization (ESI-MS), matrix-assisted laser desorption/ionization time-of-flight (MALDI-ToF-MS), nanoelectrospray time-of-flight (*n*ESI-Q-ToF-MS) mass spectrometry, are able to accurately determine the metal ion-protein stoichiometry [5], and in several cases, tandem MS/MS using top-down or bottom-up approaches, can suggest the amino acids involved in binding [6]. Spectroscopic techniques, such as X-ray crystallography [7], X-ray absorption near edge structure, extended X-ray absorption fine structure, nuclear magnetic resonance (NMR), electron paramagnetic resonance (EPR), electron spin echo envelope modulation, electron nuclear double resonance, circular dichroism, and UV-Vis can provide with three-dimensional (3D) information of the metal-bound structure of a protein but often are not applicable (i.e. lability of the drug–protein interaction under crystal conditions and electron beaming, open-shell systems, substitution of the natural metal by another one, etc.) or only provide with partial information on the binding region [6c]. Therefore, computational methods can represent a valuable complementary or alternative approach [8].

## 2.3 Computational Tools

The prediction of metal-binding sites needs to consider the molecular variables that govern metal–protein interactions. Those include long- and short-range electrostatics, what amino acids are preferential, proper coordination rules as well as protein pre- and reorganization processes. Which of these magnitudes dominates depend on the specific metal. Almost pure electrostatics drive main group metals, while coordination bonds are also crucial in transition metals (TM). Amino acids could also be specific to metal ions, and protein donors tend to be primarily amino acid side chains. Depending on the *Hard* or *Soft* character of the metal, the most common coordinating amino acids are nitrogen atoms of histidine, oxygen atoms of glutamate, aspartate, tyrosinate, oxygen, or nitrogen atoms of asparagine and glutamine, and finally, sulfur atoms of methionine and cysteine. Lysine and arginine are two amino acids that can also, less commonly, be reported to coordinate metal under deprotonation of their positively charged nitrogenated groups. Backbone atoms can also be implicated in the first coordination sphere, although to a lower extent [9]. Main chain atoms are also relevant in defining the degree of preorganization of the protein to bind metals, but the real impact is still an unanswered question. Indeed, while some systems show major rearrangements upon the binding of the metal (i.e. zinc fingers); others do not display noticeable modifications between apo- and holo-structures.

Metal-binding site predictors have been mostly developed on pattern matching so far. This means that the algorithm looks for known motifs of metal-binding sites. This could be performed at the sequence or structural level, and some approaches mix both information. These approaches allow a metal-binding site to be interpolated even in conjunction with or without crystallographic structures. However, this can only recognize metal environments previously identified. When no metallic fingerprints are known, an in-depth exploration of the structural space of the protein is necessary. In those cases, a hypothetical *ab initio* search is required, although it needs to overcome major challenges: (i) the wide conformational space to explore on the protein scaffold so as to detect metal-binding site(s) and (ii) the need to model intrinsic properties of the metal first coordination sphere during binding (e.g. the directionality of metal–ligand interactions, possible changes in the number of ligands or geometry distortion). Despite the challenge, some solutions have arisen in recent years.

In this chapter, we will explore the different sequence-based and structure-based approaches that have been successful in the prediction of metal-binding sites as well as novel approaches that could reach further than established procedures.

### 2.3.1 Pattern Recognition Algorithms

Pattern recognition algorithms rely on sequence or structural information of well-characterized motifs (fingerprints) that are systematically used to search for similar patterns in queried proteins. For metal-binding site prediction, several tools are available as web servers or standalone software. In general, two main approaches are possible: algorithms based on pure sequence motifs or structural analogies between the query apoprotein and well-characterized holo-metalloproteins [10]. Neural networks have also more recently introduced for metal ion- binding site prediction [11].

Some of the algorithms are limited to selected metals and a reduced set of amino acids are considered as potential donors. MetalDetector v2.0 web server [12] identifies potential metal-binding sites in proteins, basing its analysis only on the pure sequence information, and its usage is limited to TM (with the addition of heme and Fe/S clusters) and CYS and HIS side chains as donors. The output consists of a classification of the residues as potentially free or metal bound but no structural information or 3D models are given. MetalPredator web server [13], also based on pure sequence analysis, is aimed to predict potential Fe–S motifs in proteins finding analogies using two libraries representing Pfam domains and structural iron–sulfur clusters. Similar to MetalDetector, no structural information is given and the output defined as a potential binding site or ligand-free the different regions of the queried protein. SeqCHED Server [14] potentially enables the user to analyze a sequence for Zn, Fe, Ni, Cu, Co, Mn, Mg, and Ca metal-ion-binding sites (only Cys, His, Glu, or Asp) checking for homologies of the target sequence to Protein Data Bank (PDB) templates and finally models the target side chains in 3D structure. On the framework of machine learning, ZincFinder [11c], based on sequence information is available for the prediction of potential zinc-binding residues.

Other pieces of software utilize structural motifs for the prediction of the metal-ion-binding sites. The family of MIB predictors [15a,b] is built to predict diverse metal-ion-binding residues of a protein and to generate the predicted metal-ion-bound 3D structure. The method is based on the structural comparison with binding templates of regions that bind 12 types of metals ($Ca^{2+}$, $Cu^{2+}$, $Fe^{3+}$, $Mg^{2+}$, $Mn^{2+}$, $Zn^{2+}$, $Cd^{2+}$, $Fe^{2+}$, $Ni^{2+}$, $Hg^{2+}$, $Co^{2+}$, and $Cu^{+}$). MIB2 actually includes the $(PS)^2$ modeling method [16], and the AlphaFold protein structure database [17]. TEMSP [18] software bases its prediction on the structural comparison of the $C_\alpha$ and $C_\beta$ relative position of the analyzed protein with a template library searching for Zn-binding motifs that matches in terms of structural criteria. The standalone script FINDSITE-metal [11a], is based on the structural superposition of metal-binding templates detected on the query protein with a template dataset of 860 structures containing Ca, Co, Cu, Fe, Mg, Mn, Ni, and Zn. Template-bound metal ions are clustered and ranked. The script requires two input files: the target protein structure in PDB format and the list of identified binding regions. As well as the binding site regions and ranking information a 3D structure of the final site is given. The standalone mFASD [19] software, starting from a protein structure, could predict if it contains binding regions for different types of metals (Fe, Zn, Cu, Ca, Mn, and Mg) and discriminate between them. The algorithm is based on a structural comparison between the query and a reference motifs dataset. The server FEATURE [20], individuates potential zinc-binding sites in a given protein structure. Similarly to previous commented methods, FEATURE does not define the metal-binding site as a set of residues neither provide 3D models of the sites, the output basically consists in the xyz coordinates of the probes that are compatible with Zn binding.

Other algorithms, such as MetSite [21], 3DLigandSite [22], fragment transformation method [23] implement the combination of both, structural and sequence information to predict possible binding residues of a target protein. MetSite web server, for example, uses neural networks to predict six metal ion-binding sites ($Ca^{2+}$, $Cu^{2+}$, $Fe^{3+}$, $Mg^{2+}$, $Mn^{2+}$, and $Zn^{2+}$). The method is developed for the identification of the potential residues that can coordinate the metal in protein super-families by combining sequence and structural analysis. Since the algorithm is based only on the distribution of individual residues, it presents some limitation in the identification of the binding site location. In 3DLigandSite web server, a protein sequence or a 3D structure can be analyzed and compared with a structural library to identify homologies with metal-bound proteins. Only a few methods are devoted to individuating metal-binding sites without recurring to a template-based approach. For example, the recent IonCom server [24], combines *ab initio* training and template-based algorithms. It focuses on binding site prediction of $Zn^{2+}$, $Cu^{2+}$, $Fe^{2+}$, $Fe^{3+}$, $Ca^{2+}$, $Mg^{2+}$, $Mn^{2+}$, $Na^{+}$, and $K^{+}$).

As in any field of molecular modeling, the application of machine learning is rapidly expanding across various topics, including metal-binding site predictions. One example of this can be seen in Metal3D, a software published in a 2023 study by Rothlisberger et al. [25], which uses a convolutional neural network to train their model. While it has proven efficient for predicting zinc-binding sites, the authors claim that the model is valid for any metal. Additionally, the newest version of AlphaFold, AlphaFold 3, also considers the location of metal ions [26].

In most of the methods exposed so far, the main assumption is that well-characterized metal-ion-binding sites could be used to infer the binding residues of a general target protein or peptide. This presents several limitations when it comes to cases for which no or little templates are available or in the case of metal-induced folding. However, when there is a template, successful applications can be found in *de novo* enzyme design. Ward et al. proposed an artificial metallo-peroxidase upon the addition of Cu(II) ions to a 6-phosphogluconolactonase [27]. The biological scaffold was identified applying the STAMPS algorithm (search for 3D atom motifs in protein structure) [28] which allows to identify candidates by a systematic search in the PDB for apo-structures with potential binding regions containing the facial two histidines one-carboxylate binding motif (either aspartate or glutamate, *fac*-$[M(His)_2(O_2CR)]$, facial two-histidines one-carboxylate motif [FTM], Figure 2.1a) [29].

The STAMPS algorithm, based on the web-server RASMOT-3D PRO [28], was programmed to perform a structural comparison between the target protein and a dataset of well-characterized FTM motifs. As the first step, distances between the $C_\alpha$ and $C_\beta$ of the histidine residues and glutamate or the aspartate residues are compared with those of the FTMs dataset and screened using distance thresholds. Subsequently, after alignment, the identified structures are filtered by root mean square deviation (RMSD) with respect to the FTMs. As a result, among the 114 protein identified by a visual inspection, seven scaffolds, for which the FTM motif was reorganized to bind metals were selected applying the criteria reported in Refs. [28, 29]. Furthermore, scaffolds with mutated aspartate or glutamate for asparagine or glutamine were considered. Among the 13 cloned proteins, six were overexpressed in *Escherichia coli* and the wild-type **6-PGLac** after addition of $CuSO_4$ salt demonstrated peroxidase activity with comparable efficiency to that of naturally or artificially well-known peroxidases [30]. The crystal structure of **Cu-6-PGLac** (PDB code 4TM7) was determined showing the T-shaped geometry for the catalytic Cu(II) ion bonded by two *trans* histidine residues (His67 and His104) and a water molecule in the third coordination position (Figure 2.1b). However, the computationally predicted FTM motif was not reflected in the crystal structure. This may be due to some experimental difficulties but also to

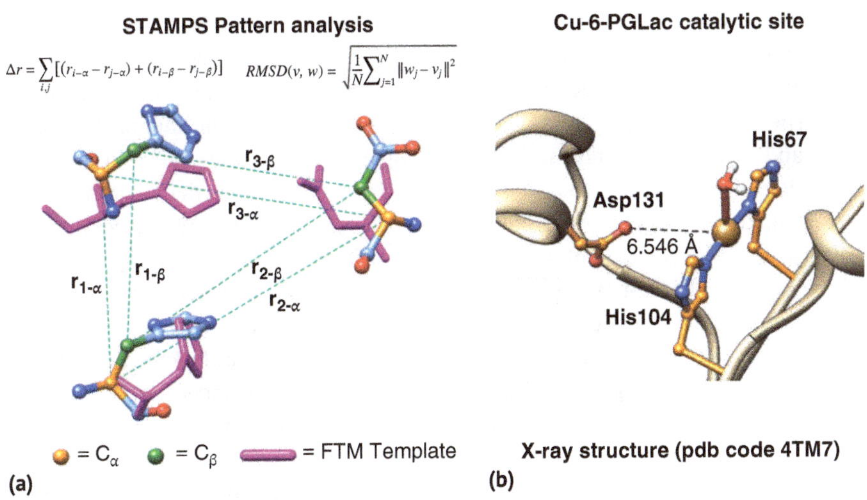


**Figure 2.1** Schematic representation of (a) the STAMPS template-based parameters used for search FTM motifs, and (b) representation of the Cu(II)-binding site in Cu-6-PGLac (PDB code 4TM7) artificial metalloenzyme.

limitations in the STAMPS algorithm, like the need to account for major conformational changes of the protein scaffold as well as side-chain flexibility in solution.

### 2.3.2 The Backbone Preorganization Hypothesis

Whereas sequence-based algorithms do not consider geometric constraints at all, the vast majority of the structure-based methods described above are based on geometric features related to the first coordination sphere of the metal (for example, distance from the metal to the donor atom) and/or a very general description of the geometry around it in terms of the location of the amino acids. This imposes the need to account for structures where the side chains are well organized to bind the metal. However, one could imagine that side chains can change their orientation with a relative facility due to their high flexibility. In this framework, the exact configuration of the side chains could be ignored to assess whether a region of the protein can accept a metal. The primary constraint would then be the presence of suitable donor amino acids with a correct preorganization of their backbone atoms.

Some works have taken the profit of the backbone preorganization hypothesis. For example, Muñoz Robles et al. [31] studied an artificial metalloenzyme obtained by inserting a Fe moiety into the heme oxygenase of *Corynebacterium diphtheriae*. They utilized a computational protocol that includes a filter of those residues with a distance $C_{\alpha}$-metal of a maximum of 7 Å to identify the candidates to coordinate the metal. A posterior study proposed a method to locate potential metal-binding sites by probing the protein's 3D space for regions where donor-containing residues feature a β-carbon within a threshold distance respect the position of the queried point [32]. The same STAMPS method described above, evaluates the matches between the query structure and the templates using metrics based on the distances between the α-carbons and the β-carbons of the possible coordinating amino acids, although ignoring the position of the metal (Figure 2.1a).

In a recent study [33], several geometric features related to the backbone preorganization hypothesis were analyzed for the whole set of structures contained in the MetalPDB [34], a publicly accessible database of metal-containing proteins from the PDB. The dataset includes around 400 000 bonds, implying a coordination between a metal and a side-chain donor, and the following descriptors were studied in that case: the distance from the α-carbon to the metal; the distance from the β-carbon to the metal; and the angle between the metal, the α-carbon, and the β-carbon (Figure 2.2).

The statistical study did not reveal significant differences when grouping the results by metal: all of them presented similar ranges for each of the geometric descriptors, meaning that it is not possible to differentiate which metal could bind a residue through the direct use of these descriptors. However, the grouping by type of amino acid allowed to establish different ranges for each feature, as a characteristic fingerprint of each type of amino acid. For example, whereas cysteine has a range from 3.25 to 4.45 Å for the distance from metal to α-carbon, lysine has a range from 5.13 to 9.28 Å for this same feature. Thus, in agreement with the backbone preorganization hypothesis, it seemed that those geometric descriptors encode sufficient information to decide whether a metal in a given position has the possibility of binding to a certain set of amino acids in its vicinity. Furthermore, one could take advantage of the preferences for certain amino

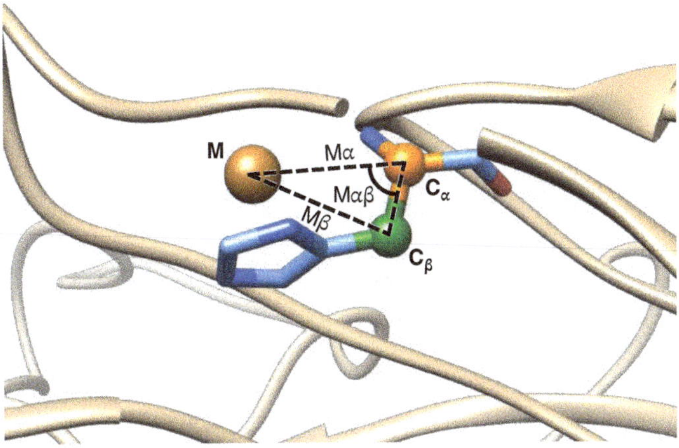


**Figure 2.2** Schematic representation of the geometric features studied for the case of a metal coordination bond with a side-chain donor. Metal, α-carbon, and β-carbon are depicted in brown, orange, and green balls, and labeled $M$, $C_\alpha$, and $C_\beta$.

acids of each metal (e.g. copper ions have a high affinity for histidine residues) to indirectly discriminate which metals could bind at that position.

The conclusions of that study crystallized in a new piece of open-source software, called BioMetAll, which allows identifying those regions of a protein where a metal could bind. It takes as an input the 3D structure file of the protein (in.pdb format) and the user-defined constraints: (i) the minimum number of amino acids that should coordinate the metal; (ii) the list of potential types of amino acids, or a specific motif that could coordinate the metal; and (iii) whether to consider coordination with backbone oxygen atoms or not. An interesting feature of BioMetAll is that the user can also request possible mutations to complete an already present motif, which can be used as a starting point in an enzyme-design protocol. Once the parameters are set, the program probes the 3D space of the protein by searching for points where the user-defined constraints and the geometric descriptors are accomplished. Those points that share the same potential coordinator amino acids are grouped and constitute the areas where a metal could bind that set of amino acids, also called "coordinating environments." As an output, the user obtains the list of coordinating environments, along with a.pdb file ready to open with any visualization program.

One of the advantages of the backbone-based approach is the possibility to determine how a conformational change of the protein could affect the formation of metal-binding sites. Another advantage, as the side chains are not explicitly taken into account, is that the method can detect not only sites with well-positioned rotamers but also secondary or transient-binding sites. In the BioMetAll article, both benefits are illustrated using the human serum albumin (HSA) as a showcase.

As the most abundant protein in the blood plasma and first carrier for most of the essential metal ions [35], HSA is known for containing two main metal-binding sites. They are called NTS and MBS, and both are composed of a two-histidine, one-carboxylate motif [36]. Along with these two primary sites, other several secondary sites appeared when the structure of HSA was determined under the presence of high concentrations of $Zn^{2+}$ [37]. Interestingly, the binding of medium- and long-chain fatty acids (FAs) to the site FA2 (located at the same region of MBS) induces a change on the scaffold of HSA, which hinders the metal coordination at MBS [36b,c, 38].

Two HSA structures were used to test the discriminative capacities of the program, one with an FA bound to the site FA2 (fatted conformation, PDB code 1bj5), and the other without the presence of FAs (defatted conformation, PDB code 5ijf). BioMetAll was instructed to search for two-histidines one-carboxylate binding motifs and, in correspondence with the experimental background, was able to detect both NTS and MBS in the defatted conformation (Figure 2.3a) and only the NTS in the fatted one (Figure 2.3b). Furthermore, when instructed to search looser considering any combination of His, Asp, and Glu (with a minimum of three amino acids to form the binding site), BioMetAll was able to identify other secondary binding sites reported in the literature [37], which demonstrates the versatility and utility of the backbone preorganization approach to better understand biometallic interactions.

Pattern recognition and backbone preorganization approaches to predict metal-binding sites either based on sequence, coordination rules, template fitting or a combination of them are definitely promising tools. However, they remain the first line and fine molecular structural prediction requires modeling tools which models are closer to the atomic reality. Although certain predictors provide 3D models of the metal-bound complexes, in certain circumstances it could be more interesting to explore a wider space of solutions. For this task, the protein–ligand docking framework could be considered. However, these should be optimized for dealing with metalloligands.

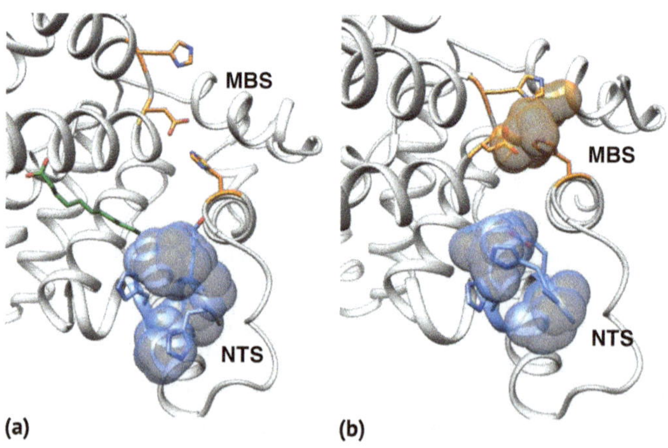


**Figure 2.3** Representation of the primary metal-binding sites of albumin (MBS and NTS, in orange and blue sticks, respectively), along with the binding sites predicted by BioMetAll (in surface representation). (a) Fatted conformation. The MBS binding site is hindered by the binding of a FA (in green sticks), and not detected by the BioMetAll calculation. (b) Defatted conformation. Both binding sites are well organized and detected by the BioMetAll calculation. *Source*: Adapted from [33].

### 2.3.3 Protein–Ligand Dockings

Protein–ligand docking methods were designed for fast and accurate prediction of the binding sites and modes of small organic molecules to proteins and could offer a valuable alternative to MM or MD methodologies. Despite their high performances in terms of computational time and wide system exploration, the method considers the macromolecule as a rigid entity neglecting potential conformational changes in solution or induced by metal binding.

The binding energy calculation is based on simplified MM force fields (*Scoring Functions*) that consider nonbonded interactions between the *ligand* (in docking terminology the small molecule) and the protein host. The docking problem is limited to finding, generally with a genetic algorithm (GA) or a Monte Carlo (MC) algorithm, a series of protein–*ligand* relative positions (*poses*), evaluating their interaction energy throughout the *Scoring Function* and rank them by scoring. Several high-quality reviews about *Scoring Functions* and their performances have been published in the recent years [39]. A *Scoring Functions* is generally built as a sum of terms representing different interaction types: hydrogen bond (*Hbond*), electrostatic (*elec*), or van der Waals (*vdW*), such term could be multiplied by an empirical parameter to fine-tune its relative weight. Additionally, other terms may also be incorporated, such as desolvation (*desolv*), interatomic contacts (*Clashes*), ligand torsions (*tors*), intermolecular ligand interactions, or relative to restraints defined (i.e. covalent bonds, coordination interactions, etc.). For example, GoldScore implemented in GOLD software has the following form [40]

$$\text{Fitness}\,(F) = \alpha \cdot S_{\text{hbond}}^{\text{ext}} + \beta \cdot S_{vdW}^{\text{ext}} + \gamma \cdot S_{\text{hbond}}^{\text{int}} + \delta \cdot \left(S_{vdW}^{\text{int}} - S_{\text{tors}}\right) \tag{2.1}$$

$S_{\text{hbond}}^{\text{int}}$ and $S_{vdW}^{\text{int}}$ the intramolecular interactions described by LJ-like potentials, $S_{\text{tors}}$ the energy change related to the molecular torsions, and $\alpha$, $\beta$, $\gamma$, and $\delta$ are empirical parameters.

The background of the docking is widely discussed in Refs. [41, 42], while the historical development of the techniques in Refs. [43–45].

Unfortunately, most docking suites, including commercial ones, are mainly focused on organic species [41a,c,d] and how to deal with the coordination bond formation between the metal ion and a donor of an amino acid side chain [41e, 46] is still an open question in computational bioinorganics. Some attempts to apply protein–ligand dockings to the prediction of metal species binding sites have reached only partial success [8f, 47]. Some stand on pure electrostatic models, which, despite their step forward, fail in respecting coordination rules. If the metal is part of the protein and the coordination bond takes place with the organic *ligand*, most of the software offers several solutions [8d,e, 48]. However, for systems in which the metal ion is part of or represents *per se* the *ligand*, the coordination bonds can be only described through *covalent docking* approaches nowadays implemented in many programs such as GOLD [48] or AutoDock [47a], or source code modification of commercial software, for example, CovalentDock [47b] or DOCKTITE [49]. For all these approaches, the specific protein–*ligand* bond needs to be *a priori* defined and forced during the simulations imposing energy restraints and/or geometrical constraints. Therefore, their applicability is limited to systems for which the specific amino acid donors involved in coordination are already known, which significantly limits the predictive capabilities of the method. Moreover, coordination bonds parametrization is totally lacking in the scoring functions, excluding the correct applicability of covalent docking to metal ions as *ligands*.

The relatively simple physical chemical representation of the energy under a *Scoring Function* scheme opens some opportunities for metal interaction implementations. Recently, a new approach has been proposed on the GOLD software framework consisting of defining the coordination bond using the *Hbond* function included in GoldScore without source code modifications. The metal is virtually described as a hydrogen bond donor liable to interact with the *Hbond* acceptors of protein by means of dummy atoms positioned in the coordination vacancies. In terms of Lewis acid and base theory, the acid center is transferred from the metal to the fictitious proton located at the virtual coordination bond axis (Figure 2.4). The $S_{\text{Hbond}}$ LJ-like function [50], is able to describe coordination bond lengths and angles by the multiplication of its potential for a weight block function (wt = distance$_{\text{wt}}$ + angle$_{\text{wt}}$), as reported below [50]

$$x_{\text{wt}}(x, x_{\text{ideal}}, x_{\text{max}}) = \begin{cases} 1, & \text{if } x \leq x_{\text{ideal}} \\ 1.0 - \dfrac{x - x_{\text{ideal}}}{x_{\text{max}} - x_{\text{ideal}}} & \text{if } x_{\text{ideal}} \leq x \leq x_{\text{max}} \\ 0, & \text{if } x > x_{\text{max}} \end{cases} \tag{2.2}$$

The strength of the coordination-like interactions between 15 metals (Mg, V, Cr, Mn, Fe, Co, Ni, Cu, Zn, Ru, Rh, Re, Os, Pt, and Au) and all the amino acidic acceptors are defined by a series of empirical parameters implemented in *GoldScore* that takes into account their relative affinity to different metals [50].

In a recent study, the methodology was applied to characterize the $V^{IV}O^{2+}$ binding site with carboxypeptidase [51, 52]. The analysis was carried out on the X-ray structure of the protein (1cpx [53]) while the geometry of $[V^{IV}O(H_2O)_4]^{2+}$ ion was optimized at the DFT theory level. The four equatorial water molecules were replaced by dummy hydrogen using a bond length of 0.74 Å as described in Ref. [50]. During the docking simulation, the His69, Glu72, and His196 side chains which, as suggested by

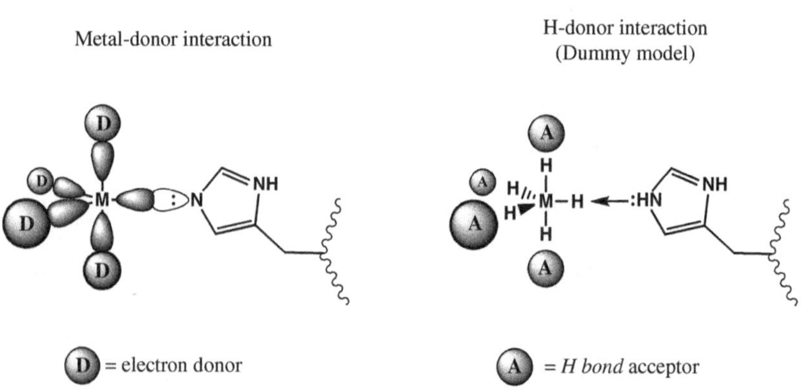


**Figure 2.4** Description of the dummy hydrogen approach to metal ions docking in proteins.

experimental evidences [51, 52] could bind $V^{IV}O^{2+}$, were considered flexible applying the GOLD rotamer libraries [54]. The dockings of the $V^{IV}O^{2+}$ ion on the carboxypeptidase show a single full-populated cluster in which the highest scoring poses show the $V^{IV}O^{2+}$ coordination by three different amino acidic donors, two histidine, His69, and His196, and glutamate Glu72. The identified binding site is analogous to that of Zn-carboxypeptidase [55] in agreement with the $Zn^{2+}$ and $V^{IV}O^{2+}$ ions similarity. The coordination geometry predicted for the adduct $V^{IV}O^{2+}$–carboxypeptidase results in a square pyramid with a vacancy in the equatorial position, liable to interact with a water solvent molecule in the solvated system (Figure 2.5). Concerning the bond length, the method appears solid predicting 2.322 Å for V—N(His196), 2.295 Å for V—N(His69), and 2.259 Å for V—COO⁻(Glu72) coordination bonds, coherent with the experimental values for $V^{IV}$—N(1-Me-Imidazole) and $V^{IV}$—COO⁻(acetate) distances [56]. The O=V—Donor angles of about 105° highlight a square pyramidal coordination geometry.

### 2.3.4 GaudiMM Platform

GaudiMM platform is a *multiobjective* molecular modeling suite based on the NSGA-II [57] GA that allows evaluating the system with diverse molecular descriptors (*objectives*) aimed to solve several chemical problems [58]. Among them, a recent objective was developed to reproduce key features of metal moieties [32]. In contrast with the pattern recognition algorithms described previously, the coordination objective does not rely on structural templates extracted

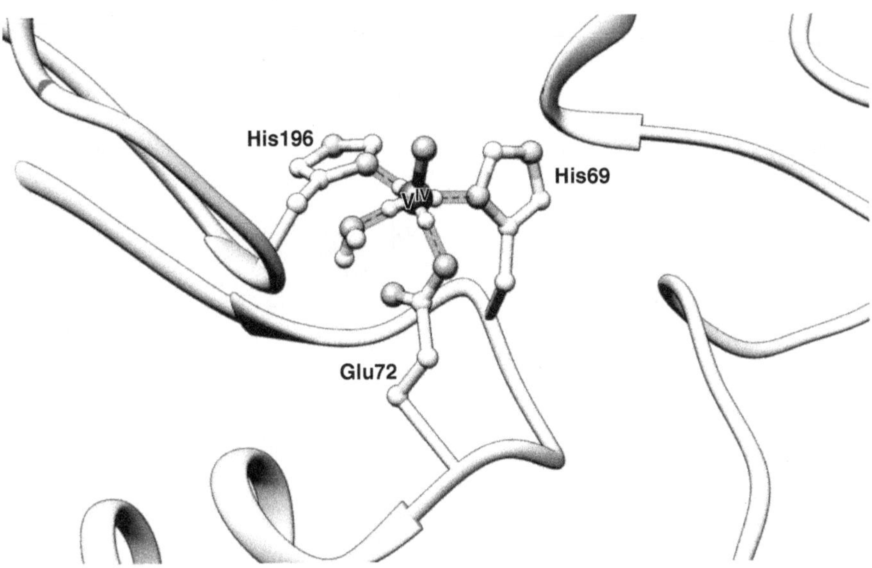


**Figure 2.5** Best docking proposal for the interaction of $[V^{IV}O(H_2O)_4]^{2+}$ to carboxypeptidase after the replacement of three $H_2O$ molecules by His69, Glu72, and His196 donors. The proposed solution results in agreement with DeKoch et al. [51] and Sanna et al. [52], on the basis of EPR data and DFT calculated $^{51}V^{1}$hyperfine coupling constant. *Source*: Adapted from [51].

from protein datasets. The strategy consists of find, through a GA exploration, protein regions in which the donor-containing side chains orientation is compatible with the coordination of a metal ion. For each candidate, the presence of potential donor atom types is first analyzed, and if the number is sufficient to fulfill the coordination vacancies of the metal ion, a series of geometrical requirements are then evaluated: (i) the RMSD between the ideal polyhedron and the one built on top of the identified donors, (ii) to account for donors orientation with respect to the central metal, deviations of angles and dihedrals with respect to the ideal values are computed (Figure 2.6). The total score (coord_fitness, CF) is extrapolated as the sum of RMSD and angle deviation terms as follows:

$$\mathrm{CF} = \mathrm{RMSD} + \left( \frac{\sum_{n}^{1} \left| R_{MX_j}^{\mathrm{Ideal}} - R_{MX_j}^{\mathrm{Formed}} \right|}{n} \right) + \left( \frac{\sum_{n}^{1} \left| \sin\left( \theta_{MX_j}^{\mathrm{Ideal}} - \theta_{MX_j}^{\mathrm{Formed}} \right) \right|}{n} \right) + \left( \frac{\sum_{n}^{1} \left| \sin\left( \vartheta_{MX_j}^{\mathrm{Ideal}} - \vartheta_{MX_j}^{\mathrm{Formed}} \right) \right|}{n} \right)$$

The method was validated on a dataset of 105 X-ray structures, obtaining up to 98% of the proposed solutions located at the correct binding site with less than 1 Å

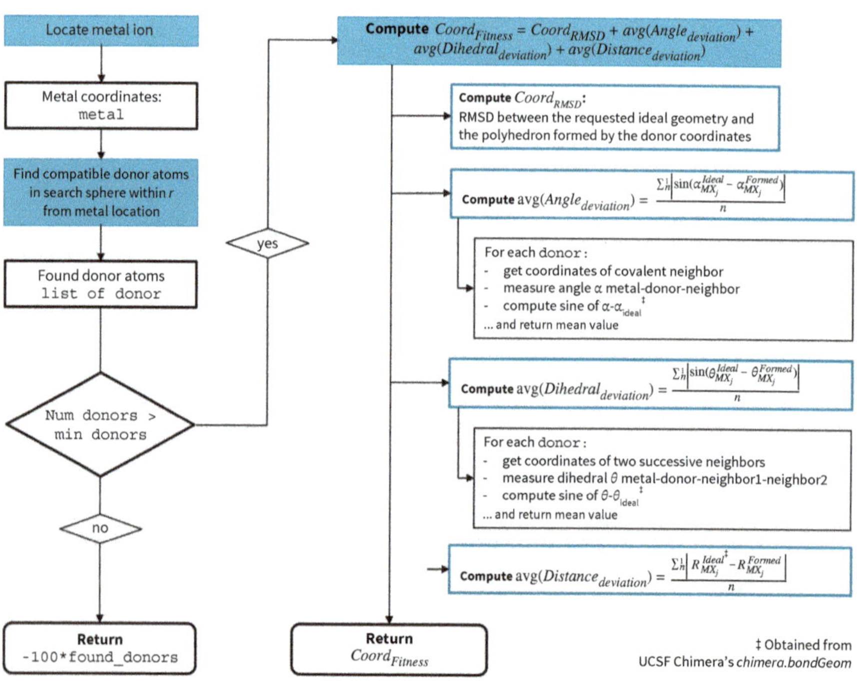


**Figure 2.6** Flowchart of the coordination objective implemented in GaudiMM. *Source*: Adapted from [32].

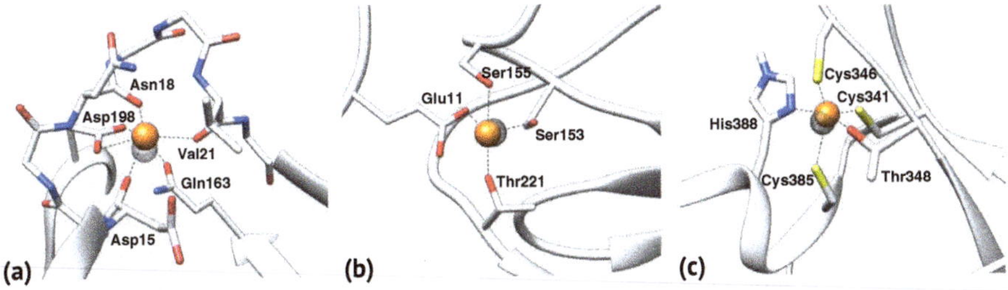


**Figure 2.7** Comparison between the GaudiMM solution (in orange) and the original XRD structure (in dots surface) for the structures (a) 1f49, (b) 1dzi, and (c) 2lq6a. The original XRD structure is also shown (in dots surface). *Source*: Adapted from [32].

distance from the crystallographic position. In Figure 2.7, as an example, a comparison is made with respect to the X-ray solution of three simulated structures of the original GaudiMM dataset.

Docking methods are excellent to grasp the first physically sound 3D model of the biometallic systems resulting from the interaction of a metallic species with a biological host. Slowly, newer methods introduce energetic and geometric terms specific to inorganic chemistry. However, when accurate prediction of the binding site in terms of first and second coordination spheres is needed, prediction may need methods of higher sensitivity.

Docking-based approaches, such as the ones described above, allow an atomistic representation of the system under analysis, primarily based on molecular mechanics. However, fine electronic effects are neglected, and scaffold changes of the receptor are challenging to take into account. If it is necessary to address such issues, higher theory-level techniques should be recommended.

### 2.3.5 Quantum Mechanical and Hybrid Approaches

In recent years, quantum mechanics (QM) approaches, in particular those based on density functional theory (DFT) have been increasingly applied to metalloproteins. They became a fundamental tool in bioinorganics chemistry. The QM approach, in contrast to MM, describes a molecular system at the electronic level. Without a doubt, the QM-approaches from which bioinorganics have mostly taken advantage of are DFT ones in which the electronic structure calculations are based on the electron density $\rho(\mathbf{r})$. It allows determining the properties of a system using a functional dependent only on $\rho(\mathbf{r})$ with relatively short computational times and is widely used in dealing with metals [59].

DFT and QM methods in general, allow computing the structure, energy, and fine molecular properties with great accuracy and are widely used in inorganic and organometallic systems because of the complexity of the metal properties, in particular TM ions for which subtle electronic effects (i.e. spin state) must to be considered [60]. Despite the increase in computing power in the last years, QM methods are computationally expensive and applicable only to a reduced number of atoms (about a maximum of 200 atoms). Considering a metalloprotein or a metalloenzyme, this is reflected in the possibility of treating a reduced part of the chemical space, typically the metal ion, its ligands, and a few amino acids. A widely used approach to deal with metalloproteins from a QM perspective consists of cutting out the metal-containing region-generating cluster models without considering the entire size of the system [61]. A valuable alternative is represented by the so-called hybrid quantum mechanics/molecular mechanics (QM/MM) methods in which part of the system, typically the metal ion, the donors of the first coordination sphere and the neighbor amino acids side chains are treated under QM formalism, and the rest of the system is at the MM level (Figure 2.8) [62]. However, for such approaches, it is vital to already now the location of the metal with a quite a high amount of certainty. It is, therefore, something more frequently used as a second step and in combination with the methods previously shown.

An important limitation of pure QM or hybrid methodologies described above is that the conformational and chemical spaces that can be explored in a reasonable amount of time are rather limited. Despite the potentiality of the methods, from a QM or QM/MM perspective, the blind characterization of where the metal ions bind to a protein is a task that is far from being achieved.

### 2.3.6 Molecular Dynamics, Integrative or Multiscale Approaches

To overcome the limitations of QM static calculations, QM theory level could be implemented in molecular dynamics via *ab initio* MD [63, 64]. However, the treatment of large systems is computationally time demanding [65], and the hybrid QM/MM/MD approximation could be a valuable alternative [66]. The approach, similarly to static QM/MM methodology, consists in partitioning the system extracting on the fly the forces of the metal ion and its environment at the

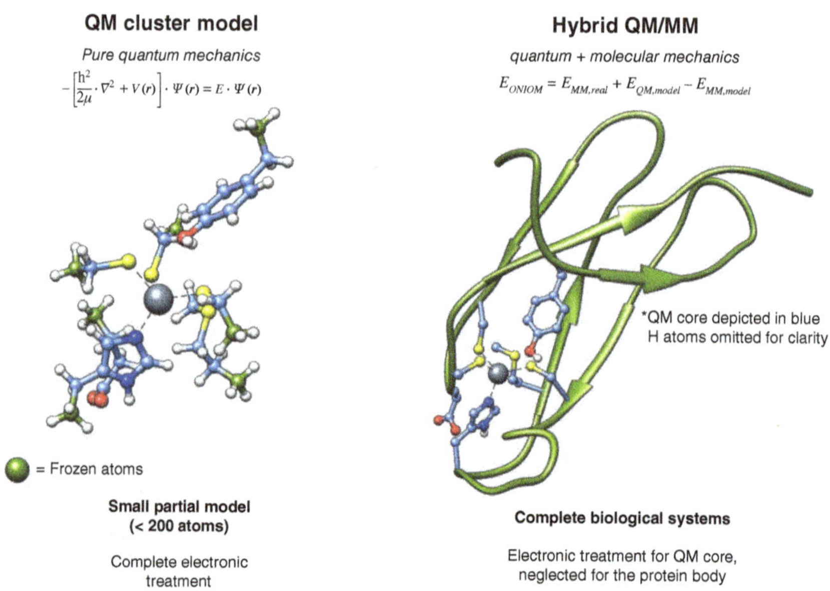


**Figure 2.8** Different approaches to deal with metal ions in quantum mechanics and hybrid QM/MM frameworks.

electronic level, while the rest of the protein and the explicit solvent are at MM theory level. The interested reader can find more details about the discussed techniques and their applications in Refs. [62a, 67]. Recently, some efforts to implement mixed QM/MM scoring functions in docking algorithms have been published [68], and in a few cases applied to metalloproteins [69].

The metal-ion-binding site prediction can be treated with multiscale strategies combining different levels of theory at different stages of the molecular modeling process. Depending on the modeling stage, the accuracy in energy calculation or the combinational and conformational space to explore could differ. Based on the same concept proposed by Warshel, Karplus, and Levitt (Nobel Prize in Chemistry in 2013), multilevel (also known as integrative or hierarchical) strategies open major opportunities.

Fast and relatively inexpensive pattern recognition or docking protocols can be combined with accurate but more time-demanding QM or MD techniques. For one side, dockings can be used for a massive exploration of the wide ion-protein relative conformational space. The major drawbacks are: (i) the inability to account for slow molecular events like long-range motions of the protein scaffold, resulting in the loss of information about conformation arrangement upon metal binding; (ii) the precision limits of the relatively simple scoring function, and (iii) the missing of electronic effects. For the other side, MD simulations can account for complete protein flexibility, including metal-induced fitting of the binding region. Moreover, the explicit solvent effect can also be considered. Finally, QM approaches reach high accuracy and deal with fine electronic phenomena related to the first coordination sphere (ligand exchange, spin-related, or RedOx processes). The combination of the techniques in hierarchical or integrated protocols, where docking gives preliminary information about potential binding sites, MD simulations are employed to explore natural or metal-induced conformational changes of the protein receptor and subsequently, QM or QM/MM to optimize, refine the structures of the final adducts, and calculate energies and molecular or spectroscopic properties; results in an interesting and rising approach in the molecular modeling landscape [41e, 70]. In a recent work by Mujika et al., they applied the combination of a recognition algorithm, QM cluster models, MD, and QM/MM approaches have been successfully applied to identify a series of plausible $Al^{3+}$–Aβ-peptide models for the aluminum interaction with β-amyloid aggregations involved in Alzheimer's disease [71].

The multistep analysis consisted of: (i) generating and evaluating the stability of a series of penta- or hexa-coordinated models for the $Al^{3+}$ with different biological donors (His, Asp, Glu, Tyr, Asn, Gln, His, terminal amine, and $H_2O$) via a QM cluster scheme; (ii) screening the PDB for potential peptide receptors compatible with the coordination environments individuated in the first step and processing it with GaudiMM, configured with a preliminary version of the coordination objective; (iii) building complete models for the $Al^{3+}$–Aβ most plausible candidates and evaluating them via MD for induced conformational changes and finally at QM/MM and QM level for binding energies. The analysis of the QM clusters showed oxygen as the most favorable donor and, in particular, the first coordination sphere with three or four carboxylates as the most stable binding site in agreement with previous studies [72]. As an example, the four most favored binding modes from a pure QM point of view are: OC—[Al($^-$OOC)$_4$;(OCNH$_2$);(H$_2$O)]$^-$ > OC—[Al($^-$OOC)$_4$;(H$_2$O)]$^-$ > OC—[Al($^-$OOC–$\kappa^2$O, O')$_2$;($^-$OOC)$_3$ ; (H$_2$O)]$^{2-}$ > TBPY—[Al($^-$OOC)$_4$;(H$_2$O)]$^-$. Starting from this information, the early version of the GaudiMM algorithm was applied for a grid-based exploration of 157 conformations taken from different NMR-resolved structures of Aβ-peptides to identify consistent binding regions. After the recognition of binding regions with at least three different carboxylic donors featuring the $C_\alpha$ within a radius of 7 Å, the early version of the coordination objective presented above was applied to evaluate bond distances, angles, and dihedrals compatible with a trigonal bipyramid or octahedral coordination geometry. In particular, searching distances about 2.5 Å, angles D-M-D as close as possible to 90º and dihedral involving the metal, the donor, and the two neighboring atoms of the ligand near to planarity. Moreover, the steric clashes were evaluated during the exploration. As a result, 13 structures were identified as possible receptors for the Al(III) involving Asp1, Glu3, Asp7, and Glu11. The other two carboxylic groups of the peptide, Glu22, and Asp23, are predicted as not coordinating. The different models were then studied in explicit solvents along 10 ns of molecular dynamics. Three representative structures of each trajectory were subsequently optimized via a QM/MM hybrid scheme, and finally the interaction energy of each binding mode was estimated at full QM theory level. The multiscale protocol was able, among the thousands of potential binding modes, to individuate five structures favored for at least 15 kcal mol$^{-1}$ up to 85 kcal mol$^{-1}$. The QM/MM most favored structure is depicted in Figure 2.9. A further discussion of these works could be found in chapter 11 of this book.

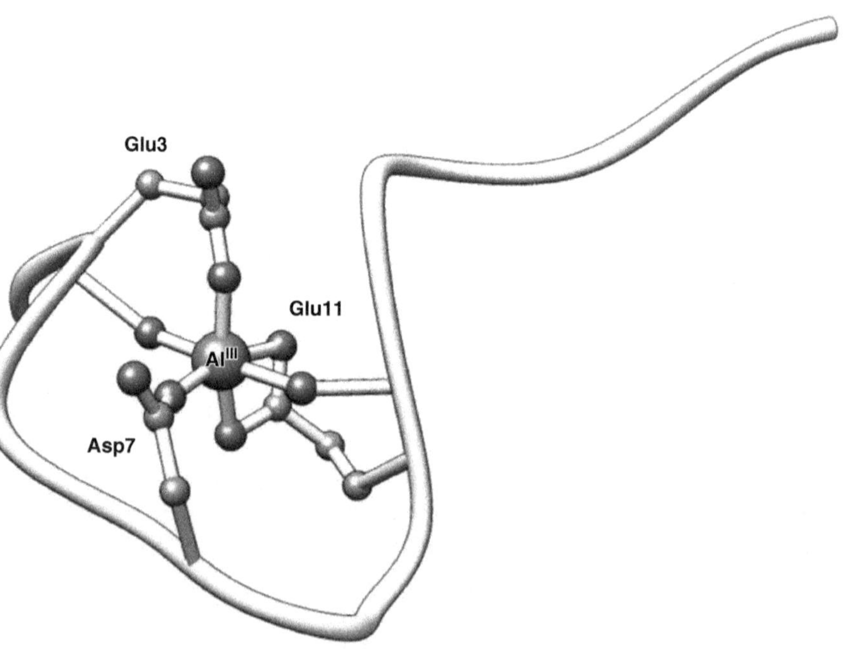


**Figure 2.9** Best predicted binding site for Al$^{III}$–Aβ-peptide showing the coordination by Asp7, Glu3, and Glu11 in an octahedral geometry.

## 2.4 Conclusions

The number of structural variables that need to be taken into account for the prediction by computational means of the interaction of metal ions in protein is large. Among the fundamental steps is the determination of the probable binding sites into the host proteins. Here, we show the main concepts that underlie the methodologies nowadays available in this area from the same prediction of the probable sites with little structural information to the refinement of different possible geometries. With the rapid expansion of structural biology and bioinformatics methods, it is foreseen that advances in the prediction of metal-binding sites in proteins will be extremely reliable in the very next years.

## References


1 Rehder, D. *Bioinorganic Chemistry*. Oxford University Press, Oxford; 2014.

2 (a) Bush, A.I. The metallobiology of Alzheimer's disease. *Trends Neurosci.* 2003, *26*, 207–214; (b) Faller, P., Hureau, C. Bioinorganic chemistry of copper and zinc ions coordinated to amyloid-β peptide. *Dalton Trans.* 2009;*7*:1080–1094; (c) Kozlowski, H., Janicka-Klos, A., Brasun, J., et al. Copper, iron, and zinc ions homeostasis and their role in neurodegenerative disorders (metal uptake, transport, distribution and regulation). *Coord. Chem. Rev.* 2009;*253*:2665–2685; (d) Bisaglia, M., Tessari, I., Mammi, S., et al. Interaction between α-synuclein and metal ions, still looking for a role in the pathogenesis of Parkinson's disease. *NeuroMol. Med.* 2009;*11*:239–251; (e) Carboni, E. and Lingor, P. Insights on the interaction of alpha-synuclein and metals in the pathophysiology of Parkinson's disease. *Metallomics* 2015;*7*:395–404; (f) Giachin, G., Mai, P.T., Tran, T.H., et al. The non-octarepeat copper binding site of the prion protein is a key regulator of prion conversion. *Sci. Rep.* 2015;*5*:15253; (g) Spagnolli, G., Rigoli, M., Orioli, S., et al. Full atomistic model of prion structure and conversion. *PLoS Pathog.* 2019;*15*:e1007864; (h) Borghesani, V., Alies, B., Hureau, C. $Cu^{II}$ binding to various forms of amyloid-β peptides: are they friends or foes? *Eur. J. Inorg. Chem.* 2018;*2018*:7–15.

3 (a) Mjos, K.D. and Orvig, C. Metallodrugs in medicinal inorganic chemistry. *Chem. Rev.* 2014;*114*:4540–4563; (b) Medici, S., Peana, M., Nurchi, V.M., et al. Noble metals in medicine: latest advances. *Coord. Chem. Rev.* 2015;*284*:329–350.

4 (a) Guo, Z. and Sadler, P.J. Metals in medicine. *Angew. Chem., Int. Ed.* 1999;*38*:1512–1531;(b) Farrell, N. *Comprehensive Coordination Chemistry II, Vol. 9* (ed. J.A. McCleverty and T.J. Meyer). Pergamon: Oxford; 2003, pp. 809–840; (c) *Metallotherapeutic Drugs and Metal-Based Diagnostic Agents. The Use of Metals in Medicine.* John Wiley & Sons Ltd, Chichester, 2005; (d) Thompson, K.H. *Encyclopedia of Inorganic and Bioinorganic Chemistry* (ed. R.B. King). John Wiley & Sons, Ltd, Chichester, 2011; (e) *Interrelations between Essential Metal Ions and Human Diseases.* Springer Science+Business Media, Dordrecht; 2013; (f) Jones, C. and Thornback, J. *Medicinal Applications of Coordination Chemistry.* The Royal

Society of Chemistry, Cambridge; 2007; (g) Dabrowiak, J.C. *Metals in Medicine*. John Wiley & Sons, Ltd, 2009; (h) *Bioinorganic Medicinal Chemistry*. Wiley-VCH Verlag GmbH & Co. KGaA, 2011; (i) Lewis, J.C. Artificial metalloenzymes and metallopeptide catalysts for organic synthesis. *ACS Catal.* 2013;*3*:2954–2975; (j) Yu, F., Cangelosi, V.M., Zastrow, M.L., et al. Protein design: toward functional metalloenzymes. *Chem. Rev.* 2014;*114*:3495–3578; (k) Bos, J., Roelfes, G. Artificial metalloenzymes for enantioselective catalysis. *Curr. Opin. Chem. Biol.* 2014; *19*:135–143; (l) Pàmies, O., Diéguez, M., Bäckvall, J.-E. Artificial metalloenzymes in asymmetric catalysis: key developments and future directions. *Adv. Synth. Catal.* 2015;*357*:1567–1586.

**5** Naylor, S. and Kumar, R. *Adv. Protein Chem., Vol. 65*. Amsterdam: Elsevier; 2003, pp. 217–248.

**6** (a) Møller, C., Sprenger, R.R., Stürup, S., et al. Determination of the binding sites for oxaliplatin on insulin using mass spectrometry-based approaches. *Anal. Bioanal. Chem.* 2011;*401*:1619–1629; (b) Wenzel, M. and Casini, A. Mass spectrometry as a powerful tool to study therapeutic metallodrugs speciation mechanisms: current frontiers and perspectives. *Coord. Chem. Rev.* 2017;*352*:432–460; (c) Shi, W. and Chance, M.R. Metalloproteomics: forward and reverse approaches in metalloprotein structural and functional characterization. *Curr. Opin. Chem. Biol.* 2011;*15*:144–148.

**7** Ward, J., Ollmann, E., Maxey, E., et al. *Metalloproteins*. Berlin: Springer; 2014, pp. 171–187.

**8** (a) Neese, F. Quantum chemical calculations of spectroscopic properties of metalloproteins and model compounds: EPR and Mössbauer properties. *Curr. Opin. Chem. Biol.* 2003;*7*:125–135; (b) Neese, F. A critical evaluation of DFT, including time-dependent DFT, applied to bioinorganic chemistry. *JBIC J. Biol. Inorg. Chem.* 2006;*11*:702–711; (c) Solomon, E.I., Scott, R.A., King, R.B. *Computational Inorganic and Bioinorganic Chemistry*. John Wiley & Sons, Ltd, Chichester; 2009; (d) Farid, R., Day, T., Friesner, R.A., et al. New insights about HERG blockade obtained from protein modeling, potential energy mapping, and docking studies. *Bioorg. Med. Chem.* 2006;*14*:3160–3173; (e) Rarey, M., Kramer, B., Lengauer, T., et al. A fast flexible docking method using an incremental construction algorithm. *J. Mol. Biol.* 1996;*261*:470–489; (f) Jain, A.N. Scoring functions for protein-ligand docking. *Curr. Protein Pept. Sci.* 2006;*7*:407–420.

**9** Gray, H.B., Valentine, J.S., Stiefel, E.I. *Biological inorganic Chemistry: Structure and Reactivity*. Sausalito: University Science Books; 2007.

**10** Akcapinar, G.B. and Sezerman, O.U. Computational approaches for de novo design and redesign of metal-binding sites on proteins. *Biosci. Rep.* 2017;*37*: BSR20160179.

**11** (a) Brylinski, M., Skolnick, J. FINDSITE-metal: integrating evolutionary information and machine learning for structure-based metal-binding site prediction at the proteome level. *Proteins* 2011;*79*:735–751; (b) Liu, T. and Altman, R.B. Prediction of calcium-binding sites by combining loop-modeling with machine learning. *BMC Struct. Biol.* 2009;*9*:72–72; (c) Passerini, C. Andreini, S. Menchetti, A. et al. Predicting zinc binding at the proteome level. *BMC Bioinform.* 2007;*8*:39; (d) Levy, R., Edelman, M., Sobolev, V. Prediction of 3D metal binding sites from translated gene sequences based on remote-homology templates. *Proteins Struct. Funct. Bioinf.* 2009;*76*:365–374; (e) Cai, C.Z., Han, L.Y., Ji, Z.L., et al. SVM-Prot: web-based support vector machine software for functional classification of a protein from its primary sequence. *Nucleic Acids Res.* 2003;*31*:3692–3697; (f) Haberal, İ. and Oğul, H. Prediction of protein metal binding sites using deep neural networks. *Mol. Inf.* 2019;*38*:1800169.

**12** Passerini, A., Lippi, M., Frasconi, P. MetalDetector v2.0: predicting the geometry of metal binding sites from protein sequence. *Nucleic Acids Res.* 2011;*39*: W288–W292.

**13** Valasatava, Y., Rosato, A., Banci, L., et al. MetalPredator: a web server to predict iron–sulfur cluster binding proteomes. *Bioinformatics* 2016;*32*:2850–2852.

**14** Levy, R., Edelman, M., Sobolev, V. Prediction of 3D metal binding sites from translated gene sequences based on remote-homology templates. *Proteins Struct. Funct. Bioinf.* 2009;*76*:365–374.

**15** (a) Lin, Y.-F., Cheng, C.-W., Shih, C.-S., et al. MIB: metal ion-binding site prediction and docking server. *J. Chem. Inf. Model.* 2016;*56*:2287–2291; (b) Lu, C.-H., Chen, C.-C., Yu, C.-S., et al. MIB2: metal ion-binding site prediction and modeling server. *Bioinformatics* 2022;*38*:4428–4429.

**16** Chen, C.-C., Hwang, J.-K., Yang, J.-M. $(PS)^2$-v2: template-based protein structure prediction server. *BMC Bioinform.* 2009;*10*:366.

**17** Varadi, M., Anyango, S., Deshpande, M., et al. AlphaFold protein structure Database: massively expanding the structural coverage of protein-sequence space with high-accuracy models. *Nucleic Acids Res.* 2021;*50*:D439–D444.

**18** Zhao, W., Xu, M., Liang, Z., et al. Structure-based *de novo* prediction of zinc-binding sites in proteins of unknown function. *Bioinformatics* 2011;*27*:1262–1268.

**19** He, W., Liang, Z., Teng, M., et al. mFASD: a structure-based algorithm for discriminating different types of metal-binding sites. *Bioinformatics* 2015;*31*:1938–1944.

**20** Ebert, J.C. and Altman, R.B. Robust recognition of zinc binding sites in proteins. *Protein Sci. Publ. Protein Soc.* 2008;*17*:54–65.

**21** Sodhi, J.S., Bryson, K., McGuffin, L.J., et al. Predicting metal-binding site residues in low-resolution structural models. *J. Mol. Biol.* 2004;*342*:307–320.

**22** Wass, M.N., Kelley, L.A., Sternberg, M.J.E. 3DLigandSite: predicting ligand-binding sites using similar structures. *Nucleic Acids Res.* 2010;*38*:W469–W473.

**23** Lu, C.-H., Lin, Y.-F., Lin, J.-J., et al. Prediction of metal ion–binding sites in proteins using the fragment transformation method. *PLoS One* 2012;*7*:e39252.

**24** Hu, X., Dong, Q., Yang, J., et al. Recognizing metal and acid radical ion-binding sites by integrating ab initio modeling with template-based transferals. *Bioinformatics* 2016;*32*:3260–3269.

**25** Dürr, S.L., Levy, A., Rothlisberger, U. Metal3D: a general deep learning framework for accurate metal ion location prediction in proteins. *Nat. Commun.* 2023;*14*: 2713.

**26** Abramson, J., Adler, J., Dunger, J., et al. Accurate structure prediction of biomolecular interactions with AlphaFold 3. *Nature* 2024;*630*:493–500.

**27** Fujieda, N., Schätti, J., Stuttfeld, E., et al. Enzyme repurposing of a hydrolase as an emergent peroxidase upon metal binding. *Chem. Sci.* 2015;*6*:4060–4065.

**28** Debret, G., Martel, A., Cuniasse, P. RASMOT-3D PRO: a 3D motif search webserver. *Nucleic Acids Res.* 2009;*37*:W459–W464.

**29** Amrein, B., Schmid, M., Collet, G., et al. Identification of two-histidines one-carboxylate binding motifs in proteins amenable to facial coordination to metals. *Metallomics* 2012;*4*:379–388.

**30** (a) Ugarova, N.N., Lebedeva, O.V., Berezin, I.V. Horseradish peroxidase catalysis I. Steady-state kinetics of peroxidase-catalyzed individual and co-oxidation of potassium ferrocyanide and *o*-dianisidine by hydrogen peroxide. *J. Mol. Catal.* 1981;*13*:215–225; (b) Kawamura-Konishi, Y., Asano, A., Yamazaki, M., et al. Peroxidase activity of an antibody–ferric porphyrin complex. *J. Mol. Catal. B Enzym.* 1998;*4*:181–190; (c) Ricoux, R., Dubuc, R., Dupont, C., et al. Hemozymes peroxidase activity of artificial hemoproteins constructed from the *Streptomyces lividans* xylanase A and iron(III)-carboxy-substituted porphyrins. *Bioconjugate Chem.* 2008;*19*:899–910.

**31** Muñoz Robles, V., Ortega-Carrasco, E., González Fuentes, E., et al. What can molecular modelling bring to the design of artificial inorganic cofactors? *Faraday Discuss.* 2011;*148*:137–159.

**32** Sciortino, G., Garribba, E., Rodríguez-Guerra Pedregal, J., et al. Simple coordination geometry descriptors allow to accurately predict metal-binding sites in proteins. *ACS Omega* 2019;*4*:3726–3731.

**33** Sánchez-Aparicio, J.-E., Tiessler-Sala, L., Velasco-Carneros, L., et al. BioMetAll: identifying metal-binding sites in proteins from backbone preorganization. *J. Chem. Inf. Model.* 2021;*61*:311–323.

**34** Andreini, C., Cavallaro, G., Lorenzini, S., et al. MetalPDB: a database of metal sites in biological macromolecular structures. *Nucleic Acids Res.* 2012;*41*: D312–D319.

**35** Bal, W., Sokołowska, M., Kurowska, E., et al. Binding of transition metal ions to albumin: sites, affinities and rates. *Biochim. Biophys. Acta Gen. Subj.* 2013;*1830*:5444–5455.

**36** (a) Harford, C. and Sarkar, B. Amino terminal Cu (II)-and Ni (II)-binding (ATCUN) motif of proteins and peptides: metal binding, DNA cleavage, and other properties. *Acc. Chem. Res.* 1997;*30*:123–130; (b) Bal, W., Christodoulou, J., Sadler, P.J., et al. Multi-metal binding site of serum albumin. *J. Inorg. Biochem.* 1998;*70*:33–39; (c) Stewart, A.J., Blindauer, C.A., Berezenko, S., et al. Interdomain zinc site on human albumin. *Proc. Natl. Acad. Sci. U. S. A.* 2003;*100*:3701–3706; (d) Sciortino, G., Sanna, D., Lubinu, G., et al. Unveiling $V^{IV}O^{2+}$ binding modes to human serum albumins by an integrated spectroscopic–computational approach. *Chem. Eur. J.* 2020;*26*:11316–11326.

**37** Kassaar, O., Schwarz-Linek, U., Blindauer, C.A., et al. Plasma free fatty acid levels influence $Zn^{2+}$-dependent histidine-rich glycoprotein–heparin interactions via an allosteric switch on serum albumin. *J. Thromb. Haemost.* 2015;*13*:101–110.

**38** (a) Bhattacharya, A.A., Grüne, T., Curry, S. Crystallographic analysis reveals common modes of binding of medium and long-chain fatty acids to human serum albumin. *J. Mol. Biol.* 2000;*303*:721–732; (b) Curry, S., Brick, P., Franks, N.P. Fatty acid binding to human serum albumin: new insights from crystallographic studies. *Biochim. Biophys. Acta Mol. Cell. Mol. Biol. Lipids* 1999;*1441*:131–140; (c) Lu, J., Stewart, A.J., Sadler, P.J., et al. Albumin as a zinc carrier: properties of its high-affinity zinc-binding site. *Biochem. Soc. Trans.* 2008;*36*:1317–1321;

(d) Blindauer, C.A., Harvey, I., Bunyan, K.E., et al. Structure, properties, and engineering of the major zinc binding site on human albumin. *J. Biol. Chem.* 2009;*284*:23116–23124; (e) Coverdale, J.P.C., Khazaipoul, S., Arya, S., et al. Crosstalk between zinc and free fatty acids in plasma. *Biochim. Biophys. Acta Mol. Cell. Mol. Biol. Lipids* 2019;*1864*:532–542.

**39** (a) Liu, J. and Wang, R. Classification of current scoring functions. *J. Chem. Inf. Model.* 2015;*55*:475–482; (b) Huang, S.-Y., Grinter, S.Z., Zou, X. Scoring functions and their evaluation methods for protein–ligand docking: recent advances and future directions. *Phys. Chem. Chem. Phys.* 2010;*12*:12899–12908.

**40** Jones, G., Willett, P., Glen, R.C. Molecular recognition of receptor sites using a genetic algorithm with a description of desolvation. *J. Mol. Biol.* 1995;*245*:43–53.

**41** (a) Warren, G.L., Andrews, C.W., Capelli, A.-M., et al. A critical assessment of docking programs and scoring functions. *J. Med. Chem.* 2006;*49*:5912–5931; (b) Xuan-Yu, M., Hong-Xing, Z., Mihaly, M., et al. Molecular docking: a powerful approach for structure-based drug discovery. *Curr. Comput. Aided Drug Des.* 2011;*7*:146–157; (c) Guedes, I.A., de Magalhães, C.S., Dardenne, L.E. Receptor–ligand molecular docking. *Biophys. Rev.* 2014;*6*:75–87; (d) Yuriev, E., Holien, J., Ramsland, P.A., Improvements, trends, and new ideas in molecular docking: 2012–2013 in review. *J. Mol. Recognit.* 2015;*28*:581–604; (e) Muñoz Robles, V., Ortega-Carrasco, E., Alonso-Cotchico, L., et al. Toward the computational design of artificial metalloenzymes: from protein–ligand docking to multiscale approaches. *ACS Catal.* 2015;*5*:2469–2480; (f) Akhter, M. Challenges in docking: mini review. *JSM Chem.* 2016;*4*:1025; (g) Agarwal, S. and Mehrotra, R. An overview of molecular docking. *JSM Chem.* 2016;*4*:1024; (h) Wang, G. and Zhu, W. Molecular docking for drug discovery and development: a widely used approach but far from perfect. *Future Med. Chem.* 2016;*8*:1707–1710.

**42** (a) Wang, Z., Sun, H., Yao, X., et al. Comprehensive evaluation of ten docking programs on a diverse set of protein–ligand complexes: the prediction accuracy of sampling power and scoring power. *Phys. Chem. Chem. Phys.* 2016;*18*:12964–12975; (b) Wang, Z., Kang, Y., Li, D., et al. Benchmark study based on $2P2I_{DB}$ to gain insights into the discovery of small-molecule PPI inhibitors. *J. Phys. Chem. B* 2018;*122*:2544–2555.

**43** Sousa, S.F., Ribeiro, A.J.M., Coimbra, J.T.S., et al. Protein-ligand docking in the new millennium–a retrospective of 10 years in the field. *Curr. Med. Chem.* 2013;*20*:2296–2314.

**44** Pagadala, N.S., Syed, K., Tuszynski, J. Software for molecular docking: a review. *Biophys. Rev.* 2017;*9*:91–102.

**45** Riccardi, L., Genna, V., De Vivo, M. Metal–ligand interactions in drug design. *Nat. Rev. Chem.* 2018;*2*:100–112.

**46** (a) Timerbaev, A.R., Hartinger, C.G., Aleksenko, S.S., et al. Interactions of antitumor metallodrugs with serum proteins: advances in characterization using modern analytical methodology. *Chem. Rev.* 2006;*106*:2224–2248; (b) Vincent, J.B., Love, S., The binding and transport of alternative metals by transferrin. *Biochim. Biophys. Acta, Gen. Subj.* 2012;*1820*:362–378; (c) Casini, A. and Reedijk, J., ChemInform Abstract: interactions of anticancer Pt compounds with proteins: an overlooked topic in medicinal inorganic chemistry? *Chem. Sci.* 2012;*3*:3135–3144;

(d) Sanna, D., Micera, G., Garribba, E. Interaction of insulin-enhancing vanadium compounds with human serum holo-transferrin. *Inorg. Chem.* 2013;*52*:11975–11985; (e) Costa Pessoa, J., Garribba, E., Santos, M.F.A., et al. Vanadium and proteins: uptake, transport, structure, activity and function. *Coord. Chem. Rev.* 2015;*301–302*:49–86.

**47** (a) Morris, G.M., Huey, R., Lindstrom, W., et al. AutoDock4 and AutoDockTools4: automated docking with selective receptor flexibility. *J. Comput. Chem.* 2009;*30*:2785–2791; (b) Glen, R.C. A fast empirical method for the calculation of molecular polarizability. *J. Comput.-Aided Mol. Des.* 1994;*8*:457–466; (c) Scholz, Knorr, S., Hamacher, K., et al. DOCKTITE–a highly versatile step-by-step workflow for covalent docking and virtual screening in the molecular operating environment. *J. Chem. Inf. Model.* 2015;*55*:398–406.

**48** Jain, N. Scoring functions for protein-ligand docking. *Curr. Protein Pept. Sci.* 2006;*7*:407–420.

**49** Scholz, Knorr, S., Hamacher, K., Schmidt, B. DOCKTITE—a highly versatile step-by-step workflow for covalent docking and virtual screening in the molecular operating environment. *J. Chem. Inf. Model.* 2015;*55*:398–406.

**50** (a) Sciortino, G., Garribba, E., Maréchal, J.-D. Validation and applications of protein–ligand docking approaches improved for metalloligands with multiple vacant sites. *Inorg. Chem.* 2018;*58*:294; (b) Sciortino, G., Sanna, D., Ugone, V., et al. Elucidation of binding site and chiral specificity of oxidovanadium drugs with lysozyme through theoretical calculations. *Inorg. Chem.* 2017;*56*:12938–12951; (c) Sciortino, G., Rodríguez-Guerra Pedregal, J., Lledós, A., et al. Prediction of the interaction of metallic moieties with proteins: an update for protein-ligand docking techniques. *J. Comput. Chem.* 2018;*39*:42–51.

**51** DeKoch, R.J., West, D.J., Cannon, J.C., et al. Kinetics and electron paramagnetic resonance spectra of vanadyl (IV) carboxypeptidase A. *Biochemistry* 1974;*13*:4347–4354.

**52** Sanna, D., Pecoraro, V.L., Micera, G., et al. Application of DFT methods to the study of the coordination environment of the $VO^{2+}$ ion in V proteins. *J. Biol. Inorg. Chem.* 2012;*17*:773–790.

**53** Bukrinsky, J.T., Bjerrum, M.J., Kadziola, A. Native carboxypeptidase A in a new crystal environment reveals a different conformation of the important tyrosine 248. *Biochemistry* 1998;*37*:16555–16564.

**54** Lovell, S.C., Word, J.M., Richardson, J.S., Richardson, D.C. The penultimate rotamer library. *Proteins Struct. Funct. Bioinf.* 2000;*40*:389–408.

**55** Rees, D.C., Lewis, M., Lipscomb, W.N. Refined crystal structure of carboxypeptidase A at 1.54 Å resolution. *J. Mol. Biol.* 1983;*168*:367–387.

**56** Smith II, T.S., Root, C.A., Kampf, J.W., et al. Reevaluation of the additivity relationship for vanadyl–imidazole complexes: correlation of the EPR hyperfine constant with ring orientation. *J. Am. Chem. Soc.* 2000;*122*:767–775; (b) Vilas Boas, L.F. and Costa Pessoa, J. In *Comprehensive Coordination Chemistry, Vol. 3* (ed. G. Wilkinson, R.D. Gillard, J.A. McCleverty). Oxford: Pergamon Press; 1985, pp. 453–583.

**57** Deb, K., Pratap, A., Agarwal, S., et al. A fast and elitist multiobjective genetic algorithm: NSGA-II. *IEEE Trans. Evolut. Comput.* 2002;*6*:182–197.

**58** Rodríguez-Guerra Pedregal, J., Sciortino, G., Guasp, J., et al. GaudiMM: a modular multi-objective platform for molecular modeling. *J. Comput. Chem.* 2017;*38*:2118–2126.

**59** Hohenberg, P. and Kohn, W. Density functional theory (DFT). *Phys. Rev.* 1964;*136*:B864–B871.

**60** (a) Sameera, W.M.C. and Maseras, F. Transition metal catalysis by density functional theory and density functional theory/molecular mechanics. *Wiley Interdiscip. Rev. Comput. Mol. Sci.* 2012;*2*:375–385; (b) Sperger, T., Sanhueza, I.A., Kalvet, et al. Computational studies of synthetically relevant homogeneous organometallic catalysis involving Ni, Pd, Ir, and Rh: an overview of commonly employed DFT methods and mechanistic insights. *Chem. Rev.* 2015;*115*:9532–9586; (c) Santoro, S., Kalek, M., Huang, G., et al. Elucidation of mechanisms and selectivities of metal-catalyzed reactions using quantum chemical methodology. *Acc. Chem. Res.* 2016;*49*:1006–1018; (d) Balcells, D., Clot, E., Eisenstein, O., et al. Deciphering selectivity in organic reactions: a multifaceted problem. *Acc. Chem. Res.* 2016;*49*:1070–1078; (e) Cramer, C.J. and Truhlar, D.G. Density functional theory for transition metals and transition metal chemistry. *Phys. Chem. Chem. Phys.* 2009;*11*:10757–10816; (f) Neese, F. Prediction of molecular properties and molecular spectroscopy with density functional theory: from fundamental theory to exchange-coupling. *Coord. Chem. Rev.* 2009;*253*:526–563; (g) Becke, A.D. Perspective: fifty years of density-functional theory in chemical physics. *J. Phys. Chem.* 2014;*140*:18A301.

**61** Siegbahn, P.E. and Himo, F. The quantum chemical cluster approach for modeling enzyme reactions. *Wiley Interdiscip. Rev. Comput. Mol. Sci.* 2011;*1*:323–336.

**62** (a) Senn, H.M. and Thiel, W. QM/MM methods for biomolecular systems. *Angew. Chem. Int. Ed.* 2009;*48*:1198–1229; (b) Chung, L.W., Sameera, W.M.C., Ramozzi, R., et al. The ONIOM method and its applications. *Chem. Rev.* 2015;*115*:5678–5796.

**63** Marx, D. and Hutter, J. *Ab Initio Molecular Dynamics: Basic Theory and Advanced Methods*. Cambridge University Press; 2009.

**64** Car, R. and Parrinello, M. Unified approach for molecular dynamics and density-functional theory. *Phys. Rev. Lett.* 1985;*55*:2471–2474.

**65** Sulpizi, M., Raugei, S., VandeVondele, J., et al. Calculation of redox properties: Understanding short-and long-range effects in rubredoxin. *J. Phys. Chem. B* 2007;*111*:3969–3976.

**66** (a) Warshel, A. and Levitt, M. Theoretical studies of enzymic reactions: dielectric, electrostatic and steric stabilization of the carbonium ion in the reaction of lysozyme. *J. Mol. Biol.* 1976;*103*:227–249; (b) Field, M.J., Bash, P.A., Karplus, M. A combined quantum mechanical and molecular mechanical potential for molecular dynamics simulations. *J. Comput. Chem.* 1990;*11*:700–733; (c) Singh, U.C. and Kollman, P.A. A combined ab initio quantum mechanical and molecular mechanical method for carrying out simulations on complex molecular systems: Applications to the $CH_3Cl + Cl^-$ exchange reaction and gas phase protonation of polyethers. *J. Comput. Chem.* 1986;*7*:718–730.

**67** (a) Vidossich, P., Lledós, A., Ujaque, G. First-principles molecular dynamics studies of organometallic complexes and homogeneous catalytic processes.

*Acc. Chem. Res.* 2016;*49*:1271–1278; (b) Li, P. and Merz, K.M. Metal ion modeling using classical mechanics. *Chem. Rev.* 2017;*117*:1564–1686; (c) Vidossich, P. and Magistrato, A. QM/MM molecular dynamics studies of metal binding proteins. *Biomolecules* 2014;*4*:616; (d) Vidossich, P., Lledós, A., Ujaque, G. *Computational Studies in Organometallic Chemistry*. Berlin: Springer; 2015, pp. 81–106.

**68** (a) Raha, K. and Merz, K.M. Large-scale validation of a quantum mechanics based scoring function: predicting the binding affinity and the binding mode of a diverse set of protein–ligand complexes. *J. Med. Chem.* 2005;*48*:4558–4575; (b) Fong, P., McNamara, J.P., Hillier, I.H., et al. Assessment of QM/MM scoring functions for molecular docking to HIV-1 protease. *J. Chem. Inf. Model.* 2009;*49*:913–924.

**69** (a) Raha, K. and Merz, K.M. A quantum mechanics-based scoring function: study of zinc ion-mediated ligand binding. *J. Am. Chem. Soc.* 2004;*126*:1020–1021; (b) Chaskar, P., Zoete, V., Röhrig, U.F. Toward on-the-fly quantum mechanical/molecular mechanical (QM/MM) docking: development and benchmark of a scoring function. *J. Chem. Inf. Model.* 2014;*54*:3137–3152; (c) Hayik, S.A., Dunbrack, R., Merz, K.M. Mixed quantum mechanics/molecular mechanics scoring function to predict protein– ligand binding affinity. *J. Chem. Theory Comput.* 2010;*6*:3079–3091.

**70** (a) Alí-Torres, J., Maréchal, J.-D., Rodríguez-Santiago, L., et al. Three dimensional models of $Cu^{2+}$-Aβ(1–16) complexes from computational approaches. *J. Am. Chem. Soc.* 2011;*133*:15008–15014; (b) Miki, Y., Pogni, R., Acebes, S., et al. Formation of a tyrosine adduct involved in lignin degradation by *Trametopsis cervina* lignin peroxidase: a novel peroxidase activation mechanism. *Biochem. J.* 2013;*452*:575–584; (c) Shi, R., Li, W., Liu, G., et al. Catalytic mechanism of cytochrome P450 2D6 for 4-hydroxylation of aripiprazole: a QM/MM study. *Chin. J. Chem.* 2013;*31*:1219–1227; (d) Piazzetta, P., Marino, T., Russo, N. Insight into the promiscuous activity of human carbonic anhydrase against the cyanic acid substrate from a combined QM and QM/MM investigation. *Phys. Chem. Chem. Phys.* 2014;*16*:16671–16676; (e) Drienovská, I., Alonso-Cotchico, L., Vidossich, P., et al. Design of an enantioselective artificial metallo-hydratase enzyme containing an unnatural metal-binding amino acid. *Chem. Sci.* 2017;*8*:7228–7235; (f) Hestericová, M., Heinisch, T., Alonso-Cotchico, L., et al. Directed evolution of an artificial imine reductase. *Angew. Chem. Int. Ed.* 2018;*57*:1863–1868; (g) Villarino, L., Splan, K.E., Reddem, E., et al. An artificial heme enzyme for cyclopropanation reactions. *Angew. Chem. Int. Ed.* 2018;*57*:7785–7789.

**71** Mujika, J.I., Rodríguez-Guerra Pedregal, J., Lopez, X., et al. Elucidating the 3D structures of Al (III)–Aβ complexes: a template free strategy based on the pre-organization hypothesis. *Chem. Sci.* 2017;*8*:5041–5049.

**72** Rezabal, E., Mercero, J.M., Lopez, X., et al. Protein side chains facilitate Mg/Al exchange in model protein binding sites. *Chem. Phys. Chem.* 2007;*8*:2119–2124.

# 3

# Quantum-mechanics Approaches in Bioinorganic Chemistry: Targeting the Oxidation State

*Marcel Swart*

*IQCC and Department of Chemistry, Universitat de Girona, Girona, Spain*
*ICREA, Barcelona, Spain*

## 3.1 Introduction

The Gold Book [1] of the International Union of Pure and Applied Chemistry (IUPAC) defines the oxidation state of an atom as "the charge of this atom after ionic approximation of its heteronuclear bonds. The underlying principle is that the ionic sign in an AB molecule is deduced from the electron allegiance in a linear combination of atomic orbitals (LCAO) – molecular orbitals (MOs) model: The bond's electrons are assigned to its main atomic contributor. Homonuclear AA bonds are divided equally." Although the definition is simple, and is accompanied by two general algorithms [1], the interpretation of the oxidation state of an atom within a compound remains subjective. An alternative approach to assign oxidation states was put forward by Salvador et al. [2], an extension of earlier work with Ramos-Cordoba et al. [3], which is based on quantum-chemical calculations (see *Quantum-chemistry approaches*) and includes a reliability estimate $R^s$ (which was chosen to go from 50 to 100%; a simple scaling can be used to transform it into the more natural 0–100% range: $R^c = 2 * R^s - 100$). Typical values for $R^s$ were observed in the 80–100% range ($R^c$: 60–100%), but also often $R^s$ values of 50–55% ($R^c$: 0–10%) were found. In particular, for mononuclear transition metal complexes, these reliability estimates were lower (e.g. $[PtO_4]^{2+}$, as studied by Truhlar et al. [4, 5], or $[Ni(S_2C_2Me_2)_2]^0$), and might be lower still for polynuclear transition metal systems and/or compounds with redox noninnocent ligands (*vide infra*). The $[Pt^{+10}(O^{2-})_4]^{2+}$ compound is a special case, as it was put forward by Truhlar et al. as a compound in a metastable state with the highest oxidation state reported so far (+10) [4, 5], larger than Pyykkö's $[Ir^{+9}(O^{2-})_4]^+$ with oxidation state +9 [6, 7], but this prediction of the +10 state has not yet been verified experimentally.

*Computational Bioinorganics: From Description to Prediction*, First Edition. Edited by Jean-Didier Maréchal.

A somewhat different approach to the oxidation states is normally used in coordination chemistry, where the formal oxidation state is defined as "the charge left on the metal after all ligands have been removed in their normal closed-shell configuration, i.e. with their electron pair" [8]. This formal oxidation state should then be contrasted with spectroscopy, from which the number of *d* electrons can often be estimated [9], leading to the "physical" oxidation state. The formal and physical oxidation states coincide for many coordination compounds, but this is not always the case. For instance, in case one of the ligands can form a stable organic (di)radical, there will be a competition between placing the radical on the metal or the ligand, and depending on the kind of metal present, and the number of *d* electrons, either a classical picture is obtained where the ligands act as an innocent bystander in a closed-shell configuration, or alternatively in a redox noninnocent radical partner with an open-shell configuration (*vide infra*); some examples of redox noninnocent ligands are shown in Figure 3.1, and examples of these in coordination compounds [10, 11] are described later on.

A special place in the set of (non-)innocent ligands is formed by Lewis acids like $Sc(OTf)_3$, which can bind to metal–oxygen [12–17] or metal–nitrogen [18] species, leading to, e.g. capped $[Fe^{III}(O)(TMC)Sc(OTf)_4(OH_2)]^0$ complexes. Originally [12], the oxidation state of iron in this crystal structure was assumed to remain in the +IV state. However, extensive computational modeling showed [19] that such an electronic structure would lead to drastically different bonds toward scandium and iron. The only viable spin/oxidation state was predicted to be high-spin Fe(III) [19], which was confirmed two years later by Münck et al. [20], Que et al. [21] in a series of papers. In the first one [20], the +III oxidation state was confirmed by a range of spectroscopic techniques, among which Mössbauer spectroscopy led to observed values for the isomer shift $\delta(0.36(3)\,\mathrm{mm\,s^{-1}})$ and quadrupole splitting $\Delta E_Q(-1.02(5)\,\mathrm{mm\,s^{-1}})$ that were in excellent agreement with the predicted [19] DFT parameters ($\delta\,0.39\,\mathrm{mm\,s^{-1}}$, $\Delta E_Q\ -0.99\,\mathrm{mm\,s^{-1}}$). In an interesting follow-up study [21], they used $Cr(OTf)_2$ to exploit the differences between $Cr^{II}$ and $Cr^{III}$, and showed that a $[(TMC)Fe^{III}$-O-$Cr^{III}(OTf)_4]$ could be formed either by $O_2$ activation or inner-sphere electron transfer. This was extended even further [22] when they obtained complexes with the oxo-$Cr(OTf)_4$ moiety bound either *syn* or *anti* with respect to the methyl groups of the TMC ligand.

The electronic structure of dinuclear complexes was very elegantly classified by Robin and Day [23] based on spectroscopy in combination with molecular orbital theory and group theory. They separated these mixed-valence complexes into three classes based on the mixing of wavefunctions in, e.g. resonance states. If this mixing is zero or low (either because of differences in symmetry in their coordination

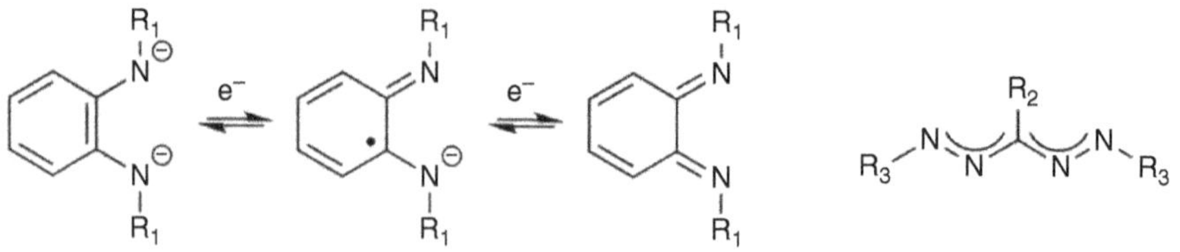

**Figure 3.1** Examples of noninnocent ligands.

spheres, or because of large separation in space between the two metals), the class I mixed-valence compounds are obtained where valences are firmly trapped, leading to clear spectra of individual constituent ions in IR/UV-Vis/Mössbauer spectroscopy. The other extreme is observed in class III when the mixing is maximal, leading to indistinguishable metals and complete delocalization, and absence of spectra of individual constituent ions. The third class, class II, falls in between these two extremes, in which the valences are distinguishable but with slight delocalization, and spectra of constituent ions that are still close to their normal frequencies/peaks.

## 3.2 Experimental Observations

Two of the most widely used experimental techniques [24, 25] for determining oxidation states are Mössbauer spectroscopy (limited to a small number of elements, including iron) and X-ray absorption spectroscopy (XAS). For the latter, in particular, the metal K-edge X-ray absorption near edge structure (XANES) – defined as the region within 50 eV of the absorption edge – was observed to show changes that for iron/manganese could be correlated to the oxidation state [26–30]. These spectra result from weak 1*s*–3*d* excitations, which are dipole-forbidden but gain intensity because of a quadrupole mechanism or mixing in of 4*p* orbitals. Directly correlated to this is the pre-edge intensity, which might give more information on the symmetry of the coordination environment around the metal: centrosymmetric environments have smaller 3*d*–4*p* mixing, and hence a lower intensity of the pre-edge. Many reports are available in the literature on the use of XANES for determining oxidation states [26–30], which was, in general, successful for the oxidation of iron (see Figure 3.2, left) [26, 27], but the changes were less pronounced for nickel (see Figure 3.2, right) [31].

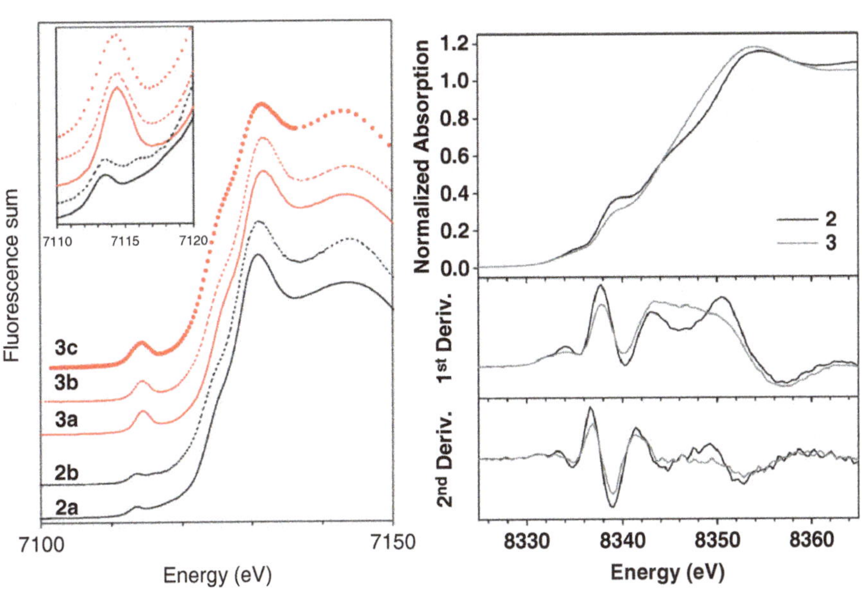


**Figure 3.2** XANES spectra for Fe (left) and Ni (right). *Source*: Reproduced from Rohde et al. (2004)/American Chemical Society and Pirovano et al. (2015)/John Wiley & Sons.

By a combination of XAS, X-ray emission spectroscopy (XES), and theory, DeBeer and Meyer [32] evaluated oxidation states of iron–sulfur clusters $[L_2Fe_2S_2]^n$, in three different forms (diferric $Fe^{III}$:$Fe^{III}$, mixed-valence $Fe^{II}$:$Fe^{III}$, diferrous $Fe^{II}$:$Fe^{II}$), and found that Fe K-edge rising edges do not necessarily shift upon reduction due to canceling effects of spin-state changes and covalency, which may lead to superimposable Fe K$\beta$ mainlines for iron–sulfur complexes in different redox states [32].

Recently, together with Browne and Tromp we studied [33] the treatment of tris-$\mu$-halido dinuclear nickel complexes $\left[(L)_2(Ni_2^{II}(\mu - X)_3\right]^{1+}$ (L = 1,4,7-trimethyl-1,4,7-triazacyclononane, TMTACN) by NaOCl/NaOBr, which led to the formation of $[(L)_2Ni_2(O)_3]^{2+}$ as shown by mass spectrometry (Figure 3.3). A combination of different spectroscopies was applied, including resonance Raman and XAS, which in combination with theory showed clearly that the assignment of the spectra belonged to tris-$\mu$-oxo structure **3**, and not oxo-peroxo structure **3a**, or the di-Ni(IV)-oxo structure **3b** (see Figure 3.3 and Table 3.1).

We also explored the use of XAS to determine the oxidation state of nickel, which in principle might help distinguish between the formal Ni(IV) oxidation state of **3** versus the formal Ni(III) oxidation state of **3a**. However, as already mentioned above, although this is normally working well for iron, in the case of nickel, it usually provides negligible changes in the XANES spectra; indeed, this was confirmed in this case as well.

Finally, for metalloporphyrins (such as the heme group [34, 35]), several reports [36–38] have shown resonance Raman spectra where specific bands were shifting depending on the formal oxidation state. Although this is a useful marker, it is limited to the porphyrin structure and analogs and not generally applicable.

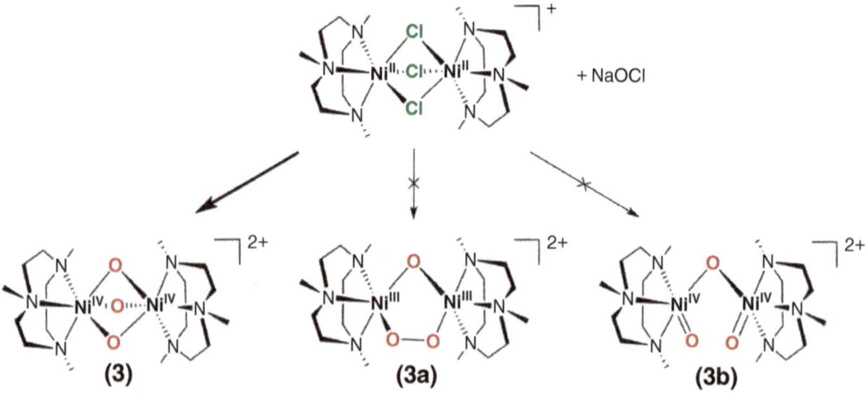


**Figure 3.3** Possible isomers after treatment of $[(L)_2Ni_2(X_3)]^{1+}$ with NaOCl or NaOBr.

**Table 3.1** Stability and spectral properties for di-nickel complexes **3**, **3a**, and **3b**.

| Exp. | 3 | 3a | 3b |
|---|---|---|---|
| | Relative energy (kcal $mol^{-1}$) | | |
| | 0.0 | 4.46 | 42.20 |
| | Vibrational frequencies ($cm^{-1}$) involving Ni (in brackets the [18] O isotope effect) | | |
| | 472 [−2] | 427 [−1] | 477 |
| | 497 [−19] | 498 [−1] | 487 |
| 521 [0] | 518 [−2] | 536 [−20] | |
| | 540 [−21] | 556 [−13] | |
| | 565 [−11] | 560 [−10] | |
| | | 585 [−15] | |
| | | 590 [−9] | |
| | 604 [−7] | 601 [−7] | |
| *631 [−29]* | 638 [−29] | 663 [−30] | 653 |
| | | | 700 |
| | | | 742 |
| *801* [0] | 773 [0] | 892 [−52] | |

## 3.3 Quantum-chemistry Approaches

There are two main branches of theoretical chemistry [39] that try to solve the Schrödinger equation $\widehat{H}\Psi = E\Psi$: (i) one focusing on improving the wavefunction $\Psi$ in wavefunction theory (WFT), (ii) the other focusing on improving the Hamiltonian $\widehat{H}$ in density functional theory (DFT). In the end, both should lead to the exact energy $E^{exact}$ for any given system made up of electrons and nuclei. However, the route to success is quite different between the two branches. In WFT, the road to the best possible wavefunction with any given basis set should lead to full configuration interaction (Full CI), which in all but trivial cases is in practice unreachable, and hence limited in scope. As Bright-Wilson wrote in 1968: [40] "A theory of chemistry that is limited to extremely small numbers of electrons is useful but, of course, does not scratch the surface of what chemistry really is and cannot even begin to present models of most of the phenomena of chemistry." In contrast, DFT is a very efficient method (which in principle depends on only three coordinates: $x$, $y$, and $z$), which can easily handle compounds of 100–150 atoms. Furthermore, the theory is exact and it was proven that it provides the exact energy [41, 42]; only the method/formulation for obtaining it is unknown [43], and approximations need to be made [44–48]. Therefore, in both cases extensive benchmarking is needed [49–52] in order to prove the worth of a given WFT/DFT method, for a number of systems for which the answers are known. Typically, these benchmarks have been biased toward organic chemistry, with less emphasis on (bio)inorganic chemistry. Only in the past decade or so has this been changed [53, 54], and performance for, e.g. spin states [55–61], has come into the spotlight. A dedicated COST Action was granted (ECOSTBio, 2014–2018), which included >100 research groups from >30 countries that collaborated on the study of transition metal-containing compounds using synthesis, catalysis, spectroscopy, and theory. It was, in particular, the combination of these different fields and backgrounds, which made ECOSTBio such a big success. Also, the first textbook on spin states appeared [60], which is in the process of being reprinted.

The use of either WFT or DFT on transition metal complexes has proven to be a slippery road, in particular when the influence of placing electrons in different metal *d* orbitals was studied. Early studies with DFT were discouraging [62], while recently, WFT has shown cracks [63–67] and ad hoc fixes need to be used (such as replacing the semicore correlation of Complete Active Space with Second Order Perturbation Theory (CASPT2) with its Coupled Cluster with Single, Double excitations and perturbative Triple excitations (CCSD(T)) counterpart [68, 69]). A more complete overview on the accuracy of WFT/DFT for transition metal complexes can be found in a recent overview [70], and in recent studies on the subject [63–73].

### 3.3.1 Chemical Bonding Analyses

Ever since the start of quantum mechanics, chemists have tried to extract useful information from orbitals/densities/wavefunctions, which might be leading to new chemical insights and classifications, such as the Robin–Day classes of mixed-valence complexes [23]. The most straightforward of doing so is by looking at the distribution of charge over the molecule [23]. Counting electrons might seem like a trivial task, but it is only so for counting the total number present in a compound: each carbon contributes six electrons (of which four are valence electrons), each hydrogen 1, etc., which together with the overall charge gives the total number *N*. Distributing these over the atoms is much more difficult, and cannot be done *a priori* (except in the case of homonuclear diatomic molecules) without the help of quantum chemistry. There are simple models, like the octet rule, which sometimes may help in determining the number of chemical bonds in simple systems. For instance, in the case of $CO_2$: 2 × 6 valence electrons from the oxygens, plus 4 valence electrons from the carbon gives a total of 16 electrons that are available; all three atoms prefer to be surrounded by eight electrons (octet rule) leading to a total demand of 24 electrons; hence, 8 electrons need to be shared in 4 chemical bonds. This simple rule works well in most cases for organic chemistry, but this works only when assuming the formal existence of these bonds, e.g. single, double, or triple bonds.

Over the years, several charge analyses have been proposed, among, which is the well-known Mulliken analysis [74] that takes the density matrix $D$ – depending on the coefficients $c_{ir}$ of atomic basis functions $rs$ in the molecular orbitals $ij$: $D_{rs} = \sum_{ij} c_{ir} \bullet c_{js}$ – and distributes the corresponding charge densities equally among the atoms to which the basis functions (and atomic orbitals) belong. There are two drawbacks related to this model, (i) the electronegativity difference between two atoms is not properly taken into account, and (ii) the results are highly dependent on the size of the basis set [75]. The latter was shown by Fonseca Guerra et al. [76], while the former has been addressed in a variety of models (among which Löwdin's analysis [77] or Bickelhaupt's Modified Mulliken model [78]). Other researchers have followed different strategies [79] through either intrinsic bond orbitals [80], through constraining multipoles [81, 82] or simply by empirical relationships (correlating, e.g. a Mulliken charge to a formal oxidation state of iron; *vide infra*).

In the end, all research groups aim for the same final goal: understanding how the electrons flow through the molecule and how this flow varies based on changes made in the molecule itself (substituent effects) or through its interaction with other molecules (intermolecular weak interactions, chemical reactions). In the past couple of years, through the Iboview program (Figure 3.4) [80] one can nowadays also visualize directly this flow of electrons along a reaction and see the orbitals (and bonds) forming and breaking [83, 84].

Decades of research into the quantification of chemical bonds have shown that putting a number to the strength of a bond is far from trivial (e.g. the uranium dimer [85, 86], the carbon–carbon bond in $C_2$ [87, 88], or the rotational profile [89, 90] for sandwich compounds $[Fe(Cp)_2]^0$ versus $[Ti(P_5)_2]^{2-}$) and is prone to interpretations that may lead to fierce discussions [91, 92]. The field of chemical bonding is too large and too diverse to be summarized here, but some reviews [93, 94] are provided, which may serve as a starting point for the interested reader. In our group, we have focused [56, 90, 95, 96] on understanding metal–ligand bonding through an energy decomposition analysis [93], where the total compound is divided into fragments, which then interact to form the full compound.

An interesting discussion was brought to the forefront by Lancaster et al. in 2016 [97], who showed for $[Cu(CF_3)_4]^{1-}$ spectroscopic evidence for a $3d^{10}$ ($Cu^{I}$) ground-state electronic configuration, instead of the classical $3d^8$ $Cu^{III}$ formal oxidation state. This conclusion by Lancaster et al. was based on UV-Vis/near-IR spectroscopy, Cu K-edge XAS, and 1*s*2*p* resonant inelastic X-ray scattering, in combination with DFT, multiplet

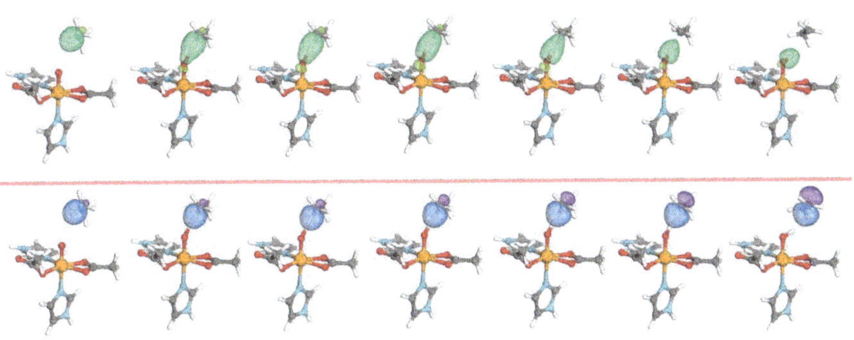

**Figure 3.4** Iboview picture of orbitals along C-H activation by a small TauD-J model complex. *Source*: Picture modified with permission from Klein and Knizia, *Angew. Chem. Int. Ed.* 2018, 57, 1191384.

theory, and multireference calculations. It followed a prediction by Snyder in 1995 based on theory [98], which showed that all five $d$ orbitals of the metal were fully occupied. The lowest unoccupied orbital is of predominantly trifluoromethyl character, which implies that all five metal-localized MOs are of lower energy relative to the trifluoromethyl-centered $\sigma$ orbitals, i.e. an inverted ligand-field (see Figure 3.5). It was reviewed by Hoffmann et al. [99], who confirmed the inverted ligand field, and they extended the discussion by including the dianion $[Cu(CF_3)_4]^{2-}$ and trianion $[Cu(CF_3)_4]^{3-}$, showing the gradual movement from a near-square planar (monoanion) to a pure tetrahedral (trianion) coordination sphere around the metal.

The inclusion of the dianion $[Cu(CF_3)_4]^{2-}$ might be able to help understand the electronic structure of the monoanion $[Cu(CF_3)_4]^{1-}$, since the oxidation of the former leads to the latter. However, ferrocene is a good example that does not necessarily mean that the occupations of the irreps remain the same during oxidation. i.e. ferrocene itself has $a_{1g}$ as the highest occupied orbital, yet after oxidation to ferrocenium, the unpaired electron is found in an $e_{2g}$ orbital [100]. The latter was shown clearly by spectroscopy, and had been used to argue for the $e_{2g}$ character of the highest occupied orbitals of ferrocene. Somewhat related to this is the spin state of a bare Fe(II) porphyrin, which was expected to have a triplet ground state based on empirical considerations [101], confirmed by DFT [102], but CASPT2 predicted a high-spin ground state. Only recently was it shown that it this is related to the limited size of the active space, i.e. by being able to correlate 34 electrons in 32 orbitals in stochastic CASSCF [65, 66], the preference for the triplet state was reinstated (and corroborated by subsequent coupled cluster calculations [66]).

Returning to the $[Cu(CF_3)_4]^{1-/2-/3-}$ systems, the question remains [99]: what is oxidized, the metal or the ligands? By comparing the NPA charges for the trianion with those of the monoanion [99], one could see that one electron was added on the copper (+0.74 monoanion, −0.27 trianion) and one electron went to the total of the four $CF_3$ ligands (−1.74 monoanion, −2.73 trianion). For the trianion, one would expect a classical formal oxidation state of $Cu^I$ and all four $CF_3$ ligands negatively charged, which would be consistent with the tetrahedral coordination sphere

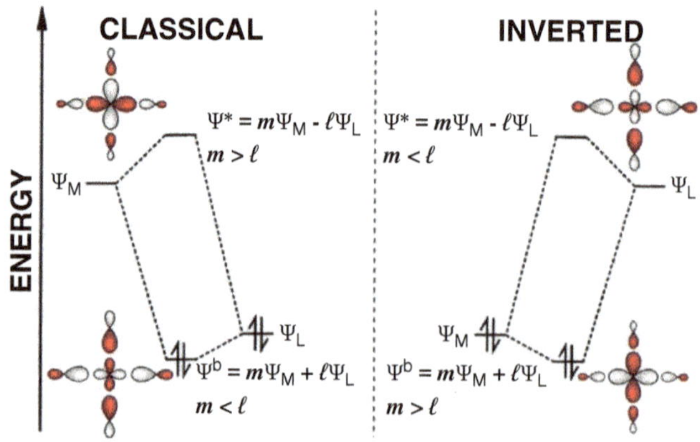


**Figure 3.5** Classical versus inverted ligand field. *Source*: Reproduced from Walroth et al. (2016)/American Chemical Society.

around copper in that case. However, by looking back to the monoanion, one would then expect a $Cu^{II}$ coupled to a $[(CF_3)_4]$-radical (since both had gained an electron). Neither theory nor spectroscopy shows this, so one cannot rely on charge analyses alone (*vide infra*). Arguments in favor of a $Cu^{III}$ formal oxidation state [99] would be the preference for a (near) square planar coordination sphere instead of tetrahedral, the expected lengthening of the Cu–C bonds. Probably, the $[Cu(CF_3)_4]^{1-}$ system should therefore be considered an interesting case study that exhibits the classical nonclassical characteristics of both a formal oxidation state of $Cu^{III}$ and a physical oxidation state of $Cu^{I}$. Without a doubt, follow-up studies will pursue that quest.

### 3.3.2 Oxidation States from Wavefunctions or Densities

Let's consider a simple example to show the effect of the different approaches for obtaining the oxidation state of a metal, i.e. $[Fe(O)]^q$ (see Figure 3.6) with the overall charge $q$ varying from −2 ("Fe(0)") to +3 ("Fe(V)"); for each of these $q$, we obtained the different spin states (low-spin LS, intermediate-spin IS, and high-spin HS). The geometries were obtained with S12g/TZ2P, including acetonitrile as solvent through the COSMO dielectric continuum model, and including scalar relativistic ZORA, and some data were obtained from NWChem with S12g/def2-tzvp/COSMO. The data in Table 3.2 show only for the most stable spin state per charge $q$, the corresponding Fe–O distance, and the expectation value of $\langle S^2 \rangle$. More importantly, also given are the atomic charges on Fe resulting from some charge analyses, the population of the $d$-shell according to the Mulliken analysis, and the Sit analysis [103] for the occupation of iron's $d$ orbitals.

In the classical view of the formal oxidation state of iron in this $[Fe(O)]^q$ system, it would go from 0 ($q = -2$) to + V ($q = 3$). In most cases, a high-spin state is observed as the ground state, and although there is some effect of the spin state on the values shown in Table 3.2, in general, the data are not drastically different between different spin states of one and the same overall charge $q$. If one looks at the values obtained for Fe from the different charge analyses, one rapidly observes that these are far from the formal oxidation state in all cases (as noted before [79]). For instance, for the $Fe^0$ FOS in $[Fe(O)]^{-2}$, the values range from −1.1 to −0.5, while for the $Fe^{IV}$ FOS in $[Fe(O)]^2$ values between +1.7 and +2.2 are observed. The relative change of 5 units (from 0 to +V) is only partially reflected in the charge analyses, which show changes between 3.0 and 3.6 units. Based on these trends, one might argue that, if at all, one should use the MDC-$d$ [82] charges (see Table 3.2).

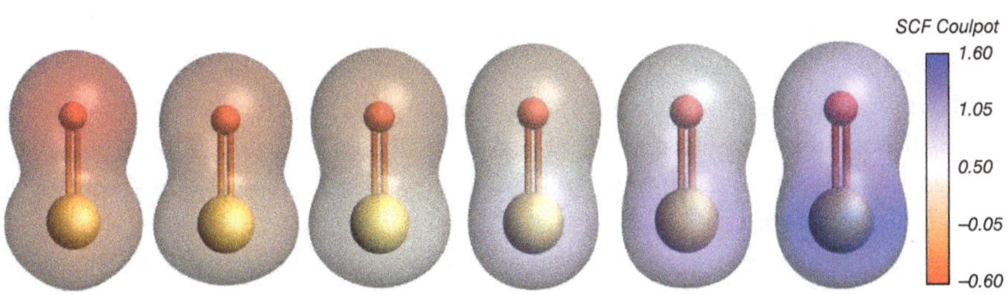


**Figure 3.6** $[Fe(O)]^q$ ($q = -2, -1, 0, 1, 2$).

**Table 3.2** Analysis of electronic structure of $[Fe(O)]^q$.

| | $^5[FeO]^{-2}$ | $^4[FeO]^{-1}$ | $^5[FeO]^0$ | $^6[FeO]^1$ | $^5[FeO]^2$ | $^4[FeO]^3$ | *Diff.*[a] |
|---|---|---|---|---|---|---|---|
| r(Fe–O) (Å) | 1.777 | 1.680 | 1.666 | 1.646 | 1.565 | 1.576 | *−0.199* |
| $\langle S^2 \rangle$ | 6.02 | 3.89 | 6.01 | 8.76 | 6.08 | 4.39 | |
| $OS(Fe)^{formal}$ | 0 | I | II | III | IV | V | |
| *Charge analyses* | | | | | | | |
| Mulliken[a] | −0.668 | 0.146 | 1.029 | 1.655 | 2.126 | 2.595 | *3.263* |
| Mulliken[b] | −0.774 | 0.060 | 0.934 | 1.545 | 2.013 | 2.463 | *3.237* |
| Hirshfeld[a] | −1.047 | −0.270 | 0.604 | 1.294 | 1.849 | 2.345 | *3.392* |
| Voronoi[a] | −0.853 | −0.132 | 0.668 | 1.282 | 1.746 | 2.171 | *3.024* |
| MDC-m[a] | −0.461 | 0.295 | 1.079 | 1.697 | 2.193 | 2.618 | *3.079* |
| MDC-d[a] | −1.097 | 0.066 | 1.053 | 1.680 | 2.113 | 2.533 | *3.630* |
| MDC-q[a] | −0.497 | 0.099 | 1.006 | 1.699 | 2.240 | 2.630 | *3.127* |
| *Population d shell* | | | | | | | |
| Mulliken[a] | 6.472 | 6.487 | 6.457 | 6.018 | 5.830 | 5.411 | *−1.061* |
| Mulliken[b] | 6.670 | 6.646 | 6.556 | 6.095 | 5.870 | 5.459 | *−1.211* |
| LCAOs[a] | 6.439 | 6.458 | 6.432 | 6.017 | 5.857 | 5.442 | *−0.997* |
| LCAOs[b] | 6.591 | 6.633 | 6.707 | 6.364 | 6.126 | 5.547 | *−1.044* |
| Sit[b] | II $(\alpha^5\beta^1)$ | III $(\alpha^4\beta^1)$ | II $(\alpha^5\beta^1)$ | III $(\alpha^5\beta^0)$ | IV $(\alpha^4\beta^0)$ | IV $(\alpha^4\beta^0)$ | |

a. Obtained with ADF.
b. Obtained with NW Chem.

Also, as indicated in Table 3.2 are the population analyses for the $d$ shell, for which different approaches have been used: directly from the Mulliken charge analysis, or by summing the weight of Fe atomic $d$ orbitals in the final molecular orbitals. Overall, the observed trend is similar, with only a small change in these populations along the series; i.e. (as an example) the Mulliken population of the $d$ shell remains constant for $q = -2$ to $q = 0$, and then gets smaller very slowly to a final value of 5.4 for $q = 3$ (corresponding to an overall change of −1.1 along the whole series from $q = -2$ to $q = 3$). Hence, the total population of the $d$ orbitals is even less reliable/useful as an indicator for the oxidation state of metals.

In our group, we have been using a slightly different approach, where we look not at the total population of the $d$ shell, but at the individual $d$ orbitals (see Table 3.3). Because of the issue of $\pi$ bonding, which is often accompanied by $\pi$ backbonding to some extent, the participation of $d$ orbitals does not just come from bonding situations. Therefore, it would be more meaningful to look at the occupation of metal $d$ orbitals beyond an arbitrary cutoff to indicate whether they are occupied or nonoccupied. A convenient cutoff (from years of experience) is 75%; anything below that number is

**Table 3.3** Population analysis (%) of individual Fe *d* orbitals.

| *q* | −2 | | −1 | | 0 | | 1 | | 2 | | 3 | |
|---|---|---|---|---|---|---|---|---|---|---|---|---|
| **Orbital** | $\alpha$ | $\beta$ | $\alpha$ | $\beta$ | $\alpha$ | $\beta$ | $\alpha$ | $\beta$ | $\alpha$ | $\beta$ | $\alpha$ | $\beta$ |
| $z^2$ | **94** | 33 | **94** | 30 | **85** | 21 | **92** | 29 | 58 | 32 | 67 | 26 |
| $x^2-y^2$ | **100** | **99** | **100** | 64 | **100** | **100** | **100** | 0 | **100** | 0 | **100** | 0 |
| $xy$ | **100** | 0 | **100** | 0 | **100** | 0 | **100** | 0 | **100** | 0 | **100** | 0 |
| $xz$ | **99** | 10 | **99** | 20 | **99** | 20 | **100** | 40 | **100** | 47 | **92** | 32 |
| $yz$ | **99** | 10 | **99** | 16 | **99** | 20 | **100** | 40 | **100** | 47 | **92** | 32 |
| | *5* | *1* | *5* | *0* | *5* | *1* | *5* | *0* | *4* | *0* | *4* | *0* |

unoccupied, above it occupied, which results for the $[Fe(O)]^q$ systems in an assignment of $Fe^{II}$ ($q = -2, 0$), $Fe^{III}$ ($q = -1, 1$) or $Fe^{IV}$ ($q = 2, 3$). Sit et al. generalized this [103] in a more satisfying manner by looking at the [5 × 5] projected occupation matrix $n$ of metal $d$-orbitals $\phi_p$ as a function of the occupied MOs $\psi_v^\sigma$ : $n_{pq}^\sigma = \sum_v \langle \psi_v^\sigma | \phi_p \rangle \langle \phi_q | \psi_v^\sigma \rangle$; through diagonalization of this matrix, and looking at the eigenvalues, one could assign a more trustworthy oxidation state to the metal. In this particular case of $[Fe(O)]^q$, with high symmetry, the results of our analysis and Sit's approach are identical. However, Sit's approach [103] will probably be more reliable, and less dependent on the arbitrariness of the cutoff parameter or orientation in space.

To finish off this section on how to obtain an estimate for the oxidation state from quantum-chemical data, a singular example has to be mentioned [104]. It involves a Mn(III)-complex that is active for the catalase reaction, i.e. it transforms two molecules of hydrogen peroxide into two molecules of water and an oxygen molecule. The reaction proceeds through two phases: in the first phase the Mn(III) binds $H_2O_2$, breaks the O—O bond to form a Mn(IV)-OH complex, and subsequently the free $OH^{\bullet}$ radical abstracts the OH-hydrogen to form water and a Mn(V) = O complex. Because of the spin-state of the different complexes (quintet at the start, then switching over to triplet once the Mn(IV)-OH is formed), the oxidation state of the metal is in sync with the spin-density MDC-$d$ charges that go down from −3.8 for Mn(III), to −2.8 for Mn(IV), to −2.0 for Mn(V) (the minus sign here because it involves electrons); note that no beta $d$-orbitals were observed to be occupied along this reaction profile. In the second phase, a second $H_2O_2$ molecule binds, which sequentially gives up its two Hs to become a triplet dioxygen with the Mn(V)-oxo complex gradually going towards a Mn(IV)-hydroxo and finally Mn(III)-aqua complex.

## 3.4 Classical Examples

In the past couple of years, we have collaborated with experimental labs on combining spectroscopy with electronic structure theory, in order to understand these spectra and give additional insights on reaction mechanisms. The first example where theory played a major role was for the $Ni^{II/III}$ complexes of McDonald [31], for which X-ray absorption spectroscopy showed ambiguous results. However, the computational chemistry study showed that upon oxidation of the $Ni^{II}$-complex to the presumably $Ni^{III}$-complex, the coordination sphere around the metal did not change much (as anticipated based on the XAS data), and a particular occupation of the $d_{xz}$-orbital (**o$_a$**) versus the $d_z^2$-orbital (**o$_b$**). These two occupations showed completely different spectroscopies, with e.g. the EPR g-values related to the $xz$-orbital (**o$_a$**) showing an axial EPR signal with $g_\perp < g_\parallel$; in contrast, occupying the $z^2$-orbital (**o$_b$**) showed $g_\parallel < g_\perp$ (similar to experiment: $g_\parallel$ 2.02, $g_\perp$ 2.25). In the case of the UV-Vis spectra (see Figure 3.7), quite different computed spectra were observed. Interestingly enough, the one with the $z^2$-orbital occupied (**o$_b$**) showed excellent agreement with experiment, while the one with $xz$-orbital occupied (**o$_a$**) did not.

Energetically, these two occupations were quite similar in stability, and only through the combination with spectroscopy were we able to pinpoint the electronic structure that was responsible for the experimental observations. Our $d$-orbital population analysis was consistent with the oxidation state of $Ni^{II}$, with 8 $d$-spin orbitals occupied >87% (the other two at 39%); for $Ni^{III}$, the situation was less straightforward. If the same cutoff of 75% (*vide supra*) was used, for both **o$_a$** and **o$_b$** it should be lowered to 70% to arrive at the expected $d^7$ oxidation state. The next highest $d$-spin orbital occupations were, however, 61%, so the choice for lowering it to 70% (instead of 75%) seems arbitrary. Sit's analysis probably would work better for these cases.

The situation was quite different for the di-$Ni^{III/IV}$ complex of Browne [33], where the most stable isomer **3** was preferred substantially over isomers **3a–3b** (see Figure 3.3); however, in this case, the most stable isomer was also clearly responsible for the experimentally observed spectra. It should be noted that *a priori* it was unclear which one of the two possible isomers would prevail, since there are precedents for either the tris-$\mu$-oxo form with manganese(IV) [105] (see Figure 3.8, left), while iron prefers to form the oxo-peroxo diiron(III) structure [14] (Figure 3.8, right). Interestingly enough, although it was mentioned at some conferences that nickel is the new iron, in fact it is not; instead, it is the new manganese (see Figure 3.8, middle). The occupation

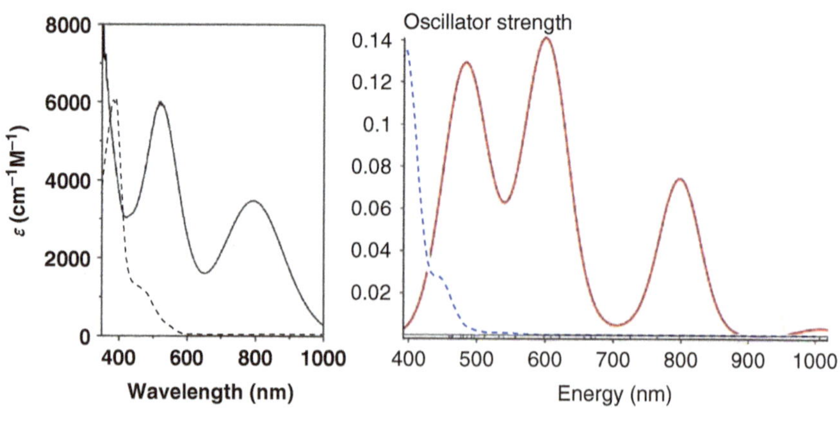


**Figure 3.7** UV-Vis spectra for $[Ni^{II/III}(OC(O)OH]$ from the experiment (left) and DFT (right, $Ni^{II}$ in blue, $Ni^{III}$ in red).

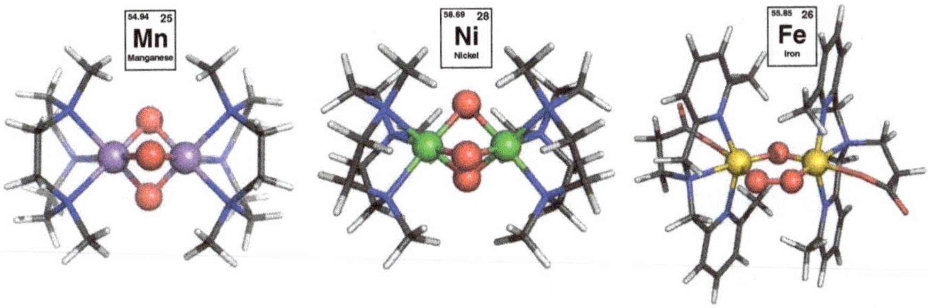


**Figure 3.8** Different possible isomers for $[(L)_2M_2(O_3)]^q$ complexes with Mn (left; *Source*: Adapted from [105]), Ni (middle; *Source*: Adapted from [33]), and Fe (right; *Source*: Adapted from [14]).

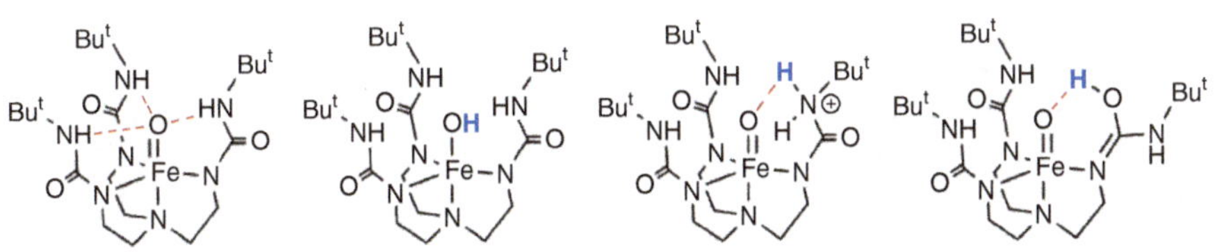

**Figure 3.9** $Fe^{III/IV}$-oxo (left) and "$Fe^{IV}$-hydroxo" complexes (a–c) [106–109] with the buea ligand.

of the *d* orbitals in these cases shows that **3** could indeed be described as $d^6$ ($Ni^{IV}$), with the other orbitals occupied, however, by approximately 60%, so a $Ni^{II}$ description would also be possible. Similar conclusions could be reached for **3a** and **3b**, where our current analysis based on *d* orbitals occupations is inconclusive.

Increasing the oxidation state in metal-oxo complexes, e.g. from Fe(III)-oxo to Fe(IV)-oxo, has a direct effect on the Fe–O vibration, which gets stronger, leading to a shorter Fe–O distance and a higher Fe–O vibrational frequency. This is extensively reported in the literature for a variety of metal-complexes, but one of the nice examples where the oxo-complex was reported for both Fe(III) and Fe(IV) (and other metals like Mn [106]) is posed by Borovik's buea-ligand [107, 108] (see Figure 3.9). This ligand contains urea units to model the secondary coordination sphere that stabilizes the oxo, and hence plays an important role in the stabilization of both the $Fe^{III}$ and $Fe^{IV}$ oxo complexes. The experimental observation of lowering of the Fe–O vibration from approximately 800 $cm^{-1}$ ($Fe^{IV}$) to approximately 670 $cm^{-1}$ ($Fe^{III}$), was confirmed [109] by the computations at PBE-$D_2$/TZ2P. However, adding a proton to the $Fe^{IV}$-oxo could lead to a multitude of different configurations, with as main possibilities [109] the anticipated $Fe^{IV}$–OH complex (**a**), and two alternatives where the proton is located on one of the nitrogens of the urea moiety (**b**), or on the oxygen of a rotated urea group (**c**; see Figure 3.9). Only Mössbauer spectra were available, and no X-ray structure; hence, theory was needed to distinguish between the different possibilities. Our study at S12g/TZ2P for spin states at PBE-D/TZ2P geometries showed conclusively that the proton was located on the rotated urea moiety (**c**), which was not only the most stable configuration, but also showed Mössbauer parameters that were in excellent agreement with experiment.

The *d* occupations again showed the expected behavior with complex **a** having a $d^5$-like ($Fe^{III}$) electronic structure, and complexes **b–c** a $d^4$ description where the $Fe^{IV}$ state remains.

## 3.5 Intriguing Examples

As described above, adding $Sc(OTF)_3$ to a Fe(IV)-oxo complex leads to a number of changes [110]: scandium picks up an additional triflate and a water as fifth ligand, and then caps the Fe–O moiety, where the Fe–O distance increases by approximately 0.10 Å (to 1.765 Å), the $^{Fe}$O-Sc distance is relatively short (1.93 Å), and the Sc–O($H_2$) is large (2.19 Å). These Sc–O distances are excellent markers [19] for the total charge on the $[Sc(OTf)_4(OH_x)]^{1-/2-}$ complex and, therefore, for the iron oxidation state. i.e. if one studies $[Sc(OTf)_4(OH_2)]^{1-}$ or $[Sc(OTf)_4(OH)]^{2-}$ on their own, the $^{Fe}$O-Sc distance for $[Sc(OTf)_4(OH_2)]^{1-}$ is 1.89 Å (similar to the values observed in the X-ray structure), while for $[Sc(OTf)_4(OH)]^{2-}$ it is 2.10 Å (inconsistent with X-ray). Note that the presence of the scandium moiety as a $[Sc(OTf)_4(OH_2)]^{1-}$ complex leads directly to a formal oxidation state of +III in the $[Fe^{III}(O)(TMC)Sc(OTf)_4(OH_2)]^0$ complexes [19]. A detailed analysis on the formation of the $Sc^{3+}$-capped iron–oxygen complex showed recently that the formation of the Fe(IV) species would be endothermic, while the Fe(III) analog would indeed be formed exothermically [110]. The charge analyses, *d*-shell population, and MDC-*d* charges confirm the notions observed for $[Fe(O)]^q$, in that the population of the *d* shell according to Mulliken's analysis is 6.05 ($Fe^{III}$) and 6.10 ($Fe^{IV}$), in both cases accompanied with a *d*-shell population of 1.01 on $Sc^{3+}$; Mulliken [MDC-*d*] charge of 1.3 [0.99] ($Fe^{III}$) and 1.28 [0.97] ($Fe^{IV}$); and hence are not indicative of the oxidation state. In contrast, inspecting the *d*-orbital occupations (similar to Table 3.3) shows direct occupations of $d^5$ ($Fe^{III}$), $d^4$ ($Fe^{IV}$), and in both cases, $d^0$ for $Sc^{3+}$ (as expected).

In collaboration with Garcia-Bosch, we investigated diamido ligands in conjunction with copper, which in principle would allow for seven oxidation states, going from $[Cu^I(L^{2-})_2]^{3-}$ (**O.S.1**) to $[Cu^{III}(L)_2]^{3+}$ (**O.S.7**) (see Figure 3.10). Through cyclic voltammetry, we were able to reach **O.S.5**, and found different preferences of the coordination sphere around the copper, depending on the oxidation and spin state [10]. The oxidation state of copper along the series **O.S.1–O.S.7** showed intriguing patterns, which were not always consistent with expectations. The end points were

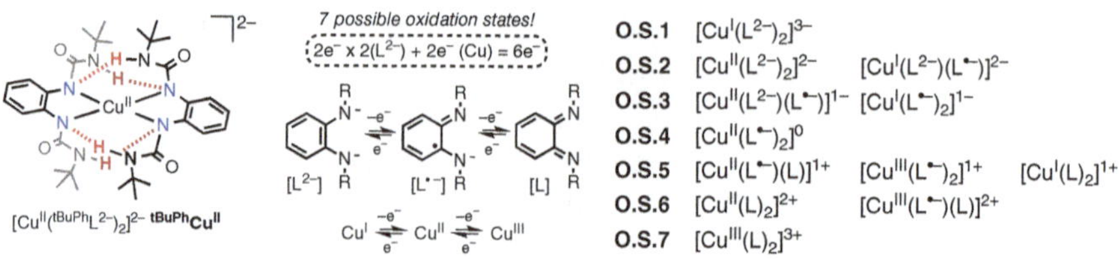


**Figure 3.10** Copper complexes with redox-active ligands and H-bonding interactions and possible description of oxidation states. *Source*: Adapted from [10].

clearly characterized as $Cu^{I}$ and $Cu^{III}$, based on the *d*-orbital occupations (similar to Table 3.3); the second and third were also clearly $Cu^{II}$, with a $d^9$ occupation.

However, the ones in the middle were mysterious, with **O.S.5** as the most intriguing one. One would expect to write it either as $Cu^{II}$ with one ligand radical, or as $Cu^{III}$ with both ligands in their radical form, leading in both cases to radicaloid species, and corresponding spin density. Nevertheless, our computational chemistry study showed no spin density at all (regardless of the density functional) [10], which will be followed up soon to clarify some aspects. The *d*-shell occupations seemed to indicate a $Cu^{I}$ description (consistent with the absence of spin density), but the cutoff used plays a major role in it being described as either $Cu^{I}$ or $Cu^{III}$, and hence, no final answer is available yet. Interestingly enough, it is reminiscent of the $[Cu(CF_3)_4]^{1-}$ system, which when extrapolated from $[Cu(CF_3)_4]^{3-}$, and removing two electrons (*vide supra*) one would expect a $Cu^{II}$ coupled to a $CF_3$-based radical [99]. Instead, a diamagnetic system was observed [97], with similar contradictory observations as found for our H-bonded copper system [10].

A range of oxidation states in one and the same complex was also reported recently by Meyer et al. [111] who showed a Fe-pincer complex to exist in four oxidation states $\left[Fe\left({}^{benzNHC}OCO\right)_2\right]^{2-/1-/0/1+}$. Although the ligand was noninnocent in the lower oxidation states, the cationic complex with a formal Fe(V) oxidation state was observed to be better described as Fe(IV) coupled antiferromagnetically to a ligand-centered radical. Their conclusions for the oxidation state were based on a combination of theory, magnetic studies (SQUID), EPR/Mössbauer/UV-Vis spectroscopy and X-ray diffraction; the theoretical part included both *ab initio* (CASSCF/NEVPT2) and DFT. Different computational methods [111] showed different descriptions of the electronic structure for $\left[Fe\left({}^{benzNHC}OCO\right)_2\right]^{1+}$, with Perdew's functionals (BP86, TPSSh) favoring a pure $Fe^{V}$ description, B3LYP a $d^4$ $Fe^{IV}$ one; both a $S = 1$ $Fe^{IV}$ coupled ferromagnetically to ligand radical, and $S = 2$ $Fe^{IV}$ coupled antiferromagnetically to ligand radical showed similar energies. This was confirmed by the state-averaged CASSCF, which indicated high multi-configuration character. However, the $S = 1$ $Fe^{IV}$ description showed Mössbauer quadrupole splitting values closer to the experiment than the $S = 2$ $Fe^{IV}$ ones.

The final case is, however, by far the most intriguing system encountered in recent years. In a series of papers [112–115], Costas and Que have focused on the formation of $Fe^{V}$-complexes based on Costas' PyNMe3 ligand and a peroxycarboxylic acid R—C(O)—OO (see Figure 3.11).

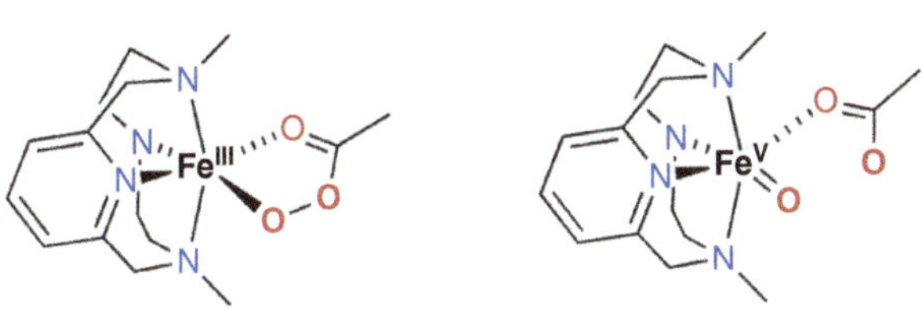


**Figure 3.11** Comparison of $Fe^{III}$/$Fe^{V}$-PyNMe3 complexes with peroxycarboxylic acid.

Breaking the **OO** bond of the peroxycarboxylic acid would lead to the $Fe^{V}$ species, and indeed, Costas et al. did observe evidence for its formation with mass spectrometry, EPR, Mössbauer spectroscopy, etc. [112–113]. However, the corresponding computational chemistry results showed a peculiar **OO** distance of approximately 2.1 Å, which would correspond to a transition state (TS) under normal circumstances. Moreover, depending on the choice of density functional they obtained either a minimum (zero imaginary frequencies) or a TS (one, *and only one*, imaginary frequency, corresponding to the breaking of the **OO** bond). Their DFT calculations were also not conclusive for distinguishing the $Fe^{III}/Fe^{V}$ complexes in terms of Mössbauer or EPR parameters. This led Ye et al. to investigate these systems with a multiconfiguration approach [116], and concluded that the electronic structure should be better described as a $Fe^{IV}$ with half an electron in the **OO** bond (i.e. corresponding to an oxidation state of 4.5). In a follow-up study [114], Costas et al. investigated the chemistry of the purported $Fe^{V}$-complex, and compared it to a genuine $Fe^{IV}$-oxo complex with the same ligand, and saw drastic differences between the "$Fe^{V}$"-complex and the pure $Fe^{IV}$-complex [114]. In our own computational study on these systems with S12g/TZ2P, we indeed found an electronic structure that based on *d*-shell occupations could be described as $d^3$ ($Fe^{V}$). However, the participation of the other *d* orbitals was certainly nonzero, with one being occupied by 60% (which could be considered to be consistent with Ye's results). The high-spin $Fe^{III}$ results (with an **OO** distance of 1.42 Å) was clearly a $d^5$ ($Fe^{III}$) description, but the S = 1/2 system (with a constrained **OO** distance of 1.42 Å) showed an electronic structure that is on the borderline and depending on choices made could be interpreted as either $d^3$, $d^4$, or $d^5$. Clearly, more investigation is needed, which is currently underway.

## 3.6 Summary and Outlook

Several computational chemistry analyses are available to interpret the oxidation state of (transition) metals in complexes, which, in simple cases, seem to work well. However, charge analyses such as Mulliken population analysis should be avoided at all costs, as they are unable to pinpoint directly what is the oxidation state. As shown here for a simple example, $[Fe(O)]^q$, these population analyses cover at best three out of the five electrons when going from $q = -2$ to $q = +3$; when looking at the *d*-shell population, it fares even worse and shows a maximum change of approximately one electron. Some interesting examples of noninnocent ligands have been reported in recent years, among which the $[Cu(CF_3)_4]^{1-}$ and our diamido-copper(II) complex, which showed contradictory observations, where based on comparison with a state where the classical picture holds (i.e. $[Cu(CF_3)_4]^{3-}$ or the **O.S.7** of our complex), one would expect to find radicals, while no spin density is observed in computational chemistry studies. Together with Costas' $Fe^{V}/Fe^{IV}/Fe^{4.5}$ complex, these systems show complex electronic structures that warrant further study and/or the development (cq. extension) of new computational analyses for gaining insights of how these exotic electronic structures affect their chemistry.

## Acknowledgments

MINECO (CTQ2014-59212-P, CTQ2017-87392-P, PID2020-114548GB-I00), FEDER (UNGI10-4E-801), and the COST Association (CM1305, ECOSTBio) are gratefully acknowledged for financial support, and Barcelona Supercomputing Center/Red Española de Supercomputación/Consorci de Serveis Universitaris de Catalunya for extensive computer time.

## References

1 IUPAC, Compendium of Chemical Terminology – the Gold Book (version 2.3.3). 2014. http://goldbook.iupac.org/index.html (accessed May 8, 2020).
2 Postils, V., Delgado-Alonso, C. Luis, J.M. et al. An objective alternative to IUPAC. *Angew. Chem. Int. Ed.* 2018;*57*:10525–10529.
3 Ramos-Cordoba, E., Postils, V., and Salvador, P. Oxidation states from wave function analysis. *J. Chem. Theory Comp.* 2015;*11*:1501–1508.
4 Yu, H.S. and Truhlar, D.G. Oxidation state 10 exists. *Angew. Chem. Int. Ed.* 2016;*55*:9004–9006.
5 Hu, S.-X., Li, W.-L., Lu, J.-B. et al. On the upper limits of oxidation states in chemistry. *Angew. Chem. Int. Ed.* 2018;*57*:3242–3245.
6 Pyykkö, P. and Xu, W.-H. On the extreme oxidation states of iridium. *Chem. Eur. J.* 2015;*21*:9468–9473.
7 Pyykkö, P. and Xu, W.-H. The formal oxidation states of iridium now run from *Angew. Chem. Int. Ed.* 2015;*54*:1080–1081.
8 Heims, F. and Ray, K. Multiple spin scenarios in transition-metal complexes involving redox non-innocent ligands. In *Spin States in Biochemistry and Inorganic Chemistry: Influence on Structure and Reactivity* (ed. M. Swart and M.Costas). Chichester, UK: Wiley; 2015, pp. 229–262.
9 Jørgensen, C.K. *Oxidation Numbers and Oxidation States.* Heidelberg: Springer; 1969.
10 Rajabimoghadam, K., Darwish, Y., Bashir, U. et al. Catalytic aerobic oxidation of alcohols by copper complexes bearing redox-active ligands with tunable H-bonding groups. *J. Am. Chem. Soc.* 2018, *submitted.*
11 Travieso-Puente, R., Broekman, J.O.P., Chang, M.-C. et al. Spin-crossover in a pseudo-tetrahedral bis(formazanate) iron complex. *J. Am. Chem. Soc.* 2016;*138*:5503–5506.
12 Fukuzumi, S., Morimoto, Y., Kotani, H. et al. Crystal structure of a metal ion-bound oxoiron(IV) complex and implications for biological electron transfer. *Nat. Chem.* 2010;*2*:756–759.
13 Li, F., Van Heuvelen, K.M., Meier, K.K. et al. $Sc^{3+}$-triggered oxoiron(IV) formation from $O_2$ and its non-heme iron(II) precursor via a $Sc^{3+}$–peroxo–$Fe^{3+}$ intermediate. *J. Am. Chem. Soc.* 2013;*135*:10198–10201.
14 Banerjee, S., Draksharapu, A., Crossland, P.M. et al. $Sc^{3+}$-promoted O–O bond cleavage of a (μ-1,2-peroxo)diiron(III) species formed from an iron(II) precursor

and $O_2$ to generate a complex with an $Fe^{IV}{}_2(\mu\text{-}O)_2$ core. *J. Am. Chem. Soc.* 2020;*142*:4285–4297.

15 Liu, Y. and Lau, T.-C. Activation of metal oxo and nitrido complexes by Lewis acids. *J. Am. Chem. Soc.* 2019;*141*:3755–3766.

16 Ray, K., Pfaff, F.F., Wang, B. et al. Status of reactive non-heme metal–oxygen intermediates in chemical and enzymatic reactions. *J. Am. Chem. Soc.* 2014;*136*:13942–13958.

17 Fukuzumi, S., Ohkubo, K., Lee, Y.-M. et al. Lewis acid coupled electron transfer of metal–oxygen intermediates. *Chem. Eur. J.* 2015;*21*:17548–17559.

18 Kundu, S., Miceli, E., Farquhar, E. et al. Lewis acid trapping of an elusive copper-tosylnitrene intermediate using scandium triflate. *J. Am. Chem. Soc.* 2012;*134*(36):14710–14713.

19 Swart, M. A change in oxidation state of iron: scandium is not innocent. *Chem. Commun.* 2013;*49*:6650–6652.

20 Prakash, J., Rohde, G.T., Meier, K.K. et al. Spectroscopic identification of an $Fe^{III}$ Center, not $Fe^{IV}$, in the crystalline Sc–O–Fe adduct derived from $[Fe^{IV}(O)(TMC)]^{2+}$. *J. Am. Chem. Soc.* 2015;*137*:3478–3481.

21 Zhou, A., Kleespies, S.T., Van Heuvelen, K.M. et al. Characterization of a heterobimetallic nonheme Fe(III)–O–Cr(III) species formed by $O_2$ activation. *Chem. Commun.* 2015;*51*:14326–14329.

22 Zhou, A., Prakash, J., Rohde, G.T. et al. The two faces of tetramethylcyclam in iron chemistry: distinct Fe–O–M complexes derived from $[Fe^{IV}(O_{anti/syn})(TMC)]^{2+}$ isomers. *Inorg. Chem.* 2017:*56*:518–527.

23 Robin, M.B. and Day, P. Mixed valence chemistry-a survey and classification. *Adv. Inorg. Chem. Rad.* 1968;*10*:247–422.

24 Duboc, C. and Gennari, M. Experimental techniques for determining spin states. In *Spin States in Biochemistry and Inorganic Chemistry: Influence on Structure and Reactivity*, (ed. M. Swarta and M. Costas.). Chichester, UK: Wiley; 2015, pp. 59–83.

25 Que Jr., L. *Physical Methods in Bioinorganic Chemistry: Spectroscopy and Magnetism*. Sausalito, CA: University Science Books; 2000.

26 Rohde, J.-U., Torelli, S., Shan, X. et al. Structural insights into nonheme alkylperoxoiron(III) and oxoiron(IV) intermediates by X-ray absorption spectroscopy. *J. Am. Chem. Soc.* 2004;*126*:16750–16761.

27 Berry, A.J., O'neill, H.S.C., Jayasuriya, K.D. et al. XANES calibrations for the oxidation state of iron in a silicate glass. *Am. Miner.* 2003;*88*:967–977.

28 Berry, A.J., Danyushevsky, L.V., O'Neill, H.S.C. et al. Oxidation state of iron in komatiitic melt inclusions indicates hot Archaean mantle. *Nature* 2008;*455*:960–964.

29 Wernet, P. Chemical interactions and dynamics with femtosecond X-ray spectroscopy and the role of X-ray free-electron lasers. *Philos. Trans. R.Soc. A* 2019;*377*: 20170464.

30 Shan, X., Rohde, J.-U., Koehntop, K.D. et al. X-ray absorption spectroscopic studies of high-spin nonheme (alkylperoxo)iron(III) intermediates. *Inorg. Chem.* 2007;*46*:8410–8417.

31 Pirovano, P., Farquhar, E., Swart, M. et al. Characterization and reactivity of a terminal nickel(III)-oxygen adduct. *Chem. Eur. J.* 2015;*21*:3785–3790.

**32** Kowalska, J.K., Hahn, A.W., Albers, A. et al. X-ray absorption and emission spectroscopic studies of $[L_2Fe_2S_2]^n$ model complexes: implications for the experimental evaluation of redox states in iron–sulfur clusters. *Inorg. Chem.* 2016;*55*:4485–4497.

**33** Padamati, S.K., Angelone, D., Draksharapu, A. et al. Transient formation and reactivity of a high-valent nickel(IV) oxido complex. *J. Am. Chem. Soc.* 2017;*139*:8718–8724.

**34** Bowman, S.E. and Bren, K.L. The chemistry and biochemistry of heme c: functional bases for covalent attachment. *Nat. Prod. Rep.* 2008;*25*(6):1118–1130.

**35** Bren, K.L. NMR spectroscopy of paramagnetic heme proteins. *Curr. Inorg. Chem.* 2012;*2*:273–291.

**36** Spiro, T.G. and Burke, J.M. Protein control of porphyrin conformation. Comparison of resonance Raman spectra of heme proteins with mesoporphyrin IX analogues. *J. Am. Chem. Soc.* 1976;*98*:5482–5489.

**37** Chottard, G., Sappacher, M., Ricard, L. et al. Resonance raman studies of models for the reduced states of cytochrome $P_{450}$. *Inorg. Chim. Acta* 1983;*79*:103–104.

**38** Rana, A., Amanullah, S., Das, P.K. et al. Formally ferric heme carbon monoxide adduct. *J. Am. Chem. Soc.* 2019;*141*:5073–5077.

**39** Jensen, F. *Introduction to Computational Chemistry*. New York: Wiley & Sons; 1998.

**40** Bright Wilson Jr., E. Some remarks on quantum chemistry. In *Structural Chemistry and Molecular Biology*, (ed. A. Rich and N. Davidson), San Francisco, USA: W.H. Freeman and Company; 1968, pp. 753–760.

**41** Hohenberg, P. and Kohn, W. Inhomogeneous electron gas. *Phys. Rev.* 1964;*136*:B864–B871.

**42** Handy, N.C. The importance of Colle–Salvetti for computational density functional theory. *Theor. Chem. Acc.* 2009;*123*:165–169.

**43** Cohen, A.J., Mori-Sánchez, P., and Yang, W. Insights into current limitations of density functional theory. *Science* 2008;*321*:792–794.

**44** Becke, A.D. Perspective: fifty years of density-functional theory in chemical physics. *J. Chem. Phys.* 2014;*140*: 18A301.

**45** Perdew, J.P., Ruzsinszky, A., Tao, J. et al. Prescription for the design and selection of density functional approximations: more constraint satisfaction with fewer fits. *J. Chem. Phys.* 2005;*123*(6): 062201.

**46** Parr, R.G. and Yang, W. *Density Functional Theory of Atoms and Molecules*. New York: Oxford University Press; 1989.

**47** Dreizler, R. and Gross, E. *Density Functional Theory*. New York: Plenum Press; 1995.

**48** Koch, W. and Holthausen, M.C. *A Chemist's Guide to Density Functional Theory*. Weinheim: Wiley-VCH; 2000.

**49** Goerigk, L. and Grimme, S. A thorough benchmark of density functional methods for general main group thermochemistry, kinetics, and noncovalent interactions. *Phys. Chem. Chem. Phys.* 2011;*13*:6670–6688.

**50** Yu, H.S., He, X., Li, S.L. et al. MN15: A Kohn–Sham global-hybrid exchange–correlation density functional with broad accuracy for multi-reference and single-reference systems and noncovalent interactions. *Chem. Sci.* 2016;*7*:5032–5051.

**51** Swart, M., Solà, M., and Bickelhaupt, F.M. Energy landscapes of nucleophilic substitution reactions: a comparison of density functional theory and coupled cluster methods. *J. Comput. Chem.* 2007;*28*(9):1551–1560.

**52** Swart, M., Solà, M., and Bickelhaupt, F.M. Density functional calculations of E2 and S(N)2 reactions: effects of the choice of method, algorithm, and numerical accuracy. *J. Chem. Theory Comp.* 2010;*6*(10):3145–3152.

**53** Weymuth, T., Couzijn, E.P.A., Chen, P. et al. New benchmark set of transition-metal coordination reactions for the assessment of density functionals. *J. Chem. Theory Comp.* 2014;*10*:3092–3103.

**54** Houghton, B.J. and Deeth, R.J. Spin-state energetics of FeII complexes – the continuing voyage through the density functional minefield. *Eur. J. Inorg. Chem.* 2014;*2014*:4573–4580.

**55** Swart, M., Groenhof, A.R., Ehlers, A.W. et al. Validation of exchange-correlation functionals for spin states of iron-complexes. *J. Phys. Chem. A* 2004;*108*:5479–5483.

**56** Swart, M. Accurate spin-state energies for iron complexes. *J. Chem. Theory Comp.* 2008;*4*:2057–2066.

**57** Swart, M., Solà, M., and Bickelhaupt, F.M. A new all-round DFT functional based on spin states and $S_N2$ barriers. *J. Chem. Phys.* 2009;*131*: 094103.

**58** Swart, M. A new family of hybrid density functionals. *Chem. Phys. Lett.* 2013;*580*:166–171.

**59** Swart, M. Spin states of (bio)inorganic systems: successes and pitfalls. *Int. J. Quantum Chem.* 2013;*113*:2–7.

**60** Swart, M. and Costas, M. *Spin States in Biochemistry and Inorganic Chemistry: Influence on Structure and Reactivity*. Oxford: Wiley; 2015.

**61** Swart, M. and Gruden, M. Spinning around in transition-metal chemistry. *Acc. Chem. Res.* 2016;*49*:2690–2697.

**62** Paulsen, H., Duelund, L., Winkler, H. et al. Free energy of spin-crossover complexes calculated with density functional methods. *Inorg. Chem.* 2001;*40*:2201–2203.

**63** Feldt, M., Phung, Q.M., Pierloot, K. et al. Limits of coupled-cluster calculations for non-heme iron complexes. *J. Chem. Theory Comp.* 2019;*15*:922–937.

**64** Flöser, B.M., Guo, Y., Riplinger, C. et al. Detailed pair natural orbital-based coupled cluster studies of spin crossover energetics. *J. Chem. Theory Comp.* 2020;*16*:2224–2235.

**65** Li Manni, G. and Alavi, A. Understanding the mechanism stabilizing intermediate spin states in Fe(II)-porphyrin. *J. Phys. Chem. A* 2018;*122*:4935–4947.

**66** Li Manni, G., Kats, D., Tew, D.P. et al. Role of valence and semicore electron correlation on spin gaps in Fe(II)-porphyrins. *J. Chem. Theory Comp.* 2019;*15*:1492–1497.

**67** Zobel, J.P., Nogueira, J.J., and González, L. The IPEA dilemma in CASPT2. *Chem. Sci.* 2017;*8*:1482–1499.

**68** Phung, Q.M., Feldt, M., Harvey, J.N. et al. Toward highly accurate spin state energetics in first-row transition metal complexes: a combined CASPT2/CC approach. *J. Chem. Theory Comp.* 2018;*14*:2446–2455.

**69** Phung, Q.M., Martín-Fernańdez, C., Harvey, J.N. et al. Ab initio calculations for spin-gaps of non-heme iron complexes. *J. Chem. Theory Comp.* 2019;*15*: 4297–4304.

**70** Swart, M. Dealing with spin states in computational organometallic catalysis. *Top. Organomet. Chem.* 2020, *accepted*.

**71** Johansson, M.P., Kaila, V.R.I., and Sundholm, D. Ab initio, density functional theory, and semi-empirical calculations. In *Biomolecular simulations*, (ed. L. Monticelli and E. Salonen). Totowa, NJ: Humana Press; 2011.

**72** Cramer, C.J. and Truhlar, D.G. Density functional theory for transition metals and transition metal chemistry. *Phys. Chem. Chem. Phys.* 2009;*11*:10757–10816.

**73** Gruden, M., Stepanovic, S., and Swart, M. Spin state relaxation of iron complexes: the case for OPBE and S12g. *J. Serb. Chem. Soc.* 2015;*80*:1399–1410.

**74** Mulliken, R.S. Electronic population analysis on LCAO–MO molecular wave functions II. Overlap populations, bond orders, and covalent bond energies. *J. Chem. Phys.* 1955;*23*:1841–1846.

**75** Lehtola, S. and Jońsson, H. Pipek–Mezey orbital localization using various partial charge estimates. *J. Chem. Theory Comp.* 2014;*10*:642–649.

**76** Fonseca Guerra, C., Handgraaf, J.-W., Baerends, E.J. et al. Voronoi deformation density (VDD) charges. Assessment of the Mulliken, Bader, Hirshfeld, Weinhold and VDD methods for charge analysis. *J. Comput. Chem.* 2004;*25*:189–210.

**77** Cho, M., Sylvetsky, N., Eshafi, S. et al. The atomic partial charges arboretum: trying to see the forest for the trees. *ChemPhysChem* 2020;*21*:688–696.

**78** Bickelhaupt, F.M., Hommes, N., Guerra, C.F. et al. The carbon-lithium electron pair bond in (CH3Li)(n) (n = 1, 2, 4). *Organometallics* 1996;*15*(13):2923–2931.

**79** Aullón, G. and Alvarez, S. Oxidation states, atomic charges and orbital populations in transition metal complexes. *Theor. Chem. Acc.* 2009;*123*:67–73.

**80** Knizia, G. Intrinsic atomic orbitals: an unbiased bridge between quantum theory and chemical concepts. *J. Chem. Theory Comp.* 2013;*9*:4834–4843.

**81** Thole, B.T. and van Duijnen, P.T. A general population analysis preserving the dipole moment. *Theor. Chim. Acta* 1983;*63*:209–221.

**82** Swart, M., van Duijnen, P.T., and Snijders, J.G. A charge analysis derived from an atomic multipole expansion. *J. Comput. Chem.* 2001;*22* (1):79–88.

**83** Knizia, G. and Klein, J.E.M.N. Electron flow in reaction mechanisms-revealed from first principles. *Angew. Chem. Int. Ed.* 2015;*54*:5518–5522.

**84** Klein, J.E.M.N. and Knizia, G. cPCET versus HAT: a direct theoretical method for distinguishing X–H bond-activation mechanisms. *Angew. Chem. Int. Ed.* 2018;*57*:11913–11917.

**85** Roos, B.O. and Gagliardi, L. Quantum chemistry predicts multiply bonded diuranium compounds to be stable. *Inorg. Chem.* 2006;*45*:803–807.

**86** Gagliardi, L. and Roos, B.O. Quantum chemical calculations show that the uranium molecule $U_2$ has a quintuple bond. *Nature* 2005;*433*:848–851.

**87** Shaik, S., Danovich, D., Wu, W. et al. Quadruple bonding in $C_2$ and analogous eight-valence electron species. *Nat. Chem.* 2012;*4*:195–200.

**88** Shaik, S., Rzepa, H.S., and Hoffmann, R. One molecule, two atoms, three views, four bonds? *Angew. Chem. Int. Ed.* 2012;*52*:3020–3033.

**89** Urnežius, E., Brennessel, W.W., Cramer, C.J. et al. A carbon-free sandwich complex $[(P_5)_2Ti]^{2-}$. *Science.* 2002;*295*(5556):832–834.

**90** Vlahovic, F., Gruden, M., and Swart, M. Rotating iron and titanium sandwich complexes. *Chem. Eur. J.* 2018;*24*:5070–5073.

**91** Bickelhaupt, F.M. and Baerends, E.J. The case for steric repulsion causing the staggered conformation of ethane. *Angew. Chem.* 2003;*115*:4315–4320.

**92** Poater, J., Solà, M., and Bickelhaupt, F.M. Hydrogen–hydrogen bonding in planar biphenyl, predicted by atoms-in-molecules theory, does not exist. *Chem. Eur. J.* 2006;*12*:2889–2895.

**93** Bickelhaupt, F.M. and Baerends, E.J. Kohn-Sham density functional theory: predicting and understanding chemistry. In *Reviews in Computational Chemistry, Vol 15.* New York: Wiley-VCH; 2000, Vol. 15, pp. 1–86.

**94** Bickelhaupt, F.M. and Houk, K.N. Analyzing reaction rates with the distortion/interaction-activation strain model. *Angew. Chem. Int. Ed.* 2017;*56*:10070–10086.

**95** Swart, M. Metal-ligand bonding in metallocenes: differentiation between spin state, electrostatic and covalent bonding. *Inorg. Chim. Acta* 2007;*360*(1):179–189.

**96** Castro, A.C., Johansson, M.P., Merino, G. et al. Chemical bonding in supermolecular flowers. *Phys. Chem. Chem. Phys.* 2012;*14*:14905–14910.

**97** Walroth, R.C., Lukens, J.T., MacMillan, S.N. et al. Spectroscopic evidence for a $3d^{10}$ ground state electronic configuration and ligand field inversion in $[Cu(CF_3)_4]^{1-}$. *J. Am. Chem. Soc.* 2016;*138*:1922–1931.

**98** Snyder, J.P. Elusiveness of $Cu^{III}$ complexation; preference for trifluoromethyl oxidation in the formation of $[Cu^I(CF_3)_4]^-$ salts. *Angew. Chem. Int. Ed.* 1995;*34*:80–81.

**99** Hoffmann, R., Alvarez, S., Mealli, C. et al. From widely accepted concepts in coordination chemistry to inverted ligand fields. *Chem. Rev.* 2016;*116*:8173–8192.

**100** Elschenbroich, C. *Organometallics (Third edition).* Weinheim: Wiley-VCH; 2005.

**101** Scheidt, W.R. and Reed, C.A. Spin-state/stereochemical relationships in iron porphyrins: implications for the hemoproteins. *Chem. Rev.* 1981:*81*:543–555.

**102** Groenhof, A.R., Swart, M., Ehlers, A.W. et al. Electronic ground states of iron porphyrin and of the first species in the catalytic reaction cycle of cytochrome P450s. *J. Phys. Chem. A* 2005;*109*(15):3411–3417.

**103** Sit, P.H.-L., Car, R., Cohen, M.H. et al. Simple, unambiguous theoretical approach to oxidation state determination via first-principles calculations. *Inorg. Chem.* 2011;*50*:10259–10267.

**104** Romero-Rivera, A. and Swart, M. Study on the catalase activity of a Mn(III)-salen complex. 2020, *in preparation.*

**105** Wieghardt, K., Bossek, U., Nuber, B. et al. Synthesis, crystal structures, reactivity, and magnetochemistry of a series of binuclear complexes of manganese(11), –(111), and –(IV) of biological relevance. The crystal structure of $[L'Mn^{IV}(\mathit{mu}\text{-}O)_3Mn^{IV}L'](PF_6)_2{\cdot}H_2O$ containing an unprecedented short Mn-Mn distance of 2.296 A. *J. Am. Chem. Soc.* 1988;*110*:7398–7411.

**106** MacBeth, C.E., Gupta, R., Mitchell-Koch, K.R. et al. Utilization of hydrogen bonds to stabilize M-O(H) units: synthesis and properties of monomeric iron and manganese complexes with terminal oxo and hydroxo ligands. *J. Am. Chem. Soc.* 2004;*126*(8):2556–2567.

**107** Lacy, D.C., Gupta, R., Stone, K.L. et al. Formation, structure, and EPR detection of a high spin FeIVsOxo species derived from either an FeIIIsOxo or FeIIIsOH complex. *J. Am. Chem. Soc.* 2010;*132*:12188–12190.

**108** MacBeth, C.E., Golombek, A.P., Young Jr., V.G. et al. O2 Activation by nonheme iron complexes: a monomeric Fe(III)-Oxo complex derived from O2. *Science* 2000;*289*:938–941.

**109** Hill, E.A., Weitz, A.C., Onderko, E. et al. Reactivity of an $Fe^{IV}$-Oxo complex with protons and oxidants. *J. Am. Chem. Soc.* 2016;*138*:13143–13146.

**110** Vlahovic, F.Z., Gruden, M., Stepanovic, S. et al. Density functional approximations for consistent spin and oxidation states of oxoiron complexes. *Int. J. Quantum Chem.* 2020;*120*:e26121.

**111** Gravogl, L., Heinemann, F.W., Munz, D. et al. An iron pincer complex in four oxidation states. *Inorg. Chem.* 2020;*59*:5632–5645.

**112** Serrano-Plana, J., Oloo, W.N., Acosta-Rueda, L. et al. Trapping a highly reactive nonheme iron intermediate that oxygenates strong C–H bonds with stereoretention. *J. Am. Chem. Soc.* 2015;*137*(50):15833–15842.

**113** Fan, R., Serrano-Plana, J., Oloo, W.N. et al. Spectroscopic and DFT characterization of a highly reactive nonheme Fe(V)-Oxo intermediate. *J. Am. Chem. Soc.* 2018;*140*(11):3916–3928.

**114** Dantignana, V., Serrano-Plana, J., Draksharapu, A. et al. Spectroscopic and reactivity comparisons between nonheme oxoiron(IV) and oxoiron(V) species bearing the same ancillary ligand. *J. Am. Chem. Soc.* 2019;*141*:15078–15091.

**115** Kal, S., Xu, S., and Que Jr., L. Bio-inspired nonheme iron oxidation catalysis. Growing evidence for the involvement of oxoiron(V) oxidants in cleaving strong C–H bonds. *Angew. Chem. Int. Ed.* 2020;*59*:7332–7349.

**116** Mondal, B., Neese, F., Bill, E. et al. Electronic structure contributions of non-heme oxo-iron(V) complexes to the reactivity. *J Am Chem Soc* 2018;*140*(30):9531–9544.

4

# Computation and Spectroscopy

*Eugenio Garribba*

*Dipartimento di Medicina, Chirurgia e Farmacia, Università di Sassari, Sassari, Italy*

## 4.1 Introduction

From the birth of chemistry as a scientific discipline, the dream of chemists was to determine the structure of compounds and have some tools to predict the arrangement of the atoms in ions, molecules, and complexes.

In this context, spectroscopy plays a key role, and it is now applied in chemistry, physics, biology, and medicine. The spectroscopic techniques are the only physicochemical tools available to characterize a compound and study its properties when the resolution of the structure through X-ray diffraction analysis is not possible. Often, the combined use of different spectroscopic methods (for example, vibrational, electronic, and resonance spectroscopy) provides valuable information on the molecular systems. In bioinorganic chemistry, this means reaching information on the ligands and donors interacting with the metal in simple coordination compounds and/or regions where the metal species binds in more complicated systems such as proteins, enzymes, or DNA.

Unfortunately, interpreting a metal compound's spectroscopic response is often difficult since it depends on several factors, such as the specific donors bound to the metal, the geometry, symmetry, and size of the species formed, the oxidation number and metal spin state.

During the past decades, many computational methods have been developed to predict the structure, energy, molecular properties, and – what will be discussed here – the spectroscopic behavior of the chemical species. In particular, the computation can be used at different levels: (i) to support the spectroscopic data when the calculated spectrum for a given species overlaps with the experimental one; (ii) to predict the spectroscopic response for a new chemical compound; (iii) to get information on the origin of the transitions observed and on the metal and ligand role.

In this chapter, the most used spectroscopic techniques will be examined: electron paramagnetic resonance (EPR), electron spin echo envelope modulation (ESEEM), electron-nuclear double resonance (ENDOR), nuclear magnetic

*Computational Bioinorganics: From Description to Prediction*, First Edition. Edited by Jean-Didier Maréchal.

resonance (NMR), UV-Vis spectroscopy, circular dichroism (CD), magnetic circular dichroism (MCD), infrared (IR), Raman, and, finally, Mössbauer spectroscopy. For each of them, the background, the computational approach, and some case studies will be discussed, giving particular emphasis to the data in the literature on bioinorganic systems. The essential elements of the computational methods available for each spectroscopy will also be presented.

## 4.2 Software, Computational Techniques, and Statistical Indicators

Commercially or free of charge, several packages are available for computation on a chemical compound, allowing one to treat small or large molecules. Among them, Gaussian [1], ORCA [2], Gamess [3], AMS [4], Spartan [5], Turbomole [6], and Jaguar [7] can be cited. They allow the users to perform several types of calculations. Often, the programs are or can be combined with a graphical interface, which permits the preparation of the input and visualize the output of the simulations (the chemical structure, molecular orbitals (MOs), predicted spectra, etc.). These graphical suites, such as GaussView [8] for Gaussian, can be specific for the software, or not specific, like Molekel [9] or Gabedit [10].

Usually, in these packages, *ab initio*, semi-empirical, molecular mechanics (MM), quantum mechanics (QM)/MM, and density functional theory (DFT) methods are available [11–13]. *Ab initio* and semi-empirical methods are based on the solution of Schrödinger's equation. In an *ab initio* simulation, the solution for a multi-electron system is obtained using self-consistent field (SCF), or Hartree–Fock (HF) method [14, 15]; on one hand, the calculations are very accurate, on the other hand, they are slow, computationally expensive and – generally – not suitable for large molecules. Calculations can be carried out on closed and open shell systems: in the closed-shell simulations, all the electrons are coupled, and each wavefunction is occupied by two electrons (restricted Hartree–Fock); in the open shell calculations, a different wavefunction is obtained for each electron (unrestricted Hartree–Fock). A well-known limitation of HF method is that the *electron correlation* is not taken into account, but it is treated only in an average way; in other words, the instantaneous interaction between electrons and the effect on a specific electron of all the other ones is not explicitly considered.

In the semi-empirical simulations, a simplified form of Hamiltonian is assumed, and one or more empirical parameters, derived from the experimental data, are used; the advantages are that the calculations are faster than those based on *ab initio* methods and large species (mainly organic compounds) can be simulated efficiently, the disadvantages being that the parameters for all the atoms must be specified. Often, the results are not accurate, mainly when transition metals are considered. The wavefunctions are represented by a basis set, i.e. a set of functions (called basis functions) consisting of a linear combination of atomic orbitals, which are usually approximated by Gaussian-type or Slater-type orbitals.

Since, in many cases, the electronic structure of a molecule is not accurately described by only one configuration, the *configuration interaction* (CI), a superposition of various configurations, is used [16, 17]. It replaces one or more occupied orbitals of the HF determinant (Slater determinant) with virtual orbitals. However, mixing all the molecular electronic states is very demanding, and handling large molecules is practically impossible. So, only a limited number of combinations is considered, and the methods indicated as CIS (single excitation in the HF determinant), CID (double excitations), and CISD (single and double excitations) are valuable examples.

The *multi-configurational self-consistent field* (MCSCF) is generally used when HF is inadequate [18]. It consists of a linear combination of configuration state functions (CSF) or Slater determinants to approximate the exact electronic wavefunction of a chemical system. Often, it is presented as a combination of CI (where the expansion of the wavefunction and not the MOs are changed) and HF (where the MOs are changed, but there is only one determinant). It is advantageous when systems with degenerate configurations are taken into account. When MCSCF is applied to the excited states, the *complete active-space self-consistent field* (CASSCF) is obtained. The orbitals included in this method can be inactive (orbitals doubly occupied), virtual (unoccupied), and active (singly occupied). The spin state identifies a CASSCF wavefunction, number of electrons ($N$), and orbitals ($M$) in the active space with the notation CASSCF ($N$, $M$). The *restricted active-space self-consistent field* method is related to CASSCF but uses only selected subspaces of active orbitals. These methods are essential for predicting the excited states and, consequently, the optical spectroscopy.

In *multireference configuration interaction* (MRCI), several sets of reference configurations are generated, and from them, excited determinants are formed and used in a CI calculation. For example, one possible approach consists of performing an MCSCF calculation and selecting a set of reference configurations; subsequently, the occupied orbitals are replaced by virtual orbitals in the individual CSFs [19].

The Møller–Plesset perturbation theory (MPPT) improves the HF method and includes electron correlation effects, adding higher excitations as a non-variational correction [20]. The calculations start from an eigenfunction of HF and allow finding energies that may be lower than true ground state energy. The inclusion of the second-, third-, or fourth-order energy correction in MPPT is indicated as MP2, MP3, or MP4, respectively.

DFT has reached enormous popularity, and many reviews have been published [21, 22]. It is a theory of electronic structure based on the electron density $\rho(\mathbf{r})$ rather than on the many-electron wavefunction. It allows determining the properties of a system using a functional dependent only on $\rho(\mathbf{r})$ with relatively short computational times.[1] With an appropriate choice of the functional and basis set, it is possible to perform calculations on large molecules with high efficiency.

1 A function maps numbers to numbers. In contrast, a functional maps functions to numbers. In DFT, functionals are used to express physical quantities – such as the total energy – as a function of the total electron density $\rho(\mathbf{r})$, i.e. $E[\rho(\mathbf{r})]$. Thus, while functions operate on variables, functionals operate on functions.

Finally, some words are dedicated to the relativistic effects and models for simulating the role of the solvent. In a molecule, the relativistic contribution becomes relevant when the nuclear charge increases; in these cases, it must be included in the calculations since determines properties such as spin-orbit (SO) coupling. When this happens, the Dirac instead of Schrödinger equation must be used [23]. Since the Dirac equation is valid for systems with only one electron, the relativistic approach for many-electron systems is most often done by using the Dirac–Coulomb or Dirac–Coulomb–Breit Hamiltonians, the Coulomb term describing the Coulombian interaction between electrons and the Breit term the retardation effects of the interactions between electrons due to the finite speed of photons [24]. Another approach is based on the second-order scalar relativistic (SR) Douglas–Kroll–Hess Hamiltonian, which is compatible with any SCF methods [25].

An explicit model for the solvent is rarely used since the computational cost is very high, and implicit models, which treat the solvent as a continuum with a dielectric constant $\varepsilon$ are usually employed. The most popular models are: (i) the polarizable continuum model (PCM), which makes use of a molecular cavity that follows the real geometry of the investigated system and of a surface charge distribution to represent the polarization of the environment [26]; (ii) the continuum solvation model based on the charge density of a solute molecule (SMD), which considers the full solute electron density without defining partial atomic charges and represents the solvent as a dielectric medium with surface tension at the solute-solvent boundary [27, 28]; (iii) the conducting-like screening solvation model (COSMO), which – in contrast to the other models – predicts the polarization charges of the continuum, due to the solute polarity, from a scaled conductor approximation and, moreover, its implementation for real solvents (COSMO-RS), which allows the calculation of the molecular chemical potential in a solvent or in a mixture of solvents [29, 30].

Most of the computational protocols applied to predict the spectroscopic behavior of a chemical compound are based on their validation of a dataset formed by molecules with known behavior. The statistical dispersion of a calculated spectroscopic parameter $X\left(X^{\text{calcd}}\right)$ from the experimental $X$ $\left(X^{\text{exptl}}\right)$ for the $j$th compound of the dataset can be defined in terms of percentage deviation (PD) and absolute percentage deviation (APD). In the analysis of a large dataset, of course, it is necessary to refer to a mean value, and two types of mean deviations are relevant, the first based on the deviations with their algebraic sign and the second based on their absolute value [31]: the mean percentage deviation (MPD) and the mean absolute percentage deviation (MAPD). In particular, it has been pointed out that the value of MAPD can be chosen as a quality criterion [32]:

$$\text{MAPD} = \frac{1}{N}\sum_{j=1}^{N}\left[\left|\left(\frac{X^{\text{calcd}}(j) - X^{\text{exptl}}(j)}{X^{\text{exptl}}(j)}\right)\right| \times 100\right] \tag{4.1}$$

where $N$ is the number of examined compounds. Associated with the values of MPD and MAPD, is often reported as the standard deviation (SD), which gives an idea of the data dispersion. As an alternative to MAPD, the mean absolute deviation (MAD) can be used.

## 4.3 Electron Paramagnetic Resonance

EPR is a technique applicable, in principle, to any paramagnetic compound with one or more unpaired electrons. In an EPR experiment, the magnetic field splits the energy of the electron spin states $\alpha$ and $\beta$ (Zeeman effect) [33]. When a mononuclear paramagnetic species with only one unpaired electron is considered, the spectrum is characterized by two spin Hamiltonian parameters: the g factor and the hyperfine coupling (HFC) constant between the unpaired electron and metal nucleus ($g_{iso}$ and $A_{iso}$ in an isotropic spectrum, and $g_x$, $g_y$, $g_z$ and $A_x$, $A_y$, $A_z$ in an anisotropic spectrum) [33]. The **g** and ***A*** tensors are strongly perturbed by the metal chemical environment and their change could provide valuable information on the structure of the metal species under examination. In particular, the values of $g_z$ and $A_z$ are more sensitive to the coordination sphere variations than the *x* and *y* components. With increasing the electron donor capability of the ligands, an increase of the experimental value of $g_z$ and a decrease of $A_z$ is observed for $V^{IV}$, $d^1$ configuration, while the opposite trend is expected for $Cu^{II}$, $d^9$ configuration. The presence of a ligand in an axial position causes an increase of $g_z$ and a decrease of $A_z$ [34].

The general Hamiltonian is the sum of the Zeeman interaction, the HFC interaction, the zero-field splitting (ZFS), and the electrostatic quadrupole interaction:

$$\widehat{\mathrm{H}} = \beta\mathbf{H}\cdot\mathbf{g}\cdot\widehat{\mathbf{S}} + \boldsymbol{A}\cdot\widehat{\mathbf{I}}\cdot\widehat{\mathbf{S}} + \widehat{\mathbf{S}}\cdot\boldsymbol{D}\cdot\widehat{\mathbf{S}} + \widehat{\mathbf{I}}\cdot\mathbf{Q}\cdot\widehat{\mathbf{I}} \tag{4.2}$$

The ZFS term must be considered only if $S > 1/2$, i.e. when there is more than one unpaired electron; the term $\widehat{\mathbf{S}}\cdot\boldsymbol{D}\cdot\widehat{\mathbf{S}}$ is valid only when the SO coupling is weak.

The HFC constant in an EPR spectrum arises from the interaction between the spin angular momentum of the electron ($S = 1/2$) with the spin angular momentum of the nucleus (for example, $^{51}V$ nucleus has $I = 7/2$ and 99.8% natural abundance, while the two EPR-active nuclei for copper, $^{63}Cu$ and $^{65}Cu$, have $I = 3/2$ and abundance of 69.1 and 30.9%, respectively). The HFC tensor ***A*** has three contributions: the isotropic Fermi contact ($A^{FC}$), where FC is the abbreviation for Fermi contact), the anisotropic or dipolar (D) hyperfine interaction ($\boldsymbol{A}^{D}$), and the second-order effect that arises from SO coupling ($\boldsymbol{A}^{SO}$) [35]:

$$\boldsymbol{A} = A^{FC}\mathbf{1} + \boldsymbol{A}^{D} + \boldsymbol{A}^{SO} \tag{4.3}$$

where **1** is the unit tensor. $A^{FC}$ and the components $A^{D}_{\mu\nu}$ and $A^{SO}_{\mu\nu}$ of the tensors $\boldsymbol{A}^{D}$ and $\boldsymbol{A}^{SO}$ are given by:

$$A^{FC} = \frac{4\pi}{3} g_e g_N \beta_e \beta_N \langle S_z\rangle^{-1} \rho_N^{\alpha-\beta} \tag{4.4}$$

$$A^{D}_{\mu\nu} = \frac{1}{2} g_e g_N \beta_e \beta_N \langle S_z\rangle^{-1} \sum_{k,l} P^{\alpha-\beta}_{k,l} \left\langle \Phi_k \left| \frac{\mathbf{r}^2\delta_{\mu\nu} - 3\mathbf{r}_\mu \mathbf{r}_\nu}{\mathbf{r}^5} \right| \Phi_l \right\rangle \tag{4.5}$$

$$A^{SO}_{\mu\nu} = -\frac{1}{2S} g_e g_N \beta_e \beta_N \sum_{k,l} \frac{\partial P^{\alpha-\beta}_{k,l}}{\partial I_\mu} \langle \Phi_k | h^{SOC}_\nu | \Phi_l \rangle \tag{4.6}$$

where $g_e$ and $g_N$ are the g factors of the free electron and the nucleus, $\beta_e$ and $\beta_N$ are the electron and nuclear magnetons, $\langle S_z \rangle$ is the expectation value (EV) of the electronic spin on the $z$ axis, $\rho_N^{\alpha-\beta}$ is the spin density at the nucleus, $P_{k,l}^{\alpha-\beta}$ is the spin density matrix, $\mathbf{r}$ is the distance between the unpaired electron and the nucleus, and $h_v^{SOC}$ the spatial part of an effective one-electron SO operator.

The tensor $\boldsymbol{A}^D$ is traceless, i.e. $A_x^D + A_y^D + A_z^D = 0$. The values of the anisotropic HFC constants along the $x$, $y$, and $z$ axes are $A_i = A^{FC} + A_i^D + A_i^{SO}$ with $i = x, y, z$. In a computational study, $A_z$ calculated $\left(A_z^{calcd}\right)$ can be compared with the experimental value of $A_z$ $\left(A_z^{exptl}\right)$. The value of $A_{iso}$ is:

$$A_{iso} = \frac{1}{3}(A_x + A_y + A_z) = A^{FC} + \frac{1}{3}\left(A_x^{SO} + A_y^{SO} + A_z^{SO}\right) = A^{FC} + \frac{1}{3}A^{PC} \qquad (4.7)$$

with the term $1/3\left(A_x^{SO} + A_y^{SO} + A_z^{SO}\right)$ named pseudocontact, $A^{PC}$. If the second-order terms $A_{\mu\nu}^{SO}$ are neglected, $A_{iso}$ coincides with $A^{FC}$.

EPR spin Hamiltonian parameters were computed for most of the transition metal complexes with an odd number of electrons and the calculations were based mainly on the DFT methods. The values of $A^{FC}$, $A^D$, and $A^{SO}$ can be calculated with several packages, for example, ORCA [2, 36]. In Figure 4.1, an example of an ORCA output is shown.

Among the three contributions to $A$, the most critical parameter is the Fermi contact, which is directly proportional to the spin density at the nucleus $\left(\rho_N^{\alpha-\beta}\right)$ and, therefore, a good model must predict the spin polarization in the electron distribution of the core and valence shells induced by the unpaired $d$ electron(s) [37]. Concerning the SO contribution, Van Lenthe *et al.* [38] proposed the zero-order regular approximation (ZORA) to the Dirac equation to determine the SO coupling; this approach, however, has the limitation that the spin polarization and SO effects cannot be introduced contemporaneously and has been overcame by Neese in the coupled-perturbed Kohn–Sham implementation of SO contribution to HFC constant [35]. The SO contribution increases from left to right of the transition series and becomes important for late $d$ elements, for example, Co [39], Ni [40], and mainly, Cu [35, 39]. $A_z^{SO}$is approximately 6% compared to the sum $((A^{FC}+A_z^D)$, see Eq. (4.3)) for $V^{IV}O$ [41], while it reached approximately 20% for $Cu^{II}$ [42]. Consequently, the isotropic pseudocontact $A^{PC}$ is around 7% of $A^{FC}$ for $V^{IV}$ (Eq. (4.7)), while it is 26.4% for a $Cu^{II}$ compound like $[Cu(acac)_2]$ [39]. All these data suggest that SO effect cannot be completely neglected in an accurate prediction of HFC tensor [35, 39].

```
Raw HFC matrix (all values in MHz):
A(FC)      -209.5824     -209.5824      -209.5824
A(SD)       107.0679       76.9386      -184.0065
A(SO)       -13.0262      -15.5564       -45.8837 A(PC)  =     -24.8221
           ----------    ----------     ----------
A(Tot)     -115.5406     -148.2002      -439.4726 A(iso)=    -234.4045
```

**Figure 4.1** ORCA output for the simulation of the EPR spectrum of amavadin, a $V^{IV}$ natural occurring compound. A(FC) represents the FC, A(SD) and A(SO) the terms due to the dipolar hyperfine and SO interactions, A(PC) the pseudocontact, while A(Tot) and A(iso) the anisotropic and isotropic HFC constants.

For the first-row transition metal complexes the SR contributions is little important or negligible [43], and it must be considered only for elements of the second and third series [44].

The prediction of $A$ is strongly influenced by the functional chosen. Despite the high number of the published papers, up to now it is not available a functional working efficiently for all the systems and, for this reason, it is necessary to validate which is the best for a specific metal [45]. Some examples can be cited: for $Fe^{III}$ complexes OLYP works better than meta-GGA and hybrid functionals like B3LPY and TPSSh [46], for $Mo^{V}$ species hybrid functionals with 30–40% of exchange admixture are the most promising ones [47], while for copper the results indicate that B3LYP gives the best agreement with the experiment [35]. The validation of HFC constants on a series of 3*d* transition metal complexes performed by Neese agreed well with the experimental data, the main error being related to the underestimation of the FC [35].

The case represented by vanadium(IV) is particularly interesting. The value of $A$ for $V^{IV}O$ and for "bare" $V^{IV}$ species was calculated by fixing the basis set 6-311g(d,p) and varying the functional. The datasets consisted of 18 and 20 compounds, respectively, with different ligands, donors, electric charge, coordination number, and coordination geometry [41, 48]. $A_z$ or $A_x$ were determined from the output. The results were completely different and, to obtain a prediction in line with the experiment, for each class of compounds a different functional must be used. Generally, meta-gradient functionals such as BP86 underestimate the value of $A$ and an elevated fraction of the HF exchange in the hybrid functionals, which improves the prediction of the spin polarization, is necessary. For $V^{IV}O$ PBE0 gives an MAPD around 6%, while for "bare" $V^{IV}$ species B2PLYP gives an MAPD around 3% (Figure 4.2). An interesting datum is that for $V^{IV}O$ species the calculation of $A_z$ with the half-and-half hybrid BHandHLYP functional without the inclusion of the second-order SO term in Eq. (4.3) predicts $A_z$ with an MAPD below 4% (Figure 4.2) [41]. The good agreement obtained with BHandHLYP functional in the prediction of $A_z$ is due to a compensation between the overestimation of $A_{iso}$ and the underestimation of $A_z^D$; however, this systematic effect allows to get an excellent prediction of an EPR spectrum. The MAPD for the prediction of $A_{iso}$ for $V^{IV}O$ is 4.7% with PBE0 and 4.0% with BHandHLYP (without the SO coupling), while it is 8.2% with B2PLYP for $V^{IV}$ species.

Another instructive application concerns amavadin, the "bare" $V^{IV}$ complex synthesized in the several species of mushrooms of the genus *Amanita*, such as *A. muscaria, A. regalis,* and *A. Velatipes* to accumulate vanadium (in *A. muscaria* the vanadium content exceeds 400 times the amount detected in other species of the same genus and is independent of the V concentration in the soil) [49, 50]. Amavadin is a low molecular mass, anionic $V^{IV}$ compound formed by *N*-hydroxyimino-2,2′-diisopropionic acid, (*S,S*)-$H_3$hidpa, in its triply deprotonated form hidpa(3-) [49, 50]. The prediction of $A_{iso}$ and $A_z$ with B2PLYP functional and 6-311g(d,p) basis set gives PD of −3.5 and −4.2%, respectively, in excellent agreement with the experimental results (Table 4.1).

Concerning the prediction of **g** tensor, several calculation methods were used. Both *ab initio* and semiempirical techniques were applied, even if the latter have the limitation that empirical parameters are necessary, precluding a possible

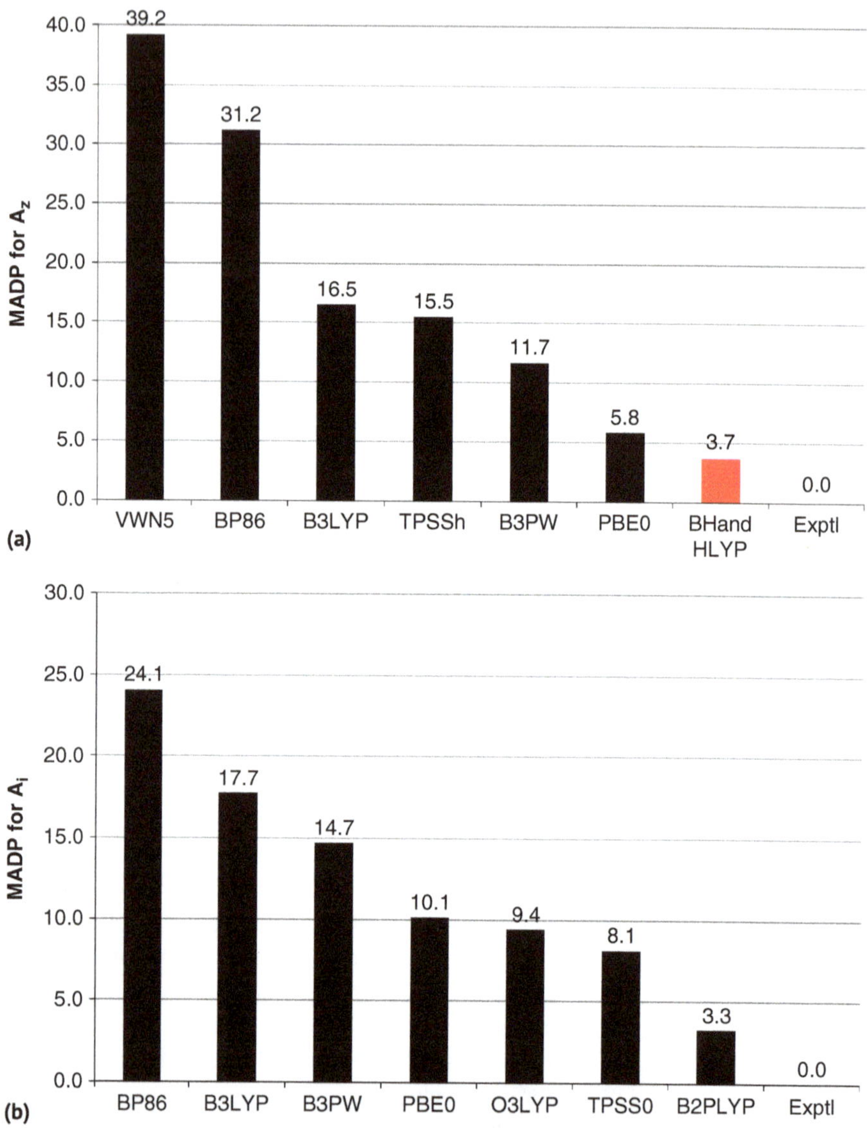


**Figure 4.2** MAPD from experimental $A_z$ (for $V^{IV}O$) and $A_i$ ($i = x$ or $z$, depending on the symmetry, for "bare" $V^{IV}$ species) calculated using several functionals and the basis set 6-311g(d,p). The bars in black refer to the calculations carried out with ORCA software taking into account the SO effect (see Eq. (4.3)), the bar in red the calculations with Gaussian software using the BHandHLYP functional without the SO term.

generalization of these methods. DFT techniques are the most used due to the low computational cost, a wide choice of functionals, and the possibility of including the SO effect in the calculations. While the results are rather encouraging for organic radicals, the main problem is the treatment of the SO coupling for transition metal ions. The best results up to now have been reached with multiconfigurational scalar-relativistic CASSCF and N-electron valence

**Table 4.1** $^{51}$V HFC constants for amavadin calculated by DFT methods.[a,b]

| Method | $A_{iso}^{calcd}$ | $A_x^{calcd}$ | $A_y^{calcd}$ | $A_z^{calcd}$ | PD ($A_{iso}$)[c] | PD ($A_z$)[c] |
|---|---|---|---|---|---|---|
| VWN[d] | −43.9 | −0.7 | −28.2 | −100.9 | −45.8 | −34.1 |
| BP86[d] | −48.0 | −6.6 | −31.2 | −106.1 | −40.7 | −30.7 |
| B3PW91[d] | −59.0 | −15.9 | −40.8 | −120.5 | −27.2 | −21.2 |
| BPW91-35[d,e] | −67.2 | −23.9 | −48.4 | −129.4 | −17.0 | −15.4 |
| PBE0[f] | −73.0 | −33.6 | −45.1 | −140.4 | −9.9 | −8.2 |
| B2PLYP[f] | −78.2 | −38.5 | −49.4 | −146.6 | −3.5 | −4.2 |

a. All the values reported in $10^{-4}$ cm$^{-1}$ units.
b. In the literature the experimental spin Hamiltonian parameters of amavadin are: $g_x = g_y = 1.982$, $g_z = 1.920$; $A_x = A_y = -45.0 \times 10^{-4}$ cm$^{-1}$, $A_z = -153.0 \times 10^{-4}$ cm$^{-1}$ [51].
c. Percentage deviation.
d. Data calculated by Kaupp et al. in Ref. [52] with the basis set (15*s*11*p*6*d*)/[9*s*7*p*4*d*] on V and DZVP on the main group atoms without polarization *p* functions for H.
e. BPW91 functional with 35% HF exchange.
f. Data calculated with 6-311g(d,p) basis set in Ref. [48].

perturbation theory (NEVPT2) methods, which improve slightly the DFT approach, even if the data are still relatively far from the experimental values and an overestimation of the g factor and large PD values are obtained. DFT calculations indicate that PBE0 and B3LYP behave generally better than other functionals, and the bad results can be related to the description of metal–ligand bonds and the presence of low-lying excited states [53].

A case study concerns the validation of DFT methods on fourteen Cu$^{II}$ complexes with several functionals and applications to the binding of Cu$^{II}$ ion to two important proteins, such as human serum albumin (HSA) and prion protein (PrP) [42]. Geometry optimization and harmonic frequencies were computed with the hybrid Becke three-parameter B3LYP functional coupled with the Grimme's D3 dispersion, while the split-valence Pople basis set 6-311g(d,p) for the main group elements and Stuttgart–Dresden implemented with *f*-functions and pseudo-potential for Cu were applied. The tested functionals can be ranked on the basis of predicting $A_z$ as follows: B3LYP > B3PW91 ~ B3P86 > PBE0 > CAM-B3LYP > TPSSh > BHandHLYP > B2PLYP > MPW1PW91 > $\omega$-B97x-D >> M06. For B3LYP, the MAPD from the experimental values of $A_z$ is 8.6% with an associated SD of 4.2%. The data can be rationalized considering the low accuracy of the functionals with a large number of empirical parameters in the prediction of the electronic density around the nucleus compared to the less parameterized functionals [54]. The ranking of the functionals in the prediction of $g_z$ is: PBE0 > BHandHLYP > B2PLYP > $\omega$-B97x-D > B3LYP ~ B3P86 ~ B3PW91 > CAM-B3LYP > TPSSh ~ MPW1PW91 >> M06. For PBE0, the MAPD is 2.9% and SD is 1.1%.

The results have been applied to the binding of Cu$^{II}$ to HSA, the most abundant protein in the blood that transports several metal ions, among which copper and nickel [55]. $Cu^{II}$ and $Ni^{II}$ are bound to the *N*-terminal one (NTS) with the donor set ($NH_2$, $N^-$, $N^-$, His-N) in a little distorted square planar geometry [55]. Computation allows to confirm this attribution through the calculation of $A_z$ and $g_z$ for the peptide formed by the first five amino acids of HSA ($AspAlaHisLysSerNHCH_3$). B3LYP and PBE0 functionals can be used to calculate **A** and **g** tensors. The optimized structure of the adduct formed by $Cu^{II}$ ion and is represented in Figure 4.3 and the results reported in Table 4.2: the PD from $A_z^{calcd}$ and $g_z^{calcd}$ is −1.1 and −1.7%, respectively.

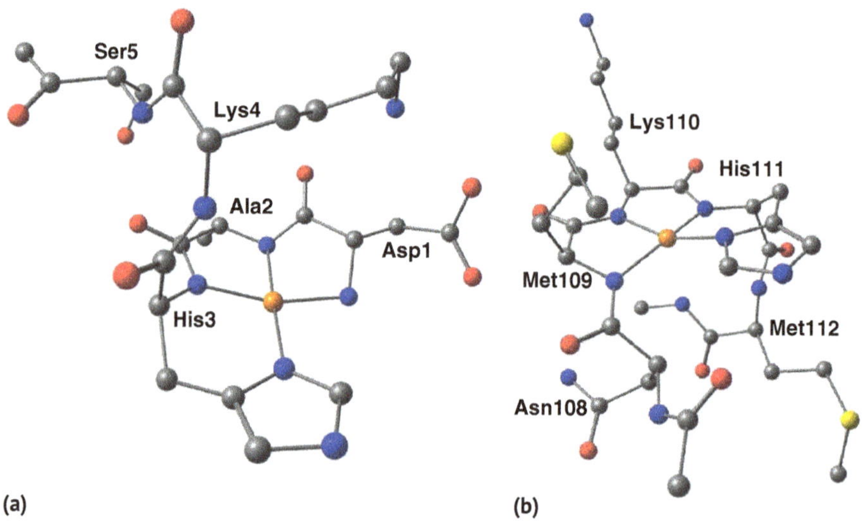


**Figure 4.3** Optimized structure at the level of theory B3LYP-D3 of the adduct formed by $Cu^{II}$ with the pentapeptide: (a) AspAlaHisLysSer–$NHCH_3$ (region 1–5 of HSA) and (b) Ac–AsnMetLysHisMet–$NHCH_3$ (region 108–112 of PrP). Hydrogen atoms are omitted for clarity. *Source*: Adapted from [42].

**Table 4.2** Experimental spin Hamiltonian parameters for the interaction of $Cu^{II}$ with HSA and fragments 106–126 of PrP, and comparison with the values calculated by DFT methods.

| Protein | $g_x$ | $g_y$ | $g_z$ | $A_x$ [a] | $A_y$ [a] | $A_z$ [a] |
|---|---|---|---|---|---|---|
| Experimental for HSA[b] | 2.045 | 2.045 | 2.173 | 22.0 | 22.0 | 206.6 |
| Calculated for AspAlaHisLysSer–$NHCH_3$ (1–5)[c] | 2.041 | 2.042 | 2.135 | 16.1 | 20.9 | 204.4 |
| Experimental for PrP (region 106–126)[d] | e | e | 2.198 | e | e | 198 |
| Calculated for Ac–AsnMetLysHisMet–$NHCH_3$ (108–112)[c] | 2.044 | 2.045 | 2.143 | 9.9 | 12.5 | 194.2 |

a. *A* measured in $10^{-4}$ $cm^{-1}$ units.
b. Experimental parameters for the system $Cu^{II}$/HSA reported in Ref. [42].
c. Calculated parameters using B3LYP functional for **A** and PBE0 for **g** tensor.
d. Experimental parameters reported in Ref. [59].
e. Values not reported.
*Source*: Adapted from [42].

The interaction of $Cu^{II}$ with PrP is another case. PrPs are involved in various transmissible spongiform encephalopathies, particularly Creutzfeldt-Jakob disease in humans, bovine spongiform encephalopathy in cattle, and scrapie in sheeps and goats [56]. In these pathologies, the α-helix in the region 128–231 undergoes a misfolding to β-sheet forming several aggregates [57]. It has suggested that Cu binding in the region 106–126 could be involved in the prion misfolding [58, 59]. The binding of $Cu^{II}$ to the pentapeptide 108–112 (Ac–AsnMetLysHisMet–$NHCH_3$) can be studied, and **g** and ***A*** tensors calculated by DFT methods (Figure 4.3). The values for $A_z^{calcd}$ and $g_z^{calcd}$ (194.2 × $10^{-4}$ $cm^{-1}$ and 2.143) can be compared with $A_z^{exptl}$ and $g_z^{exptl}$ (198 × $10^{-4}$ $cm^{-1}$ and 2.198), confirming the coordination mode proposed in the literature (Table 4.2).

## 4.4 Electron Spin Echo Envelope Modulation and Electron-nuclear Double Resonance

In most cases, the broad spectral lines observed in an EPR spectrum (usually larger than 10 Gauss) preclude the resolution of the coupling between the unpaired electron(s) and the donor nuclei directly bound to the metal. This coupling is named superhyperfine coupling constant and indicated with $A^L$, where L indicates the specific donor (for example, $^1H$, $^{14}N$, or $^{31}P$). Information on the superhyperfine coupling could be obtained by ENDOR [60, 61], and ESEEM [62, 63] techniques. In an ENDOR experiment, the magnetic field is fixed to study a specific EPR signal, and microwave radiation with sufficient power is used to saturate the transition; subsequently, a radiofrequency scan is carried out over the range of the probable resonances. ESEEM is a pulsed technique in which a series of microwave pulses, followed by a waiting period, replaces continuous microwave irradiation. If the magnetic field is set to a particular resonance and two microwave pulses are applied, the emission of a microwave signal is revealed.

Unfortunately, only a few papers are devoted to the prediction of superhyperfine coupling. Two of them concern $Cu^{II}$ and $V^{IV}O$ ions, which have been widely studied by EPR spectroscopy [64, 65]. A DFT approach suggests hybrid functionals, like B3LYP and PWP1, perform better than simple generalized gradient approximation functionals, such as BP or BLYP [64]. Using a scalar-relativistic, spin-unrestricted, open-shell Kohn–Sham (SR UKS) approach, the superhyperfine coupling with $^{14}N$ for $V^{IV}O$ can be calculated with an MAPD of 4.5% [65].

The superhyperfine coupling constants reported for $^{14}N$ are very useful in determining the type of nitrogen bound to a metal ion and, for this reason, can provide information on which donors of a specific bioligand interact with a metal in a biological matrix. Here, we will discuss some systems with $V^{IV}$, for which $A_{iso}^N$ is approximately 5.0 MHz for a $sp^3$ amine-N, in the range 6.0–7.0 for a $sp^2$ imine-N (6.0–6.5 MHz for Schiff base-N, pyridine-N and imidazole-N and ~7.0 MHz for porphyrin-N and phthalocyanin-N), and 7.5 MHz for a *sp* N [66]. This allows proving if a protein binds V with Lys-N, His-N, or Arg-N donors; for example, ESEEM suggested the coordination of His249 in the $V^{III,IV,V}$–transferrin adducts [67], and of

Lys-$NH_2$ to $V^{IV}$ in Vanabin2, one of the five proteins that bind vanadium in the cytoplasm of vanadocytes of ascidians [68].

In this case too, a good exercise is to start with the validation of the method using DFT and varying the functional, with the effect of the basis set being less critical. For a group of five $V^{IV}O$ compounds, the best functional is BHandH followed by B3PW91 and B3P86 (Figure 4.4). For BHandH the maximum PD is for complex [VO(Hhida)($H_2O$)], in which the N coordination is in the axial position, while when N donor is equatorially bound, the prediction is always rather satisfactory.

The validated protocol can be applied to $V^{IV}O$-substituted spinach chloroplast $CF_1$-APTase. Vanadium, indeed, from the 1970s, was used as a spin probe to extract specific information on the metal binding in various peptides and proteins [74]. $CF_1$-ATPase is an extrinsic membrane protein complex that synthesizes ATP, exploiting the proton gradient generated by the light reactions of photosynthesis [75]. In the metal-substituted protein, vanadium is bound at specific $Mg^{2+}$-binding sites (there are six sites for $Mg^{2+}$ nucleotides, of which three are catalytic [76]) and can provide information about the groups that act as $Mg^{2+}$ donors. The analysis of the X-ray structure for Mg environment, within 5.5 Å from the metal ion, suggests that the equatorial plane of $V^{IV}O^{2+}$ ion may be formed by Thr176, Asp269, and two water molecules, plus Lys175 in the axial position (Figure 4.5). It is possible to test the procedure optimizing the structure of a model complex with acetate, methanol, and butylamine for Asp-$COO^-$, Thr-OH, and Lys-$NH_2$ coordination. The calculated $A^N_{iso}$, 0.9 MHz, can be compared with the experimental value around 1 MHz and is compatible with a lysine residue in the axial position [73].

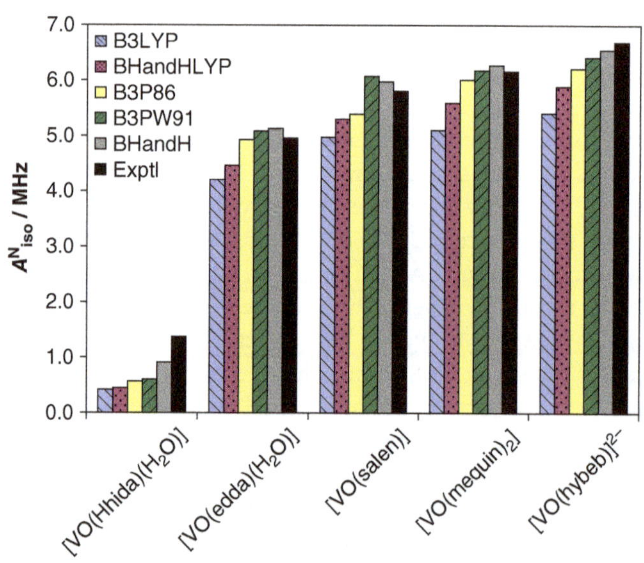


**Figure 4.4** Values of $A^N_{iso}$ calculated with varying the functional (basis set 6-311g) for five $V^{IV}O$ complexes. Experimental (exptl) are reported in Refs. [69–72]. Hhida is *N*-(2-hydroxyethyl)iminodiacetate(2-), edda is ethylene-*N,N'*-diacetate(2-), salen is *N,N'*-ethylenebis(salicylideneiminate)(2-), mequin is 2-methyl-8-hydroxyquinolinate(-), and hybeb is 1,2-bis(2-hydroxybenzamido)benzenate(4-). *Source*: Adapted from [73].

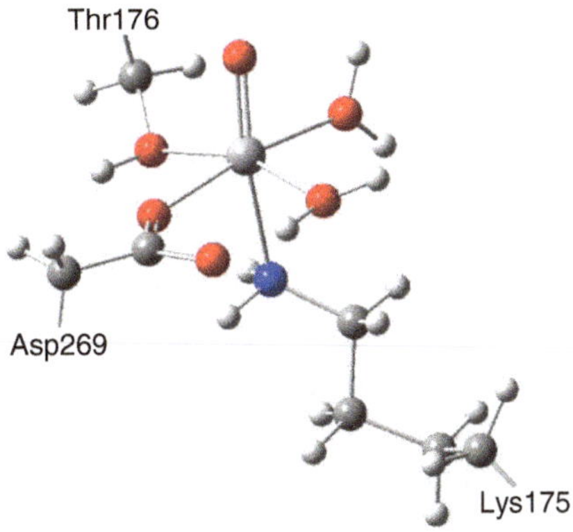


**Figure 4.5** Most probable vanadium environment in V-CF$_1$-ATPase (model optimized by DFT methods). *Source*: Adapted from [73].

## 4.5 Nuclear Magnetic Resonance

NMR is the resonance technique "twin" of EPR. In the NMR spectroscopy, the magnetic field splits the energy of nuclear spin states ($M_I = \pm 1/2$ for proton in the simplest version of NMR, $^1$H). The information usually extracted from an NMR spectrum is the chemical shift ($\delta$), the coupling constant ($J$), and – less commonly – the relaxation rate [77, 78]. Compared with EPR, the chemical shift $\delta$ corresponds to the g factor and the coupling constant $J$ to the HFC constant [79]. The difference between the two techniques is that NMR can be applied to diamagnetic and paramagnetic systems, while EPR can be applied only to paramagnetic molecules with one or more unpaired electrons.

The background of NMR is very similar to EPR, and the readers can refer to excellent books [77, 78]. The chemical shift ($\delta$) depends on the chemical environment of the nucleus of interest and can be expressed with the general formula [80, 81]:

$$\delta = \delta^{\text{orb}} + \delta^{\text{FC}} + \delta^{\text{PC}} \tag{4.8}$$

In Eq. (4.8), $\delta^{\text{orb}}$ corresponds to the orbital shift, $\delta^{\text{PC}}$ is the pseudocontact shift and arises from the dipolar interaction between the static magnetic moment due to the unpaired spin and the magnetic moment of the specific nucleus (it is proportional to the magnetic susceptibility tensor $\chi$), while $\delta^{\text{FC}}$ is the FC and derives from the isotropic hyperfine interaction between the nuclear magnetic moment and the effective local field due to the electron spin density at the nucleus [81]. Often, at a first approximation, $\delta^{\text{PC}}$ and the effect of the ZFS for species with more than one unpaired electron ($S > 1/2$) can be neglected [82–84].

In the classical theory of NMR, the orbital shift is the difference between $\sigma^{\text{ref}}$ (isotropic part of the shielding tensor $\sigma^{\text{ref}}$ for the reference) and $\sigma$ (isotropic part of the shielding tensor $\sigma$ for the species under examination) [85, 86]. The reference is often tetramethylsilane (TMS), but can be another molecule, for example, $VOCl_3$ for vanadium [49].

$$\delta = \sigma^{\text{ref}} - \sigma \tag{4.9}$$

The orbital term in Eq. (4.8) is approximately independent of the temperature and can be associated with the shift of a diamagnetic system [87]. In contrast, the contact and pseudocontact shifts are temperature-dependent and must be taken into account for paramagnetic molecules (paramagnetic NMR or pNMR). In the simplest situation, when the ZFS can be neglected, the energy of the ground state (with a $2S + 1$ degeneration) is much lower than the excited states, and the system follows the Curie law, $\sigma$ being calculated with Eq. (4.10) [88]:

$$\sigma = \frac{\beta_e}{\hbar \gamma_N} \frac{S(S+1)}{3kT} \mathbf{g} \cdot \boldsymbol{A} \tag{4.10}$$

where $\beta_e$ is the Bohr magneton, $\gamma_N$ is the gyromagnetic ratio of nucleus, $kT$ is the thermal energy, $\mathbf{g}$ is the g factor tensor, and $\boldsymbol{A}$ is the HFC tensor of the nucleus. It is important to note that, when the relativistic effects can be neglected, pNMR shift can be determined through the calculation of $\mathbf{g}$ tensor on one hand and of the HFC constant $A$ from the FC, dipolar interaction (D), and SO contribution through Eqs. (4.4)–(4.6) on the other.

The coupling constant $J$ is the sum of FC (the largest contribution), diamagnetic spin-orbit (DSO), paramagnetic spin-orbit (PSO), and spin-dipole contribution (SD) [84, 89]:

$$J = J^{FC} + J^{DSO} + J^{PSO} + J^{SD} \tag{4.11}$$

In contrast to EPR, in which the most important spin Hamiltonian parameter is the HFC constant between an unpaired electron and a metal nucleus, in the NMR the prediction of the chemical shift is more common than that of the coupling constant.

The usual approach to pNMR is based on calculating the ground-state multiplet, and the shielding tensor is predicted through the EPR Hamiltonian (Eq. (4.2)) involving the terms related to the electronic Zeeman effect, HFC, and ZFS ("EPR route"); the prediction could be improved by including the relativistic contribution, if necessary. An alternative approach is based on *ab initio* methods.

The evolution of the computational strategy to calculate the pNMR chemical shifts could provide information on the methods available. After the development of the theory in 1970 by Kurland and McGarvey with an *ab initio* approach to pNMR [80], only in 2004 a complete theory and a computational protocol was formulated [90]. In particular, the "EPR route" was suggested as a possible tool to predict the NMR shifts with the inclusion of the SO contribution [90]. The efficiency of the computation made significant progress after the addition of the ZFS effect, which initially was included as a correction in the systems when it is small [82] and subsequently was considered systematically [83]. Van den Heulen and Soncini extended the approach to the general case, including also strong SO effects [91, 92]. Concerning the relativistic effects, the first calculations were based on ZORA [93], but this approach resulted in significant errors in the prediction of the $\mathbf{g}$ tensor (approximately 25–30% for molecules containing light elements and approximately 10% for compounds with heavy elements). To the best of my knowledge, the first attempt to calculate the pNMR shifts (on a doublet system) based on a four-component relativistic DFT theory and a full Dirac–Coulomb Hamiltonian was proposed by Komorovsky et al. [87].

From 2015, four great steps have been taken. Initially, Vaara et al. proposed a hybrid *ab initio*/DFT approach, in which CASSCF and NEVPT2 methods were used to predict **g** and ZFS tensor, while DFT methods with several functionals were used to calculate the HFC tensor ***A*** and orbital shielding [94]. Subsequently, Autschbach et al. determined the pNMR shifts from first principles without "EPR route" through the Eq. (4.12) [95]:

$$\sigma^{\text{ref}} - \sigma = \frac{1}{3}\sum_{i}\frac{1}{Q}\sum_{\lambda} e^{-\frac{E_\lambda}{kT}} \times \left[ 2\text{Re}\sum_{\lambda \neq \lambda'}\sum_{a,a'} \frac{\langle \lambda a | \widehat{H}_i^Z | \lambda' a' \rangle \langle \lambda' a' | \widehat{H}_{Ni}^{HF} | \lambda a \rangle}{E_{\lambda'} - E_\lambda} + \frac{1}{kT}\sum_{a,a'} \frac{\langle \lambda a | \widehat{H}_i^Z | \lambda a' \rangle \langle \lambda a' | \widehat{H}_{Ni}^{HF} | \lambda a \rangle}{E_{\lambda'} - E_\lambda} \right] \tag{4.12}$$

where the sums are over the low-energy electronic states related to the metal open shell. In 2016, Vaara et al. predicted the long-range pseudo-contact pNMR shifts in the presence of the ZFS through an *ab initio* quantum-chemical method and a modern version of the Kurland–McGarvey protocol for pNMR [96]. In 2018, finally, Neese et al. considered a benchmark of 34 $^{1}H$, $^{13}C$, $^{15}N$, $^{17}O$, $^{19}F$, and $^{31}P$ shielding constants and compared the pNMR chemical shielding at double hybrid DFT (DHDFT) level – using GIAOs – and the RI-MP2 approximation: in particular, the DSD-PBEP86 functional produced the smallest deviation from the CCSD(T) reference data with an MAPD of 1.9%, significantly lower than MP2 (MAPD = 4.1%) and meta-GGA DFs like M06L (MAPD = 5.4%) and TPSS (MAPD = 6.4%) [97]. In Figure 4.6, the variation of the MAPD between the calculated and experimental value as a function of the computational cost is graphically summarized.

As mentioned above, the most employed approach to predict the chemical shift of pNMR is the "EPR route," based on the calculation of the **g** and ***A*** tensors, which usually is carried out by DFT. With this method, the chemical shift of sixteen $Fe^{IV}{=}O$ species formed by a porphyrin or another multidentate ligand, whose importance in biochemistry is well known as intermediates in the catalytic cycle of heme enzymes as catalase and peroxidase [98, 99], were calculated [84]. For these species, the spin state $S = 1$ or $S = 2$ is possible for the ground state. It was shown that B3LYP*, derived from the popular B3LYP with a lower percentage of HF exchange [100], allows a correct prediction of the spin state and $^{1}H$ NMR spectra of all the spin states [84].

The comparison of the $^{1}H$ and $^{13}C$ chemical shifts for the series of open-shell 3*d* metallocenes, $\left[M\left(\eta^5\text{–}C_5H_5\right)_2\right]$, where $C_5H_5$ is the cyclopentadienyl radical

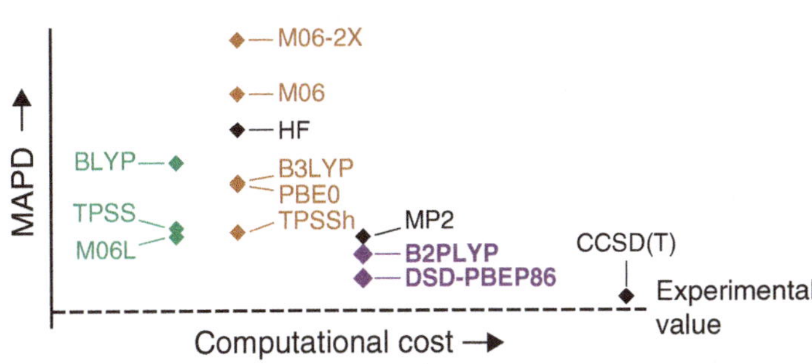


**Figure 4.6** Variation of the MAPD between the calculated and experimental value as a function of the computational cost. *Source:* Adapted from [97].

(abbreviated also with Cp) and M is V, Cr, Mn, Co, or Ni, is rather interesting. The spin multiplicities used are 4 (V), 3 (Cr), 6 (Mn), 2 (Co), and 3 (Ni). Considering the large experimental chemical shift range for these complexes, the agreement of the calculations with experimental data was good, even if the deviations were larger than those obtained for diamagnetic systems. Two different approaches to calculating the nuclear shielding can be compared, the first based on the first-order (nonrelativistic (NR)) and second-order SO contribution [82], and the second one based on the SR effects treated variationally and SO terms included as a perturbation [101]. The comparison is shown in Figure 4.7. Interestingly, these results allow relating the shifts to the features of the chemical bonds: in particular, for Ni the $^{13}$C shifts (calcd. +1540 and exptl. +1514) is larger than for V (calcd. −443 and exptl. −510), and this is caused by a different mechanism of the spin polarization: in the case of nickelocene, the Ni–Cp bond is dominated by the $\pi$ back-donation, while for vanadocene the $\sigma$ donation has the most important effect.

The results of the prediction through two different methods represented by the "EPR route" and first principles (Eq. (4.12)) can be compared (Table 4.3). From an examination of this table, two interesting facts must be pointed out: the SO terms give the largest contribution to $\sigma$ and the results obtained with the two methods are comparable.

Another interesting comparison, which can be useful for the readers, is between a simple DFT and *ab initio*/DFT calculations (with CASSCF and NEVPT2 to predict **g** and ***D*** tensors, and DFT to predict HFC tensor ***A***) of the $^1$H shifts for the complex formed by Co$^{II}$ and pyrazolylborate (Scheme 4.1) [94]. The data are listed in Table 4.4, where it can be observed that the results are very sensitive on the basis set, single or triple zeta, and that DFT performs worse than CASSCF and NEVPT2 (both as APD from the experiment and predicted sign for the shift). The calculation of $A$ depends on the choice of the DFT functional. If the effect of the variation of the exact HF exchange is taken into account (for example, PBE with 0%, PBE0 with 25%, and PBE50 with 50%), it is found that the best results are obtained with PBE0.

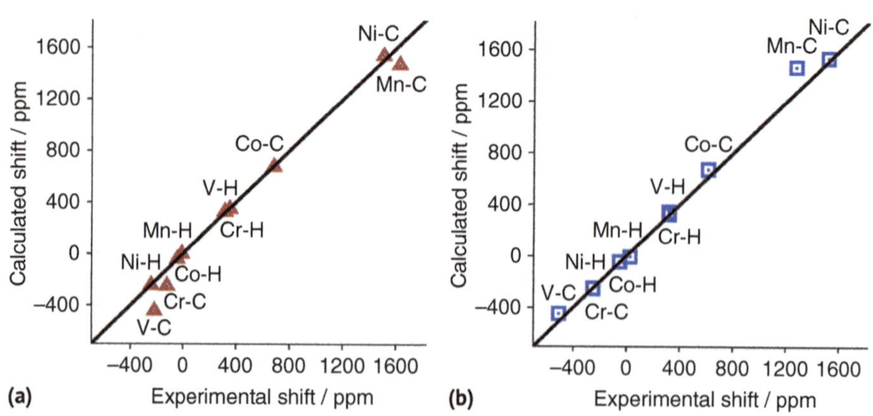


**Figure 4.7** pNMR $^{13}$C and $^1$H chemical shifts for 3*d* metallocenes: (a) results obtained with B3LYP functional reported in Ref. [82]. *Source*: Adapted from [82]; (b) results obtained with scaled ZORA and PBE0 reported in Ref. [101]. *Source*: Adapted from [101].

**Table 4.3** pNMR $^{13}C$ chemical shifts of $Np^{VI}O_2$ and $U^{V}O_2$ complexes calculated for the ground state $E_{3/2}$ and excited state $E_{1/2}$ from "EPR route" and "direct" approach.

| | | "EPR route"[a] | | | | "Direct"[b] | | | |
|---|---|---|---|---|---|---|---|---|---|
| Complex | State | FC | D | SO | **Total** | FC | D | SO | **Total** |
| $[Np^{VI}O_2(CO_3)_3]^{4-}$ [c] | $E_{3/2}$ | 22.82 | 28.46 | 58.49 | **109.78** | 22.79 | 28.43 | 58.42 | **109.65** |
| | $E_{1/2}$ | −47.99 | 34.54 | 144.21 | **130.21** | −47.94 | 32.86 | 145.14 | **130.06** |
| $[U^{V}O_2(CO_3)_3]^{5-}$ [d] | $E_{3/2}$ | 3.39 | 25.33 | 54.07 | **82.80** | 3.39 | 25.30 | 54.01 | **82.71** |
| | $E_{1/2}$ | −60.77 | 34.88 | 139.82 | **113.18** | −60.70 | 33.10 | 140.65 | **113.05** |

a. Equations (4.4)–(4.6).
b. Equation (4.12).
c. Experimental shift = 110.6 ppm.
d. Experimental shift = 106.7 ppm.
*Source*: Adapted from [95].

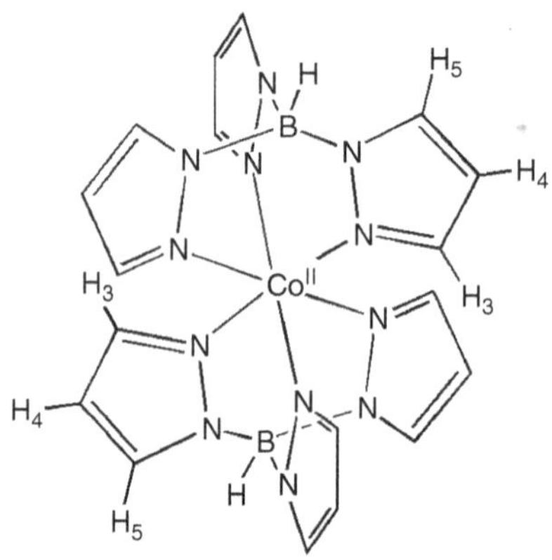


**Scheme 4.1** Structure of the $Co^{II}$ complex with pyrazolylborate.

A case study is the comparison of the prediction of $^{1}H$ chemical shift of NAMI, a ruthenium compound tested for its anticancer properties and now under clinical trials (Scheme 4.2) [102].

In 2010, Rastrelli and Bagno used NR calculations with the B3LYP hybrid and BP86 GGA functionals and a mixed basis set (cc-pVTZ for light atoms and DZV for Ru) and a relativistic approach at the ZORA SO level with the TZ2P basis set in an unrestricted calculation. They concluded that relativistic effects for $Ru^{III}$ complexes are not important (Table 4.5) [93]. The agreement was significantly improved by Autschbach *et al.* in 2011, who used a second-order perturbation theory and a relativistic ZORA Hamiltonian (Table 4.5). They performed two types of calculation: in the first type, SO coupling was introduced variationally in the ground state, and the hyperfine tensor was determined as a first-order derivative of the energy (route named "first-order" or EV), while in the second one, the SO coupling was considered as a perturbation of a NR calculation, the hyperfine tensor being calculated from double-perturbation theory as a second derivative of energy with respect to the nuclear spin angular momentum and the effective electronic spin (approach named "second-order" or "linear response" LR) [103]. In a third study, a four-component relativistic DFT protocol, including all the relativistic effects of the Dirac–Coulomb Hamiltonian theory, has been presented by Komorovsky *et al.* (mDKS) [87]. As pointed out by the authors, the chemical shifts are in good qualitative agreement with the experimental measurements and the deviation from the experimental data can be attributed to the geometry change in solution, suggesting that the vibrational and/or solvent effects should be taken into account [87].

Finally, a few words should be spent on the prediction of pNMR shift for the heavy metals. The studies are less numerous than those with $^{1}H$ and $^{13}C$ for a series of reasons: the metals have a large number of electrons that often occupy *f* orbitals, the relativistic effect must be considered, and the electronic structure is characterized by closely spaced energy levels. If the values of $\delta$ for $^{129}Xe$ and $^{181}Ta$, taken as two examples, are calculated with the GGA BP86 functional and a doubly-polarized basis sets (TZ2P), and relativistic effects treated with ZORA, the good results in Figure 4.8 are obtained [104]. Interestingly, with this protocol, the value of the coupling constant $J$ can be calculated with a good approximation in a similar manner to that used for determining $A$ in the EPR (see Eqs. (4.4)–(4.6)) [104].

**Table 4.4** Comparison of the $^{1}H$ chemical shifts calculated by DFT and *ab initio*/DFT as a function of the basis set and functional for the complex formed by $Co^{II}$ and pyrazolylborate.

| Nucleus[a] | DFT (PBE)[b] | DFT (PBE)[c] | CASSCF/ DFT (PBE)[b] | CASSCF/ DFT (PBE)[c] | NEVPT2/ DFT (PBE)[b] | NEVPT2/ DFT (PBE)[c] | NEVPT2/ DFT (PBE0)[c] | NEVPT2/ DFT (PBE50)[c] | Exptl. |
|---|---|---|---|---|---|---|---|---|---|
| $H_3$ | 45.0 | 35.0 | −74.2 | −82.9 | −74.1 | −82.5 | −92.2 | −103.2 | −111.0 |
| $H_4$ | 51.1 | 59.9 | 41.4 | 50.8 | 41.6 | 51.2 | 38.3 | 24.7 | 42.0 |
| $H_5$ | 39.9 | 52.7 | 96.6 | 110.4 | 96.9 | 110.8 | 95.5 | 88.4 | 94.2 |
| B-H | −4.0 | −4.5 | 133.4 | 130.5 | 133.5 | 130.4 | 128.5 | 133.8 | 122.2 |

a. See Scheme 4.1 for the numbering of protons.
b. Basis set SVP.
c. Basis set TZVP.
*Source*: Adapted from [94].

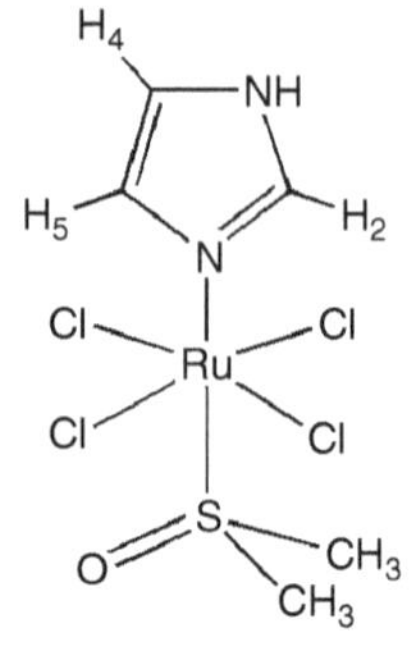


**Scheme 4.2** Structure of NAMI.

**Table 4.5** pNMR $^1H$ chemical shifts for the pharmacologically active $Ru^{III}$ complex NAMI.

| | | Ref. [93] | | Ref. [103] | | Ref. [87] |
|---|---|---|---|---|---|---|
| **Proton**[a] | **Exptl.** | **NR** | **ZORA** | **EV** | **LR** | **mDKS** |
| $CH_3$ | −14.5 | −3.6 | −5.8 | −8.9 | −9.4 | −12.1 |
| $H_2$ | −5.6 | 10.2 | 11.4 | 3.7 | −0.6 | 8.0 |
| $H_4$ | −7.8 | 6.2 | 5.6 | −2.0 | −6.8 | −2.7 |
| $H_5$ | −3.5 | 2.8 | 2.0 | −4.2 | −4.2 | −2.7 |

a. See Scheme 4.2 for the numbering of protons.

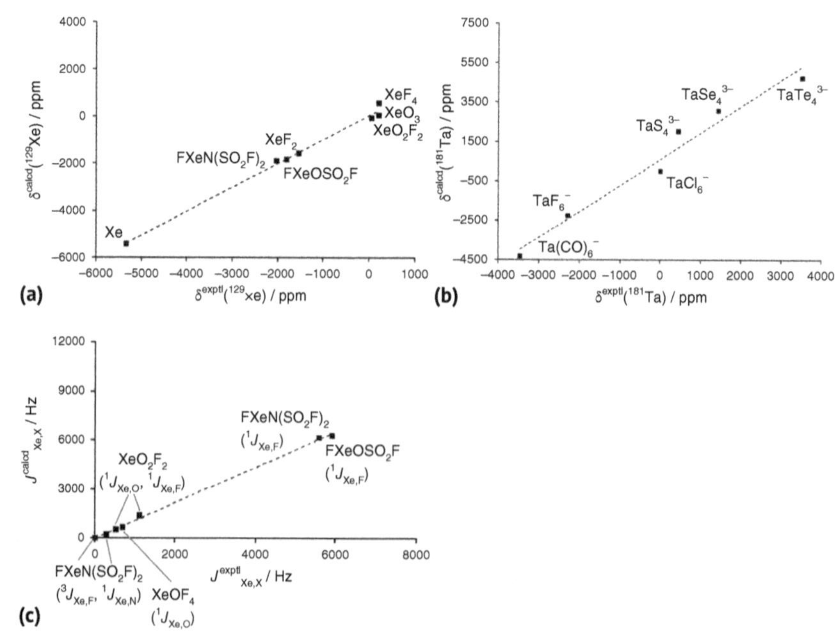


**Figure 4.8** Experimental and calculated (functional BP86, basis set TZ2P with ZORA for relativistic effect): (a) chemical shifts for Xe species; (b) chemical shifts for Ta species and (c) coupling constants for Xe species. With the dotted lines the best linear fitting is shown ($R^2$ for the fitting of the experimental data is 0.99 for graph a, 0.97 for graph b and 0.99 for graph c). *Source*: Adapted from [104].

## 4.6 Electron Absorption Spectroscopy

UV-Vis or electronic absorption spectroscopy is probably the simplest instrumental method to study the electronic transitions of organic and inorganic compounds. For a metal complex, the energy of the transitions can be related to the oxidation state, geometry, and electronic structure of a chemical species as well as the type of transition, i.e. *d-d* absorptions, metal-to-ligand charge transfers, ligand-to-metal charge transfers, and ligand-to-ligand charge transfers.

Time-dependent density functional theory (TD-DFT) [105] is an extension of DFT, and is the method most used to calculate the structure and electronic transitions of metallocompounds. TD-DFT describes a given transition in terms of a linear combination of vertical excitations from occupied to virtual MOs and, consequently, only the dominant character of each transition can be specified. TD-DFT also provides the electric dipole oscillator strength ($f$) of the transition from the ground to the excited state, which is related to the transition moment and value of $\varepsilon$. In contrast to EPR and NMR predictions, up to now only a qualitative agreement between an experimental and calculated spectrum can be reached [21, 106]. In other words, it is not possible to calculate the exact wavelength of these bands, and often considerable PDs from the experimental values are observed (see, for example, Ref. [107] for vanadium complexes). However, the results can be helpful in the transition assignment and the electronic structure description of a metal complex. The simulation of the electronic absorption spectra using TD-DFT was frequently discussed, and a number of reviews are available in the literature [21, 108].

The vertical excitations can be obtained within the linear response non-equilibrium (LRNE) [109], the linear response equilibrium [110], the state specific non-equilibrium (SSNE) [111], and state specific equilibrium (SSE) [111] frameworks (using a specific combination of functional and basis set). A Franck–Condon analysis is not always applicable for metal species: when the factor $|\langle\Psi_i|\Psi_f\rangle|^2$ is less than $10^{-4}$, the geometry of the excited state differs significantly from that of the ground state.

Often, the prediction of an electronic absorption spectrum reduces to choosing an appropriate combination of functional and basis sets. The effect of solvation as a specific correction for the excited states can also be considered, and comprehensive studies and reviews are available [108a, 112, 113].

The data discussed in the literature suggest that, at the moment, it is not possible to find, in the framework of TD-DFT calculations, a general computational condition to get a good agreement between an experimental and a calculated spectrum of a metal complex and the case of a non-oxido $V^{IV}$ compound formed by a dithiocarbazate-based tridentate Schiff-base ligand with (O, N, S) coordination may be an interesting example [107]. The spectrum has been simulated by TD-DFT methods using several functionals (VSXC, HCTH, BP86, B1LYP, B3LYP, PBE0, mPW1PW91, HSEH1PBE, BHandHLYP, APFD, WB97XD, and CAMB3LYP) and a double-$\zeta$ basis set (6-31g for V, O, C, H, and N, and 6-31+g(d) for S). In general, the order of performance of the tested functionals is: long-range corrected > half-and-half hybrid > hybrid > standalone. In Figure 4.9, the spectra calculated with one

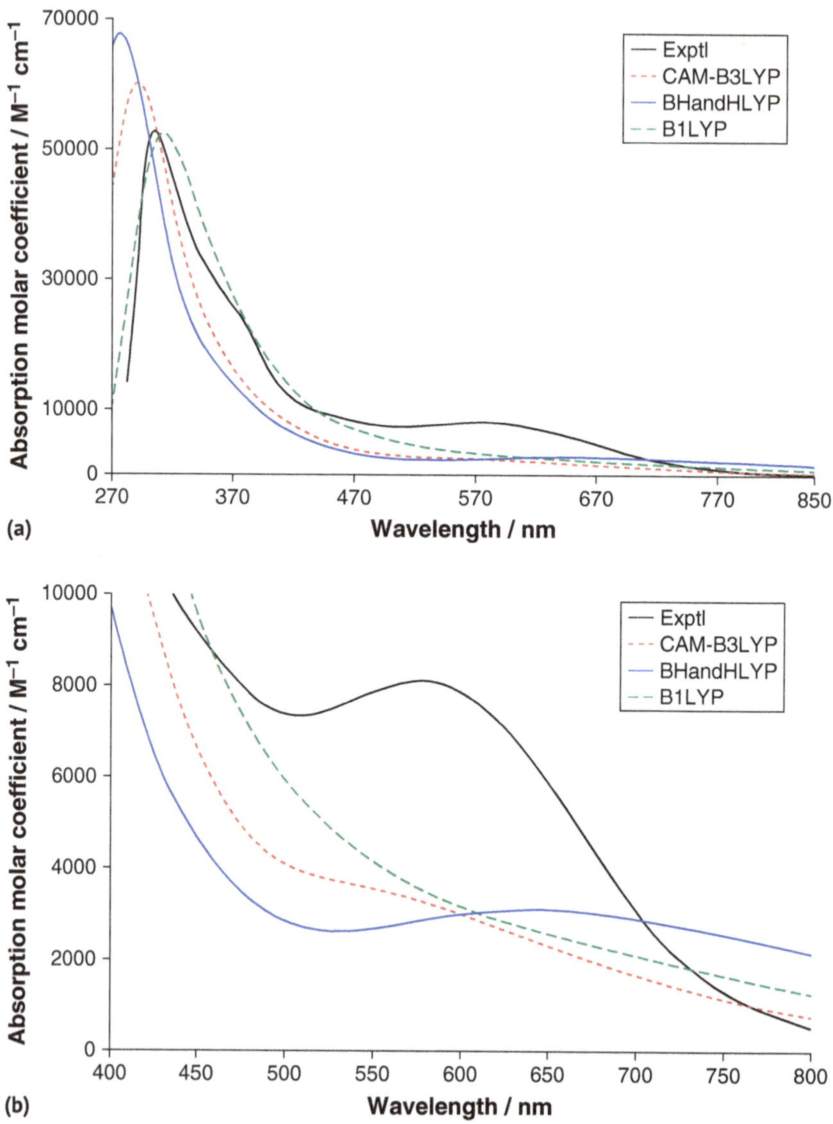


**Figure 4.9** Experimental and calculated electronic absorption spectra of a nonoxido $V^{IV}$ compound formed by a dithiocarbazate-based tridentate Schiff-base ligand: (a) full spectrum and (b) visible region. The experimental and calculated spectra with CAM-B3LYP, BHandHLYP and B1LYP functionals are shown in black, red, blue, and green, respectively. *Source*: Adapted from [107].

long-range corrected functional (CAM-B3LYP), one half-and-half hybrid functional (BHandHLYP) and one hybrid functional (B1LYP) are shown. It can be noticed that all the functionals reproduce qualitatively the experimental spectrum and, particularly, the intense absorptions in the UV region and the broad bands in the visible, but none of them is able to provide a quantitative prediction.

A study on four $Ir^{III}$ and seven $Pt^{II}$ complexes has shown that local functionals (GGA such as BP86, and meta-GGA) underestimate strongly the excitation energies with respect to the experimental values, while functionals such as CAM-B3LYP and M06-2X overestimate these energies [114]. The best performance was given from hybrid functionals with the same exchange part independently of the correlation functional; mainly, the prediction of B3LYP and B3PW91 is comparable with that of PBE0 and MPW1PW91. As observed for other spectroscopic techniques, the effect of the basis set is less relevant; for example, the triple-$\zeta$ polarized basis set *def*2-TZVP does not allow improving the prediction of the transition energies compared to a double-$\zeta$ such as LANL2DZ [114].

The TD-DFT methods were applied to the simulations of the electronic absorption spectra of $Cu^{II}$ and $Zn^{II}$ complexes through 41 exchange-correlation (XC) functionals, including pure GGA, hybrid GGA, pure meta-GGA and hybrid meta-GGA [115]. Pure and hybrid functionals and restricted and unrestricted calculations were compared. Hybrid methods were found to perform better than all pure DFT functionals with B1LYP, B97-2, B97-1, X3LYP, and B98 giving the best agreement in both restricted and unrestricted simulations. Among the functionals tested, B1LYP gave the most accurate results. The XC functional performance in spin-restricted TD-DFT calculations on the zinc complexes was found to be slightly worse and with PBE1PBE, mPW1PW91, and B1LYP the most accurate results were obtained [115].

A case study concerns the spectra of $Ni^{II}$ ion with peptides (GlyGlyGly, GlyGlyGlyGly, GlyGlyHis, and GlyGlyCys). These are important because they are models for the study of metal–protein interactions and GlyGlyHis represents the NTS (see Section 4.3) of albumin. The biological implication is that more than 90% of $Ni^{II}$ in the blood is transported by albumin-bound at NTS with the coordination mode ($NH_2$, $N^-$, $N^-$, His-N) [55]. A validation was preliminarily performed with 15 functionals and 2 basis sets [116]. On the basis of the MAPD values, the following ranking was found: HSE06 ~ MPW1PW91 ~ PBE0 > $\omega$-B97x-D ~ B3P86 ~ B3LYP ~ CAM-B3LYP > PBE ~ BLYP ~ BP86 > TPSS > TPSSh > BHandHLYP > M06 >> M06-2X. An analysis of the error dispersion based on SD allows reaching a similar conclusion, suggesting that the most accurate functional is HSE06. Between the two tested basis sets (LANL2DZ and *def2*-TZVP), the increment of the basis functions and addition of the polarization effect from LANL2DZ to *def2*-TZVP gives a progress in the quantitative prediction of the spectra. The electronic transitions were calculated at TD-DFT level of theory, using PCM for water and starting from the geometries optimized for the ground electronic state within the LRNE framework. In Figure 4.10, the comparison between the experimental and calculated spectra for $[Ni(H_{-2}GlyGlyCys)]^{2-}$ ($NH_2$, $N^-$, $N^-$, Cys-$S^-$ coordination, with $H_{-2}$ indicating the deprotonation of two amide groups that would not deprotonate in the absence of the metal ion) varying the used functionals is shown.

It can be interesting to investigate the geometry optimization's effect on the predicted spectrum's quality using $[Ni(H_{-2}GlyGlyCys)]^{2-}$ as a benchmark. The structure of the $Ni^{II}$ species was optimized with all the fifteen functionals, and the electronic transitions were computed on each geometry with the same functional applied for the optimization. The results indicate that, except for the worst functionals (TPSSh, BHandHLYP, M06, and M06-2X), the difference is very small (below 3%), and this can be related to the good performance of functionals such as

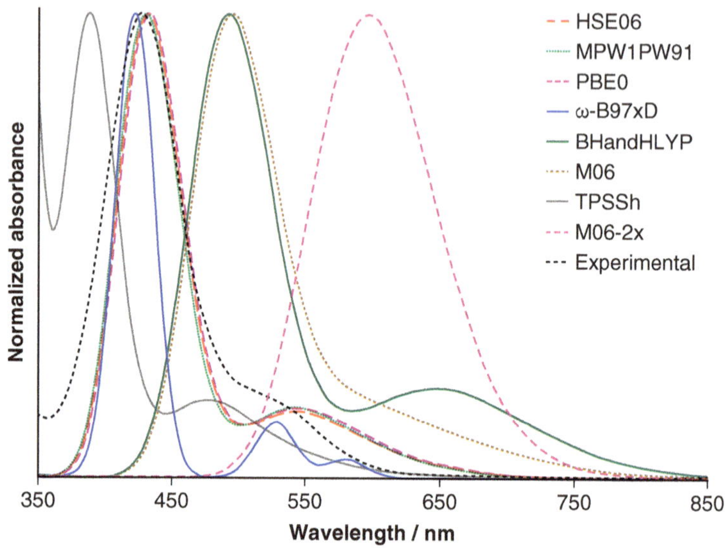


**Figure 4.10** Experimental and calculated spectra of $[Ni(H_{-2}GlyGlyCys)]^{2-}$ using *def2*-TZVP basis set and varying the functionals. The spectral absorbance was normalized for clarity. *Source*: Adapted from [116].

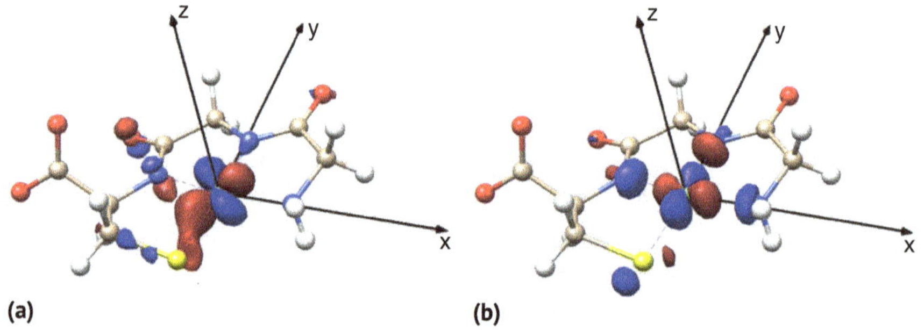


**Figure 4.11** MOs involved in the most intense transition in the visible region for $[Ni(H_{-2}GlyGlyCys)]^{2-}$, calculated by TD-DFT methods: (a) HOMO-12 and (b) LUMO. These two MOs are based mainly on the Ni-$d_{xy}$ and Ni-$d_{x^2-y^2}$ orbitals, respectively. *Source*: Adapted from [116].

MPW1PW91, PBE0, B3P86, B3LYP, BLYP, or BP86 in the structure prediction of the first-row transition metal complexes [117].

Finally, TD-DFT calculations allow attributing the dominant character of the electronic transitions of an electronic spectrum. For square planar complexes formed with $Ni^{II}$ by GlyGlyGly, GlyGlyGlyGly, GlyGlyHis, and GlyGlyCys, the main character of the most intense band in the visible region is due to the transition from the occupied MO based on Ni-$d_{xy}$ orbital to the vacant MO based on Ni-$d_{x^2-y^2}$. It can be interesting to observe that the percent amount of the Ni-*d* orbitals in the MOs involved in this transition is in the range 44–68%; the graphical representation of the MOs for $[Ni(H_{-2}GlyGlyCys)]^{2-}$is shown in Figure 4.11.

## 4.7 Circular Dichroism and Magnetic Circular Dichroism

CD is a phenomenon related to the presence of a chiral center in a chemical compound. The center of chirality may be represented by a metal or a chiral ligand. In a typical experiment, left and right circularly polarized light is generated and passed in an alternated mode through the sample compartment. A chiral compound absorbs differently the two beams, whose intensity is compared by the detector. In a CD spectrum, the difference between the molar absorption coefficients for left and right circularly polarized light, $\Delta\varepsilon = \varepsilon_L - \varepsilon_D$, is represented as a function of wavelength.

CD spectroscopy is generally used to study chiral organic and supramolecular compounds [118–120], and, in bioinorganic systems, can provide valuable information on the binding of metal ions or metal complexes with chiral ligands, for example, amino acids or proteins [121, 122].

MCD is a technique related to CD, in which a paramagnetic molecule (a metal complex, for example) is placed in a magnetic field with a direction perpendicular to that of the propagation of the polarized light. The origin of the MCD signal is due to the Zeeman splitting of the degenerate energy levels and the mixing of the electronic states by the magnetic field. MCD is a fundamental technique for the study of geometry and electronic structure of transition metal species [123]. The features that make this technique particularly useful in bioinorganic chemistry are: it allows acquiring spectra with a resolution higher than UV-Vis spectra and providing information both on the ground and excited states, revealing absorption bands in the near-IR region and studying species with an even number of electrons (difficult to be investigated by EPR) and, finally, examining the individual centers of different chromophores.

The problem of simulating a MCD spectra is to predict the *A*, *B*, and *C* terms that are related to the spectral intensity through Eq. (4.13) [124].

$$\frac{\Delta\varepsilon}{E} = \gamma B\left[-A\left(\frac{\partial f(E)}{\partial E}\right) + \left(B + \frac{C}{kT}\right)f(E)\right] \tag{4.13}$$

where $E$ is the energy of the transition, $f(E)$ is the line shape function, and the factor $\gamma$ is $(2\beta_e N_A \pi^3 \log_{10} e/250hc)$.

TD-DFT methods can be applied to calculate these terms according to the procedure proposed by Ziegler et al. [125–127]. Even if the results were rather satisfactory, the prediction of the *C* term, which is most relevant for metal and biocomplexes, is often inaccurate. A general approach includes the SO and spin–spin (SS) coupling and is based on the MRCI methods (MRCI-MCD); the ZFS derived from the interaction of SO and SS coupling and the magnetic field splitting are considered for the ground and the excited energy states. This approach is available in the ORCA software under the MRCI block and is based on the direct calculation of the transition energies and transition probabilities between the magnetic levels.

CD and MCD spectra of small models of plastocyanin and azurin, two blue copper proteins that behave as electron "shuttles" in bacteria and plants, were simulated at the TDDFT/BP86 level of theory [128]. Due to the unusual distorted geometry (tetrahedral in plastocyanin and trigonal bipyramidal in azurin) with the coordination of N, O, and S donors, the CD spectra are characterized by intense absorptions in the visible region. XRD geometries were used for the calculations since a DFT optimization would not be able to predict the distortion caused by the protein conformation and interaction of the donor groups with the residues surrounding the metal site. The results indicate a significant discrepancy in the band positions and the MOs calculations compared to the results of the ligand field theory. The blue shift of the calculated bands has been attributed to the overestimation by BP86 functional of the covalency of the metal bonds (in particular the Cu–$S_{Cys}$) [128].

A procedure that permits accounting for the ZFS due to the SO and SS coupling interactions in the ground and excited states allowed the simultaneous calculations of *A*, *B*, and *C* terms. With this approach, the MCD spectra of simple inorganic species such as OH, NH, and CH radicals, as well as coordination compounds like $[Fe^{III}(CN)_6]^{3-}$, were accurately predicted [129]. The protocol was also applied to assign the transitions and the temperature dependence of unusual mononuclear and dinuclear $Fe^{IV}O$ species: $[Fe^{IV}O(L^1)(MeCN)]^{2+}$ and $[Fe^{IV}O(L^2)(MeCN)]^{2+}$ ($L^1$ = 3,9,14,20-tetraaza-1,6,12,17-tetraazoniapentacyclohexacosane-1(23),4,6(26), 10,12(25),15,17(24),21-octane and $L^2$ = tris(3,5-dimethyl-4-methoxylpyridyl-2-methyl)amine) and $[(L^2)Fe^{IV}O(\mu\text{-}O)Fe^{IV}(OH)(L^2)]^{3+}$ [130, 131].

Very recently, a multiconfigurational restricted active space (RAS) approach was applied to the calculation of MCD spectra, through the prediction of *A*, *B*, and/or *C*-terms, for closed- and open-shell transition metal complexes such as $[Pd^{II}Cl_4]^{2-}$, $[Pd^{II}Br_4]^{2-}$, $[Au^{III}Cl_4]^-$, $[Au^{III}Br_4]^-$, $[Mn^{VII}O_4]^-$, $[Cu^{II}Cl_4]^{2-}$, $[Cu^{II}Br_4]^{2-}$, and $[Fe^{III}(CN)_6)]^{3-}$ [132]. The structure optimization was performed with KS theory using all-electron doubly-polarized triple-ζ (TZ2P) and ZORA all-electron relativistic Hamiltonian for all the species. It was found that the RAS level allows obtaining an accurate prediction of the spectra and determining unambiguously the sign and magnitude of the individual MCD terms, which can be related to a specific transition. The results indicate that: (i) in the case of closed-shell complexes such as $[Pd^{II}Cl_4]^{2-}$, $[Au^{III}Cl_4]^{2-}$, or $[Mn^{VII}O_4]^-$ the experimental spectra are well reproduced from the *A* and *B* terms (Figure 4.12a,b). For open-shell species such as $[Cu^{II}Cl_4]^{2-}$ and $[Cu^{II}Br_4]^{2-}$ the SO coupling must be included, and the *C* term must be taken into account; if this protocol is followed, the results are in good agreement with the experiment (Figure 4.12c) [132].

The spectrum of $[Fe^{III}(CN)_6)]^{3-}$ was studied for many years, but the assignment of the transitions was not straightforward. The main problem is related to the orbitally degenerate ground state that gives rise to a *C* term spectrum even without the SO coupling. Initially, the spectrum was interpreted only with the *C* term [133], but subsequently it was described with the predominant *B* term. Ganyushin and Neese satisfactorily calculated the spectrum, but without finding *A*, *B*, and *C* individually [129]. The recent RAS-level protocol of Heit *et al.* gave a very good agreement with the experimental measurement and, in addition, indicated that, due to the importance of the SO coupling, the room temperature MCD spectrum of $[Fe^{III}(CN)_6]^{3-}$ is generated both by *B* and *C* terms (Figure 4.12d) [132].

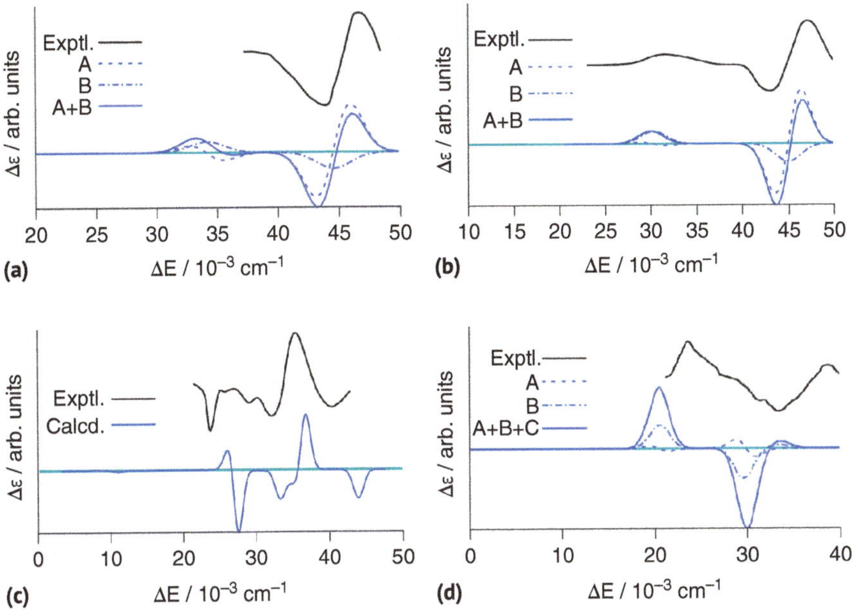


**Figure 4.12** Calculated and experimental MCD spectra for: (a) $[Pd^{II}Cl_4]^{2-}$, (b) $[Au^{III}Cl_4]^{-}$, (c) $[Cu^{II}Cl_4]^{2-}$, and (d) $[Fe^{III}(CN)_6)]^{3-}$. *Source*: Adapted from [132].

## 4.8 Vibrational Spectroscopy

Vibrational spectroscopy includes several techniques, among which IR and Raman. Both IR and Raman provide information on vibrations that are useful for elucidating the molecular structure. They are complementary techniques and allow for determining a molecule's vibrational modes completely. Even if some vibrations may be active in both Raman and IR, these techniques involve different processes and are characterized by different selection rules. In general, the maximum number of normal (or fundamental) vibrations for a linear molecule is $(3N - 5)$ and for a nonlinear molecule is $(3N - 6)$, where $N$ is the total number of atoms. Besides the normal vibrations ($\nu_1$, $\nu_2$, $\nu_3$, and so on), overtones (at $2\nu_1$, $3\nu_1$, etc.), combination (at $\nu_1 + \nu_2$, $\nu_1 + \nu_2 + \nu_3$, etc.) and difference bands (at $\nu_1 - \nu_2$, $\nu_1 + \nu_2 - \nu_3$, etc.) can appear due to the anharmonic motions of the molecules. A vibration is IR-active if there is a change in the dipole moment upon the transition, while it is Raman-active if there is a change in the polarizability of the molecule [134].

Software can calculate the vibrations of molecules in the ground and excited states and predict the wavenumber and intensity of the transitions. The molecular frequencies can be calculated from the second derivative of the energy with respect to the nuclear positions, and this is possible with HF, DFT MP2, and CASSCF methods. Graphical suites, such as GaussView with Gaussian, allow users to visualize the vibrations obtained.

The prediction of frequency is valid only at stationary points on the potential energy surface and, therefore, must be carried out on an optimized structure. For this reason, frequency calculations are used to determine the structural features of a given compound. If it corresponds to an energy minimum on the hypersurface of the potential energy all the frequencies are positive, while if it represents a saddle point, the frequency in the output is negative (imaginary), indicating that it does not correspond to a minimum. The transition states are usually characterized by only one imaginary frequency since they are first-order saddle points.

The frequencies have been predicted by DFT with contrasting results [21]. They deviate systematically from the experimental values, and must be corrected with scale factors. Several scale factors were proposed that depend on the combination of functional and basis set; the values of such factors are different for harmonic vibrational frequencies (in turn divided into low- and high-energy frequencies, <1000 and >1000 $cm^{-1}$, respectively), for vibrational contribution to enthalpy and entropy, and for zero-point vibrational energy [135]. For example, for the harmonic vibrational frequencies calculated at the level of theory B3LYP/6-31G(d), the numerical factors – that vary slightly with the database – are in the range 0.9614–0.9800 [136].

For the organic molecules, Zvereva *et al.* found that B3LYP is the best functional for the prediction of the harmonic frequencies and vibrational intensities for a dataset containing 17 small- and medium-sized compounds [137]; when the anharmonicity is considered, B3LYP and B97-1 give the best agreement with the experiment [138].

Vary recently, Katari *et al.* analyzed a benchmark containing 64 vibrational modes for 10 metallorganic compounds of Zn and Ru [136], and tested several hybrid and range-separated hybrid functionals such as CAM-B3LYP, LC-BLYP, and ωB97X-D. With the 6-31+g(d,p) basis set, the order of accuracy in predicting the vibrations is: B3LYP > $\omega$B97X-D > CAM-B3LYP > M06-2X > LC-BLYP, with deviations from the experimental values from −15 to 30 $cm^{-1}$. The use of the linear correlation analysis between the calculated and experimental data allowed finding an agreement much better than that predicted with the scaling factors [136].

The application of DFT methods to the prediction of the IR and Raman spectra of picoplatin, *cis*-$[PtCl_2(NH_3)(2\text{-picoline})]$, an anticancer agent that overcomes resistance to cell toxicity in small-cell lung cancer cells previously treated with cisplatin and carboplatin [139], was accomplished by Michalska et al. [140]. They found that PBE0 and CAM-B3LYP are the best functionals in the calculation of the whole spectrum and vibrations of the Pt–Cl and Pt–N bonds; in contrast, B3LYP gives accurate values for the stretching frequencies of the Pt–Cl bonds but underestimates the Pt–N vibrations (Figure 4.13).

## 4.9 Mossbauer Spectroscopy

Mössbauer spectroscopy is based on the Mössbauer effect, the nuclear resonance absorption without recoil of $\gamma$ radiation [141]. The effect was revealed for various nuclei, but the most common application concerns $^{57}Fe$. In a simple Mössbauer experiment, the energy of $\gamma$-ray emitted by the source is varied for the Doppler effect through the movement of the source relative to the sample (expressed in mm $s^{-1}$)

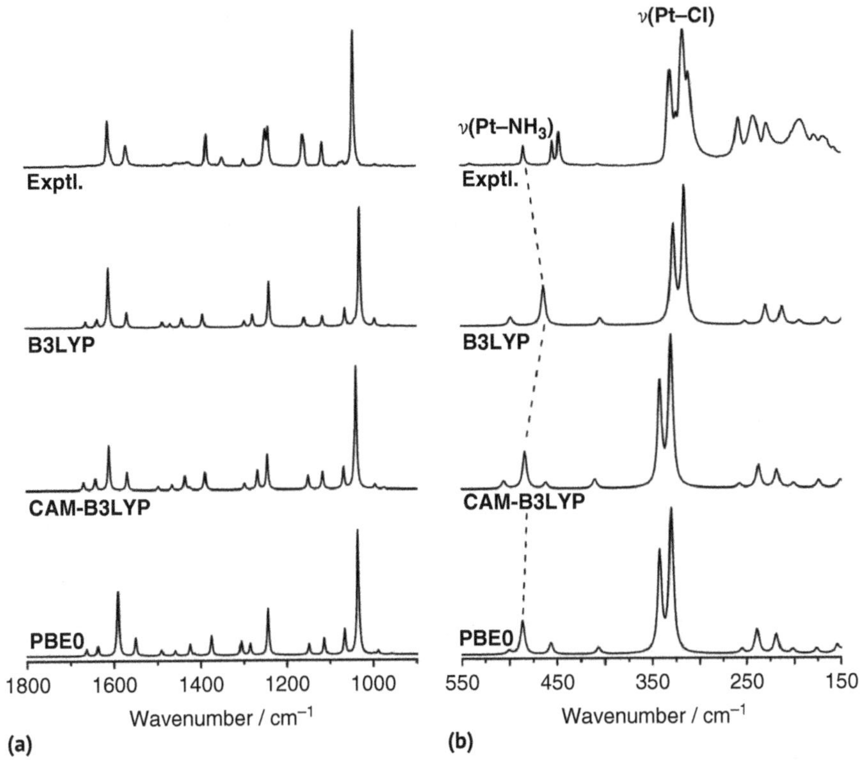


**Figure 4.13** Comparison between the experimental and calculated spectra of picoplatin: (a) Raman spectrum in the range 1800–900 $cm^{-1}$ and (b) IR spectrum in the range 550–150 $cm^{-1}$. The spectra were calculated with B3LYP, CAM-B3LYP, and PBE0 functional, and with a basis set obtained using LanL2TZ(f) on Pt, LanL2DZ on Cl and D95V(*d*,*p*) on the other elements. *Source*: Adapted from [140].

and in a spectrum the intensity of $\gamma$ radiation is plotted as a function of the source velocity (positive sign if the source moves toward the sample, negative sign in the opposite case [142]). Now, Mössbauer spectroscopy finds large use in chemistry and biology, and the two most important parameters extracted in a spectrum are the isomer shift ($\delta$) and the quadrupole splitting $(\Delta E_Q)$ [141]. The isomer shift (in mm $s^{-1}$) is influenced by the electron density at Fe nucleus and provides information on the oxidation number, spin state, and nature of the ligands bound to metal; in general, the shifts are low for high oxidation and spin states and for low coordination numbers [141]. The quadrupole splitting is the separation between the peaks (in mm $s^{-1}$) and depends on the asymmetry of the electric field around the iron center, giving information on the coordination geometry and chemical environment; for this reason, the assignment of the quadrupole splitting is straightforward for species with high symmetry (octahedral or tetrahedral complexes), while it is less easy when the symmetry decreases [141].

Since many variables influence the isomer shift and quadrupole splitting, computational methods – based mainly on DFT – were used to predict Mössbauer spectra [143]. The calculation of $\delta$ is related to the electron density of $s$ orbital at the nucleus ($\rho$) and can be found with the empirical formula [144]:

$$\delta = \alpha(\rho_0 - C) + \beta \tag{4.14}$$

where $\alpha$ is a constant which depends on the change in the distribution of the nuclear charge after the absorption, and $\rho_0$ is the electron density at the nucleus. The constants $\alpha$ and $\beta$ are determined by plotting the experimental shifts as a function of $\rho_0$ for a series of Fe complexes. Neese pointed out that, since the electron density depends on the functional and basis set employed for the simulations, the fitting must be performed for each couple of them [145]. Using the functional B3LYP and BP86 combined with a triple-$\zeta$ basis set (TZV), Neese et al. found a linear correlation between the experimental $\delta$ and $\rho_0$ that is valid for species with different electric charge, spin, iron oxidation state, and coordination number and geometry; with both the functionals, an accurate prediction of the isomer shift is obtained with an accuracy better than 0.10 mm s$^{-1}$ [144]. Bochevarov *et al.* confirmed this result, recommending also the use of functional O3LYP [146]. Subsequently, Pápai *et al.* tested the performance of six frequently used functionals (RPBE, OLYP, TPSS, TPSSh, B3LYP, and B3LYP*, where this latter is B3LYP with 15% of exact exchange) in the prediction of the isomer shift for a dataset containing 66 Fe complexes with different oxidation states, geometric isomerism, electronic structure, and ground spin state [147]; the effect of the basis set was also considered, and solvent was approximated with the COSMO model. The COSMO-B3LYP method allowed predicting $\delta$ with a MAD of 0.05 mm s$^{-1}$ and a maximum deviation of 0.12 mm s$^{-1}$ [147].

The quadrupole splitting is proportional to the largest component of the electric field gradient tensor $\boldsymbol{V}$, a traceless second-rank tensor, at the Fe nucleus:

$$\Delta E_Q = \frac{1}{2} eQV_{zz}\left(1 + \frac{\eta^2}{3}\right)^{1/2} \tag{4.15}$$

where $e$ is the electrical charge of the electron, $Q$ is the $^{57}$Fe nuclear quadrupole moment (approximately 0.16 barns), and $\eta$ is the asymmetry parameter given by the expression $\eta = (V_{xx} - V_{yy})/V_{zz}$, with $|V_{zz}| \geq |V_{yy}| \geq |V_{xx}|$.

The studies in the literature revealed larger errors in the calculation of $\Delta E_Q$ than $\delta$. It was suggested that the best functional to calculate $\Delta E_Q$ is O3LYP followed by B3LYP [146]. The results of Pápai *et al.* confirmed that an accurate quadrupole splitting prediction is still a big challenge; however, they demonstrated that to obtain a good agreement between theory and experiment, it is necessary to select appropriately the DFT method [147]. The TPSSh functional with COSMO approximation gives a satisfactory prediction for most of the Fe species, but there are three classes of Fe complexes for which a different computational protocol must be followed: for the species with $S = 3/2$ and $S = 2$ the COSMO-B3LYP and B3LYP methods, respectively, must be used, while for square-planar, $\pi$-bonded complexes with $S = 1$ and for largely distorted open-shell hexa-coordinate systems the combination of

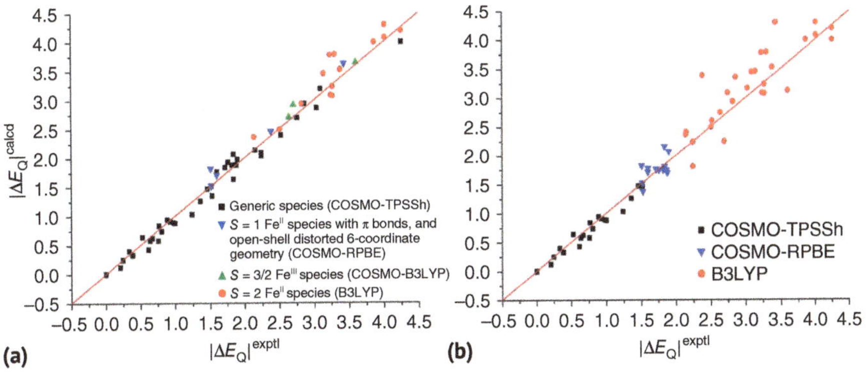


**Figure 4.14** Calculated and experimental quadrupole splitting dividing the 66 Fe complexes of the dataset in chemical classes (a) and on the basis of the $\Delta E_Q$ value (b). *Source*: Adapted from [147].

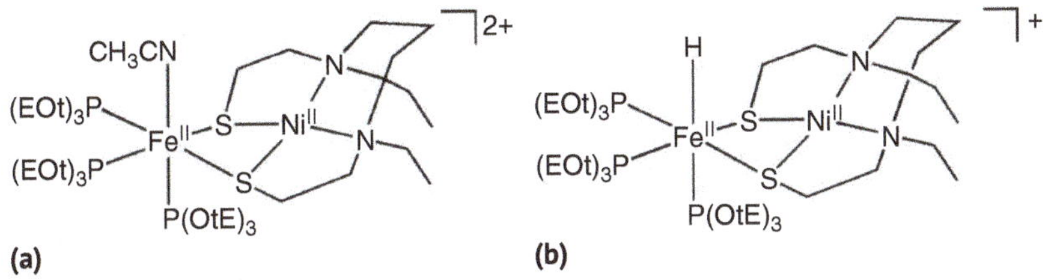


**Scheme 4.3** Structure of the two models **1** and **2** for [Ni–Fe] hydrogenase. *Source*: Adapted from [149].

COSMO with the GGA RPBE is to be preferred. If this protocol is applied to the dataset of 66 Fe species, MAD of 0.12 mm s$^{-1}$ (error 7%) and maximum deviation of 0.55 mm s$^{-1}$ (error 17%) are obtained (Figure 4.14a). The analysis of the data allows dividing the complexes into three classes characterized by the value of $\Delta E_Q$ (Figure 4.14b): (i) $\Delta E_Q \leq 1.5$ mm s$^{-1}$ (for which the method COSMO-TPSSh is the best), (ii) 1.5 mm s$^{-1}$ $< \Delta E_Q \leq 2.0$ mm s$^{-1}$ (COSMO-RPBE), (iii) $\Delta E_Q > 2.0$ mm s$^{-1}$ (B3LYP) [147]. This approach gives good results in the $\Delta E_Q \leq 2.0$ mm s$^{-1}$ region (deviations up to 0.32 mm s$^{-1}$ and errors of 17%), while in that $\Delta E_Q > 2.0$ mm s$^{-1}$ the method does not perform satisfactory (deviations up to 1.01 mm s$^{-1}$, and an error of 42%), suggesting that this approach is worse than that based on the chemical classification of iron species (presented in Figure 4.14a).

An interesting application in bioinorganic chemistry was discussed for two Ni–Fe complexes proposed as possible models for [Ni–Fe] hydrogenase [148], an extensively studied metalloenzyme involved in the transformation of molecular hydrogen. They were named **1** and **2** and are represented in Scheme 4.3. The prediction of the Mössbauer parameters was carried out on the structure optimized at the level BP86/*def*2-TZVP, using a NR method with the B3LYP functional and CP(PPP) basis set for Fe and *def*2-TZVP for the other atoms and approximating the crystal environment with COSMO, with acetonitrile as the solvent [149]. The isomer shift was calculated from the electron densities using the Eq. (4.14), with $\alpha = -0.369\,(\text{mm s}^{-1})/a_0^3$, $\beta = 2.863$ mm s$^{-1}$

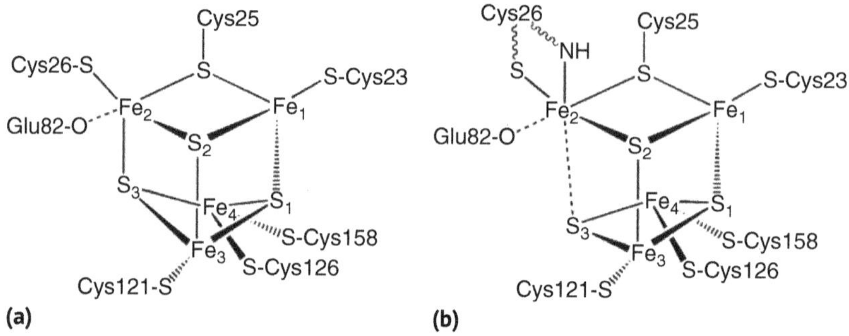


**Scheme 4.4** Structures of the [4Fe-3S] cluster proximal to the active [Ni–Fe] site of hydrogenase I from *Aquifer aeolicus*: (a) reduced form and (b) superoxidized form. *Source*: Adapted from [151].

**Table 4.6** Calculated (and experimental in parenthesis) Mössbauer parameters for the Fe sites of hydrogenase I from *Aquifer aeolicus* in the reduced and superoxidized form.

| | Reduced $[4Fe\text{-}3S]^{3+}$ form | | | Superoxidized $[4Fe\text{-}3S]^{5+}$ form | | |
|---|---|---|---|---|---|---|
| **Iron site** | $\delta$ | $\Delta E_Q$ | $\eta$ | $\delta$ | $\Delta E_Q$ | $\eta$ |
| Fe1 | 0.60 (0.71) | 1.41 (1.52) | 0.28 (0.3) | 0.39 (0.40) | −1.56 (−1.00) | 0.83 (0.7) |
| Fe2 | 0.35 (0.44) | 1.25 (1.23) | 0.99 (0.9) | 0.38 (0.46) | 2.25 (2.45) | 0.40 (0.5) |
| Fe3 | 0.36 (0.42) | 0.93 (0.84) | 0.30 (0.3) | 0.30 (0.39) | 0.81 (0.70) | 0.74 (1.0) |
| Fe4 | 0.46 (0.50) | 2.13 (2.60) | 0.51 (0.1) | 0.22 (0.28) | 0.74 (0.60) | 0.56 (0.3) |

*Source*: Adapted from [151].

and $C = 11810\ e/a_0^3$, with $a_0$ being the Bohr radius [149]. The predicted values of $\delta$, $\Delta E_Q$ and $\eta$ for **1** were 0.23 mm s$^{-1}$ (exptl: 0.25 mm s$^{-1}$), 0.29 mm s$^{-1}$ (exptl: 0.24 mm s$^{-1}$), and 0.2 (exptl: 0.3), and for **2** were 0.18 mm s$^{-1}$ (exptl: 0.18 mm s$^{-1}$), −1.75 mm s$^{-1}$ (exptl: −1.52 mm s$^{-1}$), and 0.4 (exptl: 0.4) [149].

With an analogous approach the Mössbauer parameters for the [4Fe-3S] cluster in the $O_2$ tolerant [Ni–Fe] hydrogenase I from *Aquifer aeolicus* have been predicted. It has been demonstrated that this cluster, located in proximity of the [Ni–Fe] center, protects the active site from the oxidative inactivation and can exist in two forms, reduced $[4Fe\text{-}3S]^{3+}$ and superoxidized $[4Fe\text{-}3S]^{5+}$ (shown in Scheme 4.4) [150]. The values for $\delta$, $\Delta E_Q$ and $\eta$ for the four iron atoms in the clusters are reported in Table 4.6 [151]: it is evident that the agreement is more than satisfactory.

## 4.10 Conclusions

The results discussed in this chapter emphasize the potentiality of the computational methods in bioinorganic chemistry, which allow the prediction of the parameters measured in the most used spectroscopic techniques and the formulation of valid hypotheses on the function, active site, and mechanism of action of several biomolecules. Depending on the system under examination, *ab initio*, semi-empirical, MM, QM/MM and – especially – DFT methods could be used.

Up to now, unfortunately, for most systems, there are no rules to know in advance which level of theory could give the best agreement with the experimental data, and in these cases, the best protocol must be validated on a dataset containing as many as possible metallospecies. For DFT the choice of an efficient functional is fundamental, but it depends on the molecular/spectroscopic properties and on the metal under examination, while the basis set must be large enough and include – if necessary – polarization and diffuse functions and relativistic effects. Whereas NMR and EPR can be calculated successfully for a large number of molecules, the prediction of vibrational and optical spectra is possible only qualitatively and a quantitative agreement has not been obtained yet for transition metal complexes. However, for NMR and EPR too, the choice of the best combination of functional and basis set is related to the metal ion. For example, if EPR spectroscopy is considered, for a metal ion such as $V^{IV}$, the prediction of the second-order SO contribution is negligible, and the best agreement with the experimental data is provided by two different functionals: BHandHLYP for oxidovanadium(IV) and B2PLYP for "bare" vanadium(IV). In contrast, for $Cu^{II}$ the SO effect must be included in the calculations and the functionals B3LYP and PBE0 show the best performances in the prediction of g and $A$ values.

## References


1 Frisch, M.J., Trucks, G.W., Schlegel, H.B. et al. *Gaussian 16, revision B.01*. Wallingford, CT: Gaussian, Inc.; 2016.

2 Neese, F. Software update: the ORCA program system–Version 5.0. *Wiley Interdiscip. Rev. Comput. Mol. Sci.* 2022;*12*:e1606.

3 Alexeev, Y., P. Mazanetz, M., Ichihara, O. et al. GAMESS as a free quantum-mechanical platform for drug research. *Curr. Top. Med. Chem.* 2012;*12*:2013–2033.

4 Baerends, E.J., Aguirre, N.F., Austin, N.D. et al. The Amsterdam modeling suite. *J. Chem. Phys.* 2025;*162*:162501.

5 Agarwal, M. and Frank, M.I. SPARTAN: a software tool for parallelization bottleneck analysis. In *ICSE Workshop on Multicore Software Engineering*; 2009; pp 56–63.

6 Franzke, Y.J., Holzer, C., Andersen, J.H. et al. TURBOMOLE: today and tomorrow. *J. Chem. Theory Comput.* 2023;*19*:6859–6890.

7 Bochevarov, A.D., Harder, E., Hughes, T.F. et al. Jaguar: a high-performance quantum chemistry software program with strengths in life and materials sciences. *Int. J. Quantum Chem.* 2013;*113*:2110–2142.

8 Dennington, R., Keith, T., and Millam, J. *GaussView, Version 6.1*, Semichem Inc., Shawnee Mission; 2019.

9 Portmann, S. and Lüthi, H.P. MOLEKEL: an interactive molecular graphics tool: computational chemistry column. *Chimia* 2000;*54*:766–770.

**10** Allouche, A.-R. Gabedit–A graphical user interface for computational chemistry softwares. *J. Comput. Chem.* 2011;*32*:174–182.

**11** Kohn, W. and Sham, L.J. Self-consistent equations including exchange and correlation effects. *Phys. Rev.* 1965;*140*:A1133–A1138.

**12** Cramer, C.J. *Essentials of Computational Chemistry, 2e*. Chichester, UK: Wiley; 2002.

**13** *Computational Chemistry and Molecular Modeling, Vol. 34*, Academic Press; 1999.

**14** Echenique, P. and Alonso, J.L. A mathematical and computational review of Hartree–Fock SCF methods in quantum chemistry. *Mol. Phys.* 2007;*105*: 3057–3098.

**15** Roos, B.O., Lindh, R., Malmqvist, P.Å. et al. *Multiconfigurational Quantum Chemistry*. Chichester, UK: John Wiley & Sons; 2016, pp. 43–58.

**16** Duch, W. Configuration interaction method: the past and future perspectives. *J. Mol. Struct. THEOCHEM*. 1991;*234*:27–49.

**17** Sherrill, C.D. and Schaefer, H.F. *Advances in Quantum Chemistry., Vol. 34* (ed. P.-O. Löwdin, J.R. Sabin, M.C. Zerner, et al.), Academic Press; 1999, pp. 143–269.

**18** Piela, L. *Ideas of Quantum Chemistry, 2e* (ed. L. Piela), Oxford, UK: Elsevier; 2014, pp. 577–662.

**19** Neese, F., Petrenko, T., Ganyushin, D. et al. Advanced aspects of ab initio theoretical optical spectroscopy of transition metal complexes: multiplets, spin-orbit coupling and resonance Raman intensities. *Coord. Chem. Rev.* 2007;*251*:288–327.

**20** Cremer, D. Møller–Plesset perturbation theory: from small molecule methods to methods for thousands of atoms. *Wiley Interdiscip. Rev. Comput. Mol. Sci.* 2011;*1*:509–530.

**21** Neese, F. Prediction of molecular properties and molecular spectroscopy with density functional theory: from fundamental theory to exchange-coupling. *Coord. Chem. Rev.* 2009;*253*:526–563.

**22** Mardirossian, N. and Head-Gordon, M. Thirty years of density functional theory in computational chemistry: an overview and extensive assessment of 200 density functionals. *Mol. Phys.* 2017;*115*:2315–2372.

**23** Pyykkö, P. Relativistic effects in chemistry: more common than you thought. *Annu. Rev. Phys. Chem.* 2012;*63*:45–64.

**24** Reiher, M. and Wolf, A. *Relativistic Quantum Chemistry: The Fundamental Theory of Molecular Science*. Weinheim, Germany: Wiley-VCH Verlag GmbH & Co. KGaA; 2009.

**25** Reiher, M. Relativistic Douglas–Kroll–Hess theory. *Wiley Interdiscip. Rev. Comput. Mol. Sci.* 2012;*2*:139–149.

**26** Mennucci, B. Polarizable continuum model. *Wiley Interdiscip. Rev. Comput. Mol. Sci.* 2012;*2*:386–404.

**27** Cossi, M., Barone, V., Cammi, R., et al. Ab initio study of solvated molecules: a new implementation of the polarizable continuum model. *Chem. Phys. Lett.* 1996;*255*:327–335.

**28** Marenich, A.V., Cramer, C.J., and Truhlar, D.G. Universal solvation model based on solute electron density and on a continuum model of the solvent defined by the bulk dielectric constant and atomic surface tensions. *J. Phys. Chem. B.* 2009;*113*:6378–6396.

**29** Klamt, A. Conductor-like screening model for real solvents: a new approach to the quantitative calculation of solvation phenomena. *J. Phys. Chem.* 1995; *99*:2224–2235.

**30** Klamt, A. The COSMO and COSMO-RS solvation models. *Wiley Interdiscip. Rev. Comput. Mol. Sci.* 2018;*8*:e1338.

**31** Grimme, S. Semiempirical hybrid density functional with perturbative second-order correlation. *J. Chem. Phys.* 2006;*124*:034108.

**32** (a) Schwabe, T. and Grimme, S. Towards chemical accuracy for the thermodynamics of large molecules: new hybrid density functionals including non-local correlation effects. *Phys. Chem. Chem. Phys.* 2006;*8*:4398–4401; (b) Grimme, S., Mück-Lichtenfeld, C., Würthwein, E.-U., et al. Consistent theoretical description of 1,3-dipolar cycloaddition reactions. *J. Phys. Chem. A* 2006;*110*:2583–2586; (c) Schwabe, T. and Grimme, S. Double-hybrid density functionals with long-range dispersion corrections: higher accuracy and extended applicability. *Phys. Chem. Chem. Phys.* 2007;*9*:3397–3406; (d) Schwabe, T. and Grimme, S. Theoretical thermodynamics for large molecules: walking the thin line between accuracy and computational cost. *Acc. Chem. Res.* 2008;*41*:569–579.

**33** Roessler, M.M. and Salvadori, E. Principles and applications of EPR spectroscopy in the chemical sciences. *Chem. Soc. Rev.* 2018;*47*:2534–2553.

**34** Gorelsky, S., Micera, G., and Garribba, E. Equilibrium between the octahedral and square pyramidal form and the influence of an axial ligand on the molecular properties of $V^{IV}O$ complexes: a spectroscopic and DFT study. *Chem.–Eur. J.* 2010;*16*:8167–8180.

**35** Neese, F. Metal and ligand hyperfine couplings in transition metal complexes: the effect of spin-orbit coupling as studied by coupled perturbed Kohn-Sham theory. *J. Chem. Phys.* 2003;*118*:3939–3948.

**36** Neese, F. *ORCA – An Ab Initio, DFT and Semiempirical Program Package, Version 5.0*, Mülheim an der Ruhr: Max-Planck-Institute for Chemical Energy Conversion; 2021.

**37** Munzarová, M.L., Kubáček, P., and Kaupp, M. Mechanisms of EPR hyperfine coupling in transition metal complexes. *J. Am. Chem. Soc.* 2000;*122*:11900–11913.

**38** van Lenthe, E., van der Avoird, A., and Wormer, P.E.S. Density functional calculations of molecular hyperfine interactions in the zero order regular approximation for relativistic effects. *J. Chem. Phys.* 1998;*108*:4783–4796.

**39** Remenyi, C., Reviakine, R., Arbuznikov, A.V. et al. Spin-orbit effects on hyperfine coupling tensors in transition metal complexes using hybrid density functionals and accurate spin-orbit operators. *J. Phys. Chem. A* 2004;*108*:5026–5033.

**40** Stein, M., van Lenthe, E., Baerends, E.J. et al. g- and A-tensor calculations in the zero-order approximation for relativistic effects of Ni complexes $Ni(mnt)_2$– and $Ni(CO)_3H$ as model complexes for the active center of [NiFe]-hydrogenase. *J. Phys. Chem. A* 2001;*105*:416–425.

**41** Micera, G. and Garribba, E. Is the spin-orbit coupling important in the prediction of the $^{51}V$ hyperfine coupling constants of $V^{IV}O^{2+}$ species? ORCA versus Gaussian performance and biological applications. *J. Comput. Chem.* 2011;*32*:2822–2835.

**42** Sciortino, G., Lubinu, G., Maréchal, J.-D. et al. DFT protocol for EPR prediction of paramagnetic Cu(II) complexes and application to protein binding sites. *Magnetochemistry*. 2018;*4*:55.

**43** Munzarová, M.L. *Calculation of NMR and EPR Parameters*,. (ed. M. Kaupp, M. Bühl, and V.G. Malkin). Weinheim, Germany: Wiley-VCH Verlag GmbH & Co. KGaA; 2004, pp. 461–482.

**44** Belanzoni, P., van Lenthe, E., and Baerends, E.J. An evaluation of the density functional approach in the zero order regular approximation for relativistic effects: magnetic interactions in small metal compounds. *J. Chem. Phys.* 2001;*114*:4421–4433.

**45** Pietrzyk, P., Podolska, K., and Sojka, Z. *Electron Paramagnetic Resonance, Vol. 23* (ed. B.C. Gilbert, D.M. Murphy, and V. Chechik). Cambridge, United Kingdom: The Royal Society of Chemistry; 2013, pp. 264–311.

**46** Conradie, M.M., Conradie, J., and Ghosh, A. Capturing the spin state diversity of iron(III)-aryl porphyrins: OLYP is better than TPSSh. *J. Inorg. Biochem.* 2011;*105*:84–91.

**47** Fritscher, J., Hrobárik, P., and Kaupp, M. Computational studies of electron paramagnetic resonance parameters for paramagnetic molybdenum complexes. 1. Method validation on small and medium-sized systems. *J. Phys. Chem. B* 2007;*111*:4616–4629.

**48** Sanna, D., Sciortino, G., Ugone, V. et al. Nonoxido $V^{IV}$ complexes: prediction of the EPR spectrum and electronic structure of simple coordination compounds and amavadin. *Inorg. Chem.* 2016;*55*:7373–7387.

**49** Rehder, D. *Bioinorganic Vanadium Chemistry*. Chichester, UK: John Wiley & Sons Ltd.; 2008.

**50** Silva, J.A.L., Fraústo da Silva, J.J.R., and Pombeiro, A.J.L. *Vanadium. Biochemical and Molecular Biological Approaches* (ed. H. Michibata). Dordrecht, Netherlands: Springer; 2012, pp. 35–49.

**51** Gillard, D.R. and Lancashire, J.R. Electron spin resonance of vanadium in *Amanita muscaria*. *Phytochemistry* 1984;*23*:179–180.

**52** Remenyi, C., Munzarová, M.L., and Kaupp, M. Comparative density-functional study of the electron paramagnetic resonance parameters of amavadin. *J. Phys. Chem. B* 2005;*109*:4227–4233.

**53** Singh, S.K., Atanasov, M., and Neese, F. Challenges in multireference perturbation theory for the calculations of the g-tensor of first-row transition-metal complexes. *J. Chem. Theory Comput.* 2018;*14*:4662–4677.

**54** Medvedev, M.G., Bushmarinov, I.S., Sun, J. et al. Density functional theory is straying from the path toward the exact functional. *Science* 2017;*355*:49–52.

**55** Harford, C. and Sarkar, B. Amino terminal Cu(II)- and Ni(II)-binding (ATCUN) motif of proteins and peptides: metal binding, DNA cleavage, and other properties. *Acc. Chem. Res.* 1997;*30*:123–130.

**56** Prusiner, S.B. Biology and genetics of prions causing neurodegeneration. *Annu. Rev. Genet.* 2013;*47*:601–623.

**57** Saá, P., Harris, D.A., and Cervenakova, L. Mechanisms of prion-induced neurodegeneration. *Expert Rev. Mol. Med.* 2016;*18*:e5.

**58** Giachin, G., Mai, P.T., Tran, T.H. et al. The non-octarepeat copper binding site of the prion protein is a key regulator of prion conversion. *Sci. Rep.* 2015;*5*:15253.

**59** Di Natale, G., Grasso, G., Impellizzeri, G. et al. Copper(II) interaction with unstructured prion domain outside the octarepeat region: speciation, stability, and binding details of copper(II) complexes with PrP106–126 peptides. *Inorg. Chem.* 2005;*44*:7214–7225.

**60** Smith, T.S., LoBrutto, R., and Pecoraro, V.L. Paramagnetic spectroscopy of vanadyl complexes and its applications to biological systems. *Coord. Chem. Rev.* 2002; *228*:1–18.

**61** Murphy, D.M. and Farley, R.D. Principles and applications of ENDOR spectroscopy for structure determination in solution and disordered matrices. *Chem. Soc. Rev.* 2006;*35*:249–268.

**62** Schweiger, A. Pulsed electron spin resonance spectroscopy: basic principles, techniques, and examples of applications. *Angew. Chem., Int. Ed. Engl.* 1991; *30*:265–292.

**63** Deligiannakis, Y., Louloudi, M., and Hadjiliadis, N. Electron spin echo envelope modulation (ESEEM) spectroscopy as a tool to investigate the coordination environment of metal centers. *Coord. Chem. Rev.* 2000;*204*:1–112.

**64** Neese, F. Theoretical study of ligand superhyperfine structure. Application to Cu(II) complexes. *J. Phys. Chem. A* 2001;*105*:4290–4299.

**65** Saladino, A.C. and Larsen, S.C. Density functional theory calculations of nitrogen hyperfine and quadrupole coupling constants in oxovanadium(IV) complexes. *J. Phys. Chem. A* 2003;*107*:4735–4740.

**66** Fukui, K., Ohya-Nishiguchi, H., Kamada, H. et al. Hyperfine and nuclear quadrupole couplings of the coordinated nitrogen of oxovanadium(IV) phthalocyanines studied by electron spin-echo envelope modulation spectroscopy. *Bull. Chem. Soc. Jpn.* 1998;*71*:2787–2796.

**67** Eaton, S.S., Dubach, J., More, K.M. et al. Comparison of the electron spin echo envelope modulation (ESEEM) for human lactoferrin and transferrin complexes of copper(II) and vanadyl ion. *J. Biol. Chem.* 1989;*264*:4776–4781.

**68** Fukui, K., Ueki, T., Ohya, H. et al. Vanadium-binding protein in a vanadium-rich ascidian ascidia sydneiensis samea: CW and pulsed EPR studies. *J. Am. Chem. Soc.* 2003;*125*:6352–6353.

**69** Reijerse, E.J., Shane, J., de Boer, E. et al. *Electron Magnetic Resonance of Disordered Systems*, (ed. N.D. Yordanov). Singapore: World Scientific; 1991, pp. 253–271.

**70** Fukui, K., Ohya-Nishiguchi, H., and Kamada, H. $^{14}N$ Coupling parameters in oxovanadium(IV)–amine, –imine, and –isothiocyanate complexes studied by electron spin echo envelope modulation spectroscopy. *Inorg. Chem.* 1997; *36*:5518–5529.

**71** LoBrutto, R., Hamstra, B.J., Colpas, G.J. et al. Electron spin echo envelope modulation spectroscopy reveals and distinguishes equatorial and axial nitrogen ligands bound to $VO^{2+}$. *J. Am. Chem. Soc.* 1998;*120*:4410–4416.

**72** Kôichi, F., Hiroshi, F., Hiroaki, O.-N. et al. Electron spin-echo envelope modulation spectral properties of amidate nitrogen coordinated to oxovanadium(IV) ion. *Chem. Lett.* 2000;*29*:198–199.

**73** Sanna, D., Pecoraro, V.L., Micera, G. et al. Application of DFT methods to the study of the coordination environment of the $VO^{2+}$ ion in V proteins. *JBIC, J. Biol. Inorg. Chem.* 2012;*17*:773–790.

**74** Chasteen, D.N. *Met.* Ions Biol. Syst., *Vol. 31* (ed. H. Sigel and A. Sigel). New York: Marcel Dekker; 1995, pp. 231–247.

**75** Hisabori, T., Sunamura, E.-I., Kim, Y. et al. The chloroplast ATP synthase features the characteristic redox regulation machinery. *Antioxid. Redox Signaling* 2013; *19*:1846–1854.

**76** Abrahams, J.P., Leslie, A.G.W., Lutter, R. et al. Structure at 2.8 Â resolution of F1-ATPase from bovine heart mitochondria. *Nature* 1994;*370*:621–628.

**77** Macomber, R.S. *A Complete Introduction to Modern NMR Spectroscopy*. New York: John Wiley & Sons, Inc.; 1998.

**78** Keeler, J. *Understanding NMR Spectroscopy, 1e*. Chichester, UK: Wiley; 2002.

**79** Bunce, N.J. Introduction to the interpretation of electron spin resonance spectra of organic radicals. *J. Chem. Educ.* 1987;*64*:907–914.

**80** Kurland, R.J. and McGarvey, B.R. Isotropic NMR shifts in transition metal complexes: the calculation of the fermi contact and pseudocontact terms. *J. Magn. Reson.* 1970;*2*:286–301.

**81** Bertini, I., Luchinat, C., and Parigi, G. Magnetic susceptibility in paramagnetic NMR. *Prog. Nucl. Magn. Reson. Spectrosc.* 2002;*40*:249–273.

**82** Hrobárik, P., Reviakine, R., Arbuznikov, A.V. et al. Density functional calculations of NMR shielding tensors for paramagnetic systems with arbitrary spin multiplicity: validation on 3*d* metallocenes. *J. Chem. Phys.* 2007;*126*:024107.

**83** Pennanen, T.O. and Vaara, J. Nuclear magnetic resonance chemical shift in an arbitrary electronic spin state. *Phys. Rev. Lett.* 2008;*100*:133002.

**84** Borgogno, A., Rastrelli, F., and Bagno, A. Characterization of paramagnetic reactive intermediates: predicting the NMR spectra of Iron(IV)–Oxo complexes by DFT. *Chem.–Eur. J.* 2015;*21*:12960–12970.

**85** Ramsey, N.F. Magnetic shielding of nuclei in molecules. *Phys. Rev.* 1950; *78*:699–703.

**86** Pyykkö, P. *Calculation of NMR and EPR Parameters* (ed. M. Kaupp, M. Bühl, and V.G. Malkin). Weinheim, Germany: Wiley-VCH Verlag GmbH & Co. KGaA; 2004; pp. 7–19.

**87** Komorovsky, S., Repisky, M., Ruud, K. et al. Four-component relativistic density functional theory calculations of NMR shielding tensors for paramagnetic systems. *J. Phys. Chem. A* 2013;*117*:14209–14219.

**88** McConnell, H.M. and Chesnut, D.B. Theory of isotropic hyperfine interactions in $\pi$-electron radicals. *J. Chem. Phys.* 1958;*28*:107–117.

**89** Bagno, A. and Bonchio, M. Relativistic DFT calculation of $^{99}$Ru NMR parameters: chemical shifts and spin–spin coupling constants. *Magn. Reson. Chem.* 2004; *42*:S79–S87.

**90** Moon, S. and Patchkovskii, S. *Calculation of NMR and EPR Parameters* (ed. M. Kaupp, M. Bühl, and V.G. Malkin). Weinheim, Germany: Wiley-VCH Verlag GmbH & Co. KGaA; 2004, pp. 325–338.

**91** Van den Heuvel, W. and Soncini, A. NMR chemical shift in an electronic state with arbitrary degeneracy. *Phys. Rev. Lett.* 2012;*109*:073001.

**92** Van den Heuvel, W. and Soncini, A. NMR chemical shift as analytical derivative of the Helmholtz free energy. *J. Chem. Phys.* 2013;*138*:054113.

**93** Rastrelli, F. and Bagno, A. Predicting the $^1$H and $^{13}$C NMR spectra of paramagnetic Ru(III) complexes by DFT. *Magn. Reson. Chem.* 2010;*48*:S132–S141.

**94** Rouf, S.A., Mareš, J., and Vaara, J. $^1$H chemical shifts in paramagnetic Co(II) pyrazolylborate complexes: a first-principles study. *J. Chem. Theory Comput.* 2015;*11*:1683–1691.

**95** Gendron, F., Sharkas, K., and Autschbach, J. Calculating NMR chemical shifts for paramagnetic metal complexes from first-principles. *J. Phys. Chem. Lett.* 2015; *6*:2183–2188.

**96** Benda, L., Mareš, J., Ravera, E. et al. Pseudo-contact NMR shifts over the paramagnetic metalloprotein CoMMP-12 from first principles. *Angew. Chem., Int. Ed.* 2016;*55*:14713–14717.

**97** Stoychev, G.L., Auer, A.A., and Neese, F. Efficient and accurate prediction of nuclear magnetic resonance shielding tensors with double-hybrid density functional theory. *J. Chem. Theory Comput.* 2018;*14*:4756–4771.

**98** Bertini, I., Gray, H.B., Lippard, S.J. et al. *Bioinorganic Chemistry*. Mill Valley California: University Science Books; 1994.

**99** Rehder, D. *Bioinorganic Chemistry*. Oxford, UK: Oxford University Press; 2014.

**100** Reiher, M. Theoretical Study of the $Fe(phen)_2(NCS)_2$ Spin-crossover complex with reparametrized density functionals. *Inorg. Chem.* 2002;*41*:6928–6935.

**101** Aquino, F., Pritchard, B., and Autschbach, J. Scalar relativistic computations and localized orbital analyses of nuclear hyperfine coupling and paramagnetic NMR chemical shifts. *J. Chem. Theory Comput.* 2012;*8*:598–609.

**102** (a) Leijen, S., Burgers, S.A., Baas, P. et al. Phase I/II study with ruthenium compound NAMI-A and gemcitabine in patients with non-small cell lung cancer after first line therapy. *Invest. New Drugs* 2015;*33*:201–214; (b) Rademaker-Lakhai, J.M., van den Bongard, D., Pluim, D. et al. A Phase I and pharmacological study with imidazolium-*trans*-DMSO-imidazole-tetrachlororuthenate, a novel ruthenium anticancer agent. *Clinical Cancer Research*. 2004; *10*:3717–3727.

**103** Autschbach, J., Patchkovskii, S., and Pritchard, B. Calculation of hyperfine tensors and paramagnetic NMR shifts using the relativistic zeroth-order regular approximation and density functional theory. *J. Chem. Theory Comput.* 2011; *7*:2175–2188.

**104** Bagno, A. and Saielli, G. Computational NMR spectroscopy: reversing the information flow. *Theor. Chem. Acc.* 2007;*117*:603–619.

**105** (a) Casida, M.E., Chermette, H., and Jacquemin, D. Time-dependent density-functional theory for molecules and molecular solids. *J. Mol. Struct. Theochem* 2009;*914*:1–2; (b) Runge, E., and Gross, E.K.U. Density-functional theory for time-dependent systems. *Phys. Rev. Lett.* 1984;*52*:997–1000.

**106** Rosa, A., Ricciardi, G., Gritsenko, O. et al. *Principles and Applications of Density Functional Theory in Inorganic Chemistry I*, Berlin, Heidelberg, Germany: Springer Berlin Heidelberg; 2004, pp. 49–116.

**107** Kundu, S., Mondal, D., Bhattacharya, K. et al. Nonoxido vanadium(IV) compounds involving dithiocarbazate-based tridentate ONS ligands: synthesis, electronic and molecular structure, spectroscopic and redox properties. *Inorg. Chem.* 2015;*54*:6203–6215.

**108** (a) Caricato, M., Mennucci, B., Tomasi, J. et al. Formation and relaxation of excited states in solution: a new time dependent polarizable continuum model based on time dependent density functional theory. *J. Chem. Phys.* 2006;*124*:124520; (b) Jacquemin, D., Perpète, E.A., Ciofini, I. et al. Accurate simulation of optical properties in dyes. *Acc. Chem. Res.* 2009;*42*:326–334; (c) Adamo, C. and Jacquemin, D. The calculations of excited-state properties with time-dependent density functional theory. *Chem. Soc. Rev.* 2013;*42*:845–856; (d) Laurent, A.D. and Jacquemin, D. TD-DFT benchmarks: a review. *Int. J. Quantum Chem.* 2013;*113*:2019–2039.

**109** Cammi, R., Mennucci, B., and Tomasi, J. Fast evaluation of geometries and properties of excited molecules in solution: a Tamm-Dancoff model with application to 4-dimethylaminobenzonitrile. *J. Phys. Chem. A* 2000;*104*:5631–5637.

**110** Tomasi, J., Mennucci, B., and Cammi, R. Quantum mechanical continuum solvation models. *Chem. Rev.* 2005;*105*:2999–3094.

**111** Cammi, R., Corni, S., Mennucci, B. et al. Electronic excitation energies of molecules in solution: state specific and linear response methods for nonequilibrium continuum solvation models. *J. Chem. Phys.* 2005;*122*:104513.

**112** (a) DeFusco, A., Minezawa, N., Slipchenko, L.V. et al. Modeling solvent effects on electronic excited states. *J. Phys. Chem. Lett.* 2011;*2*:2184–2192; (b) Jacquemin, D., Planchat, A., Adamo, C. et al. TD-DFT Assessment of functionals for optical 0–0 transitions in solvated dyes. *J. Chem. Theory Comput.* 2012;*8*:2359–2372.

**113** Solomon, E.I. and Lever, A.B.P. (eds.) *Inorganic Electronic Structure and Spectroscopy*. New York: Wiley; 2006.

**114** Latouche, C., Skouteris, D., Palazzetti, F. et al. TD-DFT benchmark on inorganic Pt(II) and Ir(III) complexes. *J. Chem. Theory Comput.* 2015;*11*:3281–3289.

**115** Holland, J.P. and Green, J.C. Evaluation of exchange-correlation functionals for time-dependent density functional theory calculations on metal complexes. *J. Comput. Chem.* 2010;*31*:1008–1014.

**116** Sciortino, G., Lihi, N., Czine, T. et al. Accurate prediction of vertical electronic transitions of Ni(II) coordination compounds via time dependent density functional theory. *Int. J. Quantum Chem.* 2018; e25655.

**117** (a) Bühl, M. and Kabrede, H. Geometries of transition-metal complexes from density-functional theory. *J. Chem. Theory Comput.* 2006;*2*:1282–1290; (b) Bühl, M., Reimann, C., Pantazis, D.A. et al. Geometries of third-row transition-metal complexes from density-functional theory. *J. Chem. Theory Comput.* 2008; *4*:1449–1459.

**118** Berova, N., Bari, L.D., and Pescitelli, G. Application of electronic circular dichroism in configurational and conformational analysis of organic compounds. *Chem. Soc. Rev.* 2007;*36*:914–931.

**119** Greenfield, N.J. Using circular dichroism spectra to estimate protein secondary structure. *Nat. Protoc.* 2007;*1*:2876–2890.

**120** Pescitelli, G., Di Bari, L., and Berova, N. Application of electronic circular dichroism in the study of supramolecular systems. *Chem. Soc. Rev.* 2014; *43*:5211–5233.

**121** Costa Pessoa, J., Correia, I., Gonçalves, G. et al. Circular dichroism in coordination compounds. *J. Argent. Chem. Soc.* 2009;*97*:151–165.

**122** Mehtab, S., Gonçalves, G., Roy, S. et al. Interaction of vanadium(IV) with human serum apo-transferrin. *J. Inorg. Biochem.* 2013;*121*:187–195.

**123** Solomon, E.I., Pavel, E.G., Loeb, K.E. et al. Magnetic circular dichroism spectroscopy as a probe of the geometric and electronic structure of non-heme ferrous enzymes. *Coord. Chem. Rev.* 1995;*144*:369–460.

**124** Mason, W.R. *A Pratical Guide to Magnetic Circular Dichroism Spectroscopy*. Hoboken, NJ: John Wiley & Sons, Inc.; 2007.

**125** Seth, M., Ziegler, T., Banerjee, A. et al. Calculation of the A term of magnetic circular dichroism based on time dependent-density functional theory I. Formulation and implementation. *J. Chem. Phys.* 2004;*120*:10942–10954.

**126** Seth, M. and Ziegler, T. Calculation of excitation energies of open-shell molecules with spatially degenerate ground states. II. Transformed reference via intermediate configuration Kohn-Sham time dependent density functional theory oscillator strengths and magnetic circular dichroism C terms. *J. Chem. Phys.* 2006; *124*:144105.

**127** Seth, M., Krykunov, M., Ziegler, T. et al. Application of magnetically perturbed time-dependent density functional theory to magnetic circular dichroism: calculation of B terms. *J. Chem. Phys.* 2008;*128*:144105.

**128** Solomon, E.I. and Hadt, R.G. Recent advances in understanding blue copper proteins. *Coord. Chem. Rev.* 2011;*255*:774–789.

**129** Ganyushin, D. and Neese, F. First-principles calculations of magnetic circular dichroism spectra. *J. Chem. Phys.* 2008;*128*:114117.

**130** Ye, S., Kupper, C., Meyer, S. et al. Magnetic circular dichroism Evidence for an unusual electronic structure of a tetracarbene–oxoiron(IV) complex. *J. Am. Chem. Soc.* 2016;*138*:14312–14325.

**131** Ye, S., Xue, G., Krivokapic, I. et al. Magnetic circular dichroism and computational study of mononuclear and dinuclear iron(IV) complexes. *Chem. Sci.* 2015; *6*:2909–2921.

**132** Heit, Y.N., Sergentu, D.-C., and Autschbach, J. Magnetic circular dichroism spectra of transition metal complexes calculated from restricted active space wavefunctions. *Phys. Chem. Chem. Phys.* 2019;*21*:8022–8034.

**133** Schatz, P.N., McCaffery, A.J., Suëtaka, W. et al. Faraday Effect of Charge-Transfer Transitions in $Fe(CN)_6^{3-}$, $MnO_4^-$, and $CrO_4^{2-}$. *J. Chem. Phys.* 1966;*45*:722–734.

**134** Sridharan, K. *Spectral Methods in Transition Metal Complexes*. Amsterdam, Netherlands: Elsevier; 2016.

**135** Laury, M.L., Carlson, M.J., and Wilson, A.K. Vibrational frequency scale factors for density functional theory and the polarization consistent basis sets. *J. Comput. Chem.* 2012;*33*:2380–2387.

**136** Katari, M., Nicol, E., Steinmetz, V. et al. Improved infrared spectra prediction by DFT from a new experimental database. *Chem.-Eur. J.* 2017;*23*:8414–8423.

**137** Zvereva, E.E., Shagidullin, A.R., and Katsyuba, S.A. Ab initio and DFT predictions of infrared intensities and Raman activities. *J. Phys. Chem. A* 2011;*115*:63–69.

**138** Barone, V., Biczysko, M., and Bloino, J. Fully anharmonic IR and Raman spectra of medium-size molecular systems: accuracy and interpretation. *Phys. Chem. Chem. Phys.* 2014;*16*:1759–1787.

**139** Tang, C.-H., Parham, C., Shocron, E. et al. Picoplatin overcomes resistance to cell toxicity in small-cell lung cancer cells previously treated with cisplatin and carboplatin. *Cancer Chemotherapy and Pharmacology* 2011;*67*:1389–1400.

**140** Malik, M., Wysokiński, R., Zierkiewicz, W. et al. Raman and infrared spectroscopy, DFT calculations, and vibrational assignment of the anticancer agent picoplatin: performance of long-range corrected/hybrid functionals for a platinum(II) complex. *J. Phys. Chem. A* 2014;*118*:6922–6934.

**141** Gütlich, P., Bill, E., and Trautwein, A.X. *Mössbauer Spectroscopy and Transition Metal Chemistry: Fundamentals and Applications.* Berlin, Germany: Springer-Verlag; 2011.

**142** Crichton, R.R. *Biological Inorganic Chemistry* (ed. R.R. Crichton): Amsterdam, Netherlands: Elsevier; 2008, pp. 105–116.

**143** McWilliams, S.F., Brennan-Wydra, E., MacLeod, K.C. et al. Density functional calculations for prediction of $^{57}$Fe Mössbauer isomer shifts and quadrupole splittings in $\beta$-diketiminate complexes. *ACS Omega* 2017;*2*:2594–2606 and references therein.

**144** Römelt, M., Ye, S., and Neese, F. Calibration of modern density functional theory methods for the prediction of $^{57}$Fe Mössbauer isomer shifts: meta-GGA and double-hybrid functionals. *Inorg. Chem.* 2009;*48*:784–785.

**145** Neese, F. Prediction and interpretation of the $^{57}$Fe isomer shift in Mössbauer spectra by density functional theory. *Inorg. Chim. Acta* 2002;*337*:181–192.

**146** Bochevarov, A.D., Friesner, R.A., and Lippard, S.J. Prediction of $^{57}$Fe Mössbauer parameters by density functional theory: a benchmark study. *J. Chem. Theory Comput.* 2010;*6*:3735–3749.

**147** Pápai, M. and Vankó, G. On predicting Mössbauer parameters of iron-containing molecules with density-functional theory. *J. Chem. Theory Comput.* 2013; *9*:5004–5020.

**148** Lubitz, W., Ogata, H., Rüdiger, O. et al. Hydrogenases. *Chem. Rev.* 2014;*114*: 4081–4148.

**149** Kochem, A., Bill, E., Neese, F. et al. Mössbauer and computational investigation of a functional [NiFe] hydrogenase model complex. *Chem. Commun.* 2015;*51*: 2099–2102.

**150** Shomura, Y., Yoon, K.-S., Nishihara, H. et al. Structural basis for a [4Fe-3S] cluster in the oxygen-tolerant membrane-bound [NiFe]-hydrogenase. *Nature* 2011; *479*:253.

**151** Pandelia, M.-E., Bykov, D., Izsak, R. et al. Electronic structure of the unique [4Fe-3S] cluster in $O_2$-tolerant hydrogenases characterized by $^{57}$Fe Mössbauer and EPR spectroscopy. *Proc. Natl. Acad. Sci. U. S. A.* 2013;*110*:483–488.

# 5

# Multiscale Computational Modeling for the Discovery and Characterization of New Metalloenzyme-catalyzed Reactions

*Ferran Feixas and Marc Garcia-Borràs*

*Institut de Química Computacional i Catàlisi (IQCC) and Departament de Química, Universitat de Girona, Girona, Catalonia, Spain*

## 5.1 Importance of Metalloenzymes

Transition metals (TMs) are found in the core of several biological phenomena such as catalysis, self-assembly, and biomolecular recognition, and are directly involved in a number of diseases that span from cancer to neurodegenerative disorders, microbial infections, or cardiovascular diseases. The variety of oxidation and spin states displayed by TMs makes them a versatile tool that, when incorporated into a protein scaffold, participates in a large number of biochemical processes including synthesis of bacterial cell-wall components, oxygen transport, or drug metabolism. In this chapter, we will focus on the study of biocatalytic processes involving TM-based metalloenzymes.

Enzymes are highly efficient, specific, and selective catalysts that are a consequence of billions of years of evolutionary optimization. Many enzymes evolved to incorporate TM cofactors into protein scaffolds to enhance their basic catalytic toolkit. Examples of metalloenzymes include cytochrome P450s, hydrogenases, and nitrogenases. Biochemical reactions catalyzed by metalloenzymes often present complex mechanisms that involve the formation of reactive intermediates that are stabilized and controlled by the protein environment. Characterizing biocatalytic mechanisms of metalloenzymes represents one of the current challenges of experimental and computational bioinorganic chemists and harbors essential information for metalloenzyme design involving both natural and unnatural chemical reactions, and our group has made significant contributions to this field [1, 2].

The design of metalloenzymes for catalyzing non-natural reactions represents a highly attractive alternative to synthetic catalysts. Natural metalloenzymes

*Computational Bioinorganics: From Description to Prediction*, First Edition. Edited by Jean-Didier Maréchal.

exhibit intrinsic catalytic promiscuity that can be harnessed to evolve the enzyme toward new catalytic activities by introducing specific amino acid changes (i.e. mutations) into their sequences. Catalytic promiscuity refers to the ability of an enzyme to catalyze, in addition to its native function, reactions involving different substrates or reactions that proceed via different transition states (TS) and/or reactive intermediates, i.e. different reaction pathways. Common TM biological cofactors, including Fe-heme, are potential sources of enzyme catalytic promiscuity and have been exploited to engineer new enzymes with new non-natural functionalities. Recently, a number of strategies have been developed to enable enzymes to catalyze new-to-nature reactions involving TM cofactors. However, they mainly involve experimental approaches that cannot provide a complete mechanistic description of the new chemical transformations that are taking place.

Two main strategies are commonly applied to design and engineer metalloenzymes with desired functions. The first is the rational design approach, where mutations are introduced based on the previous knowledge of the target chemical reaction, structural information on the original protein scaffold, and structure-function relationships from either previous experiments or computational modeling [3]. The second is directed evolution (DE), which consists of introducing random mutations through subsequent iterative cycles, and selecting the best variants in terms of improved enzyme activity, stability, substrate specificity, and enantioselectivity [4–7]. However, it is complex to obtain mechanistic information during the procedure, and thus it is usually hard to understand how the enzyme properties and capabilities are enhanced due to introduced mutations. For example, the laboratory evolution of P450s showed that both residues near and far from the active site can have a profound impact on P450 selectivity and function.

In this context, recent advances in experimental techniques and computational power have triggered efforts to engineer metalloenzymes combining the capabilities of laboratory DE with mechanistic information obtained from computational methods. By integrating accurate mechanistic insights from computational tools, laboratory enzyme evolution can be conducted in a more efficient and rational manner. In this chapter, we will focus on studying how different computational methodologies can be integrated to elucidate the mechanistic details of metalloenzymes and explore their potential in guiding the design of novel enzyme functionalities with particular focus on the combination of quantum mechanics (QM) and quantum mechanics/molecular mechanics (QM/MM) calculations with molecular dynamics (MD) simulations.

## 5.2 Multi-scale Computational Microscope for Modeling Metalloenzymatic Reaction Mechanisms

Enzymes are dynamic structures that exist as an ensemble of conformations that can be grouped into distinct conformational states. In general, only a fraction of the thermally accessible conformations correspond to the active state of the enzyme with a preorganized active site [8]. To accomplish function, enzymes display a variety of motions that span from femtoseconds to seconds. These motions can be linked to different aspects of enzymatic catalysis. For example, large conformational rearrangements of loops and other structural elements occurring beyond microsecond timescales are frequently required to attain the active state of an enzyme or to accommodate the substrate in the active site. On the other hand, bond breaking and forming chemical steps occurring in the femto- to picosecond timescales, and side chain rearrangements of active site residues taking place in a few nanoseconds are key for the enzymatic reaction. In metalloenzymes, reaction pathways are determined by short-lived reactive intermediates that the enzyme environment finely controls to selectively generate the reaction products. Thus, accessing this variety of time scales is required to completely reconstruct and understand enzymatic mechanisms. Reconstructing an enzyme-catalyzed reaction is challenging because it involves describing the interplay of fast and slow dynamic processes, and rearrangements of networks of inter- and intramolecular interactions that are key for the different steps of the biochemical reaction. The transient nature of all processes related to an enzyme undergoing turnover, including slow conformational rearrangements and fast formation of reactive intermediates, hampers the experimental characterization of key mechanistic states, which remain hidden for several biochemical processes.

Recent advances in structural biology have allowed the identification of key states along a biochemical reaction in metalloenzymes. Using X-ray crystallography, it has been possible to trap relevant unnatural catalytic iron-carbene intermediates in laboratory-evolved cytochrome c carbene transferase [9]. Tezcan and coworkers recently reconstructed the nitrogenase mechanism through Cryo-EM structures obtained under catalytic turnover conditions [10]. The structural analysis indicates that conformational changes in the vicinity of the iron-molybdenum cofactor are correlated with the ATP hydrolysis state of the enzyme. Additionally, NMR methods have been used to rationalize the importance of conformational changes in P450s to determine the specificity of certain reactions [11]. Despite these significant advances, additional information is essential to unravel enzymatic mechanisms completely. This includes characterizing the entire energy profiles of the reaction and conformational changes that precede and go along with the chemical reaction.

In this context, computational techniques that provide an atomistic view of the enzymatic reaction have been used to explain its molecular basis. The power of computational protocols to describe the mechanism of enzymatic processes relies on being able to access a wide range of timescales (from femtoseconds to seconds) and critically depends on the accurate description of the simulated system. Since a computational technique that provides the required level of resolution and accuracy as well as computational efficiency to cover all the aspects of enzymatic reactions still does not exist, the idea of integrating different methodologies, i.e. computational microscope, is becoming more popular [12]. These computational approaches include MD simulations, QM models, and hybrid QM/MM calculations among others (Figure 5.1).

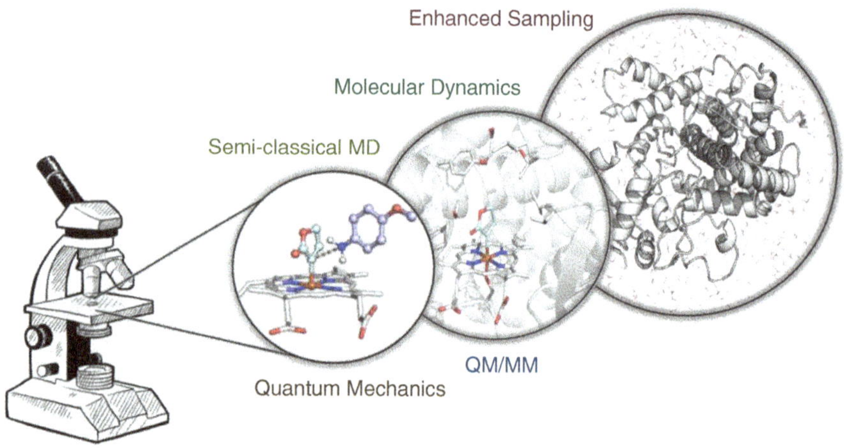


**Figure 5.1** Scheme of the computational microscope. *Source*: Adapted from [13].

### 5.2.1 Classical Molecular Dynamics Simulations

Classical MD simulations allow the description of enzyme conformational flexibility at atomic resolution. In MD simulations, classical force fields are typically employed to model the potential energy and forces within the system. From the obtained forces, Newton's second law of motion is then applied to calculate the subsequent movements of atoms within the enzyme. In this way, a trajectory is generated that illustrates the evolution of the enzyme conformation over time. This approach allows for a detailed exploration of the system motions and interactions involving the enzyme, substrate, and metal cofactors. However, conventional MD simulations are generally limited to access timescales beyond microseconds. Enhanced sampling techniques can be used to speed up MD simulations to reach timescales of biological interest [14]. MD simulations are commonly used for exploring the conformational dynamics of metalloenzymes in their *holo* state (i.e. metal cofactor bound), providing insights into global and local conformational changes such as the motion of loops or side-chain rearrangements in the active site, respectively. Additionally, these simulations have been applied to studying how both global and local conformational changes are influenced by the binding of substrates or how they stabilize key intermediate states of the reaction. Therefore, MD simulations are instrumental in deciphering some of the key steps of the enzymatic process, revealing how specific interactions and conformational shifts contribute to geometrically preorganize the enzyme for the catalytic reaction.

MD simulations and enhanced sampling techniques have been vastly used to describe the role of conformational dynamics and to evaluate the impact of mutations in laboratory-evolved metalloenzymes. For example, Dodani et al. studied a single mutation in the disordered F/G loop that modulates loop dynamics and shifts regioselectivity in nitrating cytochrome P450 TxtE [15]. They identified that the F/G loop can adopt two main conformations, one open that allows substrate binding and another closed that excludes water and facilitates catalysis. MD simulations and mutagenesis were used to identify a position on the loop that stabilizes the closed conformation and controls the regioselectivity of L-tryptophan. More recently, Rui et al. evolved a nonheme iron enzyme to catalyze an abiological C–H azidation reaction through iron-catalyzed radical relay mechanism [16]. In this work, MD simulations were used to characterize an open-to-closed transition of the C-terminal α-helix in the wild-type form of nonheme iron Sav HppD enzyme. This enzyme has been only crystallized in the open C-terminal α-helix conformation, which is not the active state. Based on substrate-bound MD simulations, it was shown that the closed conformation is the catalytically active conformation which allows properly positioning both azide and substrate for C–H enzymatic azidation via iron-catalyzed radical relay. Representative conformations of the closed state obtained from MD simulations were used as a starting points for exploring the reaction mechanism using DFT calculations with a truncated computational model (see below) offering a complete picture of the iron-catalyzed radical relay mechanism. Despite the valuable insights offered by MD simulations, quantum effects such as bond formation and breaking are still unattainable when using classical force fields, highlighting the need for QM models to elucidate the detailed reaction mechanisms of metalloenzymes.

### 5.2.2 Truncated Active Site Models at the Quantum Mechanics Level

To understand the reaction mechanism of metalloenzymes, several protocols based on QM have been proposed. The complexity and size of enzymes, and current computational power and algorithms, still prevent a full description at the QM level. Therefore, it is common to use strategies such as the *theozyme* (i.e. "theoretical enzyme") model and the cluster approach that simplify the computational treatment by considering only a few selected residues at the QM level. The simpler strategy is to perform the QM calculations on a truncated portion of the enzyme active site (*theozyme*) that includes only relevant catalytic residues, metal cofactor, and substrate. Several studies by Shaik et al. employed the *theozyme* concept to model the reaction mechanisms of P450s [17–19]. A typical *theozyme* model of a P450 includes at least a truncated Fe-oxo heme active species (together with its axial ligand) and substrate. This simple approach provides a description of the intrinsic mechanism and selectivity preference for a given metal center/metal cofactor toward a specific substrate. The cluster approach further expands the *theozyme* region by incorporating additional residues and applying positional restraints to certain atoms with the aim of mimicking the enzyme's active site shape and environment. However, these QM-based protocols exclude a large portion of the enzyme and the solvent from the calculation, limiting the representation of the reaction within the full enzymatic context.

### 5.2.3 Hybrid Quantum Mechanics/Molecular Mechanics

In this context, hybrid schemes that integrate QM with MM are commonly used. In QM/MM methods, the system is divided into distinct regions (typically two), treated with different levels of theory. The QM part usually comprises the active site of the metalloenzyme, including the metal cofactor, the substrate, and immediate surrounding residues directly involved in the catalytic process. The remaining part of the enzyme and solvent environment is modeled using the computationally more efficient MM, allowing for the complete modeling of the system. A key aspect of the computational accuracy and efficiency of QM/MM calculations is how the coupling term, which controls the interactions between the QM and MM parts, is treated (i.e. mechanical, electrostatic, or polarizable environment). For a more detailed description of QM/MM methodologies, we refer to the available literature [20]. Here, we will focus on the application of QM/MM approaches to elucidate the reaction mechanisms within metalloenzymes.

Through QM/MM simulations, it is possible to explore electron transfer, bond formation, and cleavage typical of metalloenzymes. This methodology has been key to deciphering the catalytic mechanisms of metalloenzymes, from understanding the geometric and electronic configurations of active sites to rationalizing the effects of mutations on enzyme function. In metalloenzymes, QM/MM calculations have been used, for example, to study diverse catalytic processes, such as reactivity and selectivity in cytochrome P450s [18] and electron transfer in photosynthetic oxygen evolution [21], providing a bridge between quantum chemical accuracy and the biological complexity of enzymatic reactions.

## 5.3 Practical Considerations for Computational Modeling of Metalloenzymes

### 5.3.1 Selecting the Starting Structure

When exploring the reaction mechanism of a metalloenzyme through computational modeling, choosing the correct starting structure for the simulation of the target system is essential. As mentioned before, enzymes exist as an ensemble of conformations, undergoing both global and local changes that impact their preorganization for catalysis. The accuracy of computational predictions for metalloenzyme catalysis significantly depends on selecting an enzyme conformation that accurately reflects the specific problem being addressed. For example, using conformations where the active site is not properly organized for substrate interaction in MD simulations may lead to substrate unbinding or sampling substrate orientations not relevant to the reaction. In the case of QM/MM, the initial geometry significantly influences the computational predictions of reaction barriers, influencing our understanding of the reaction mechanisms, which may lead to incorrect regio- and/or stereoselectivity patterns. Considering these aspects, the most ideal starting points for MD and QM/MM calculations would be an X-ray crystal structure (or an accurate computational prediction) of the metalloenzyme bound to the substrate or an intermediate of interest. This, in principle, ensures that the whole enzyme and the active site are properly preorganized for its interaction with the ligand, reflecting a preorganized environment to the specific reaction being studied. However, X-ray structures of the target system are often unavailable. Additionally, the transient nature of enzymes during turnover makes determining the structure of reactive states a challenge for current experimental methods, implying that the metalloenzyme is typically crystallized in the *holo* state or in the presence of substrate and TS analogs or inhibitors. Consequently, there are limited structures of metalloenzymes with bound substrates or intermediates that can be used as starting points for MD, QM, and QM/MM calculations.

In cases where an X-ray structure of the metalloenzyme is unavailable in the *apo*, *holo*, or any ligand-bound state, the first step is to generate the 3D structure of the system of interest using protein structure prediction methods. Traditional approaches include homology modeling, whose accuracy relies on the availability of structurally similar proteins. Recently, the advent of artificial intelligence-based protein structure prediction methods, exemplified by tools such as AlphaFold2 [22] and RoseTTAFold [23], has revolutionized the field by providing accurate structures for a broad range of proteins, even in the absence of closely related structural templates. Tools like AlphaFill [24] have been developed to incorporate cofactors into models generated by AlphaFold2, thus enhancing the relevance of these structures for subsequent computational modeling. More recently, Krishna et al. [25] developed RoseTTAFold All-Atom, which integrates the deep neural network-based protein structure prediction capabilities of RoseTTAFold with a graph representation of small molecules, including metal cofactors, substrates, and covalent modifications to proteins. Although we have not reached this stage yet, we anticipate that in the near future, these methods will be accurate enough to generate structures of metalloenzymes along reaction pathways. On the other hand, if X-ray structures of the metalloenzyme of interest are available, they often correspond to the *holo* state or they are co-crystallized with substrates, TS analogs, or inhibitors that are not ligands of interest for the study. Therefore, whether starting from computational predictions or X-ray structures, additional steps to incorporate the specific substrate or intermediate of interest are still required.

### 5.3.2 Initial Exploration of Conformational Dynamics

After obtaining the initial structure, either in the *holo* state or with the substrate or intermediate bound, it is recommended to explore the conformational ensemble of the metalloenzyme to identify motions that can be key for the catalytic process, e.g. substrate recognition, active site preorganization, etc. Additionally, there is a possibility that the structure, either obtained from X-ray or computational predictions, is not representative of the preorganized active conformation of the metalloenzyme. Therefore, exploring the enzyme conformational landscape of the *holo* state is advisable before introducing the substrate or intermediate. Typically, a number of MD simulations replicas, ranging from nanoseconds to microseconds, are performed to explore the conformational space and then clustered to identify relevant conformations. In addition, Monte Carlo simulations or other strategies, such as using AlphaFold2 to identify alternative conformations, can be also employed. If extensive coverage of the conformational space is required, the use of unconstrained enhanced sampling techniques, such as accelerated MD or Gaussian accelerated MD, is recommended. These techniques facilitate an extensive search of the conformational landscape without the definition of any reaction coordinate, which could be used to reveal catalytically relevant conformations. Structures obtained from MD simulations in the *holo* state can serve as the basis for positioning substrates or intermediates, especially in the absence of available X-ray structures.

When performing MD simulations of an enzyme with metal ions or cofactors, such as the heme cofactor, several considerations should be taken into account. For example, obtaining accurate force-field parameters for the metal center presents significant challenges due to the partial ionic character of metal-amino acid residue bonds. To address this aspect, various strategies for obtaining the force-field parameters for the metal center have been developed. These include the nonbonded model, which requires fine-tuning of the van der Waals term, as well as dummy and bonded models. For a more detailed description of these strategies, we refer to the existing literature [26]. Specifically, for iron-heme and nonheme enzymes, the bonded model has been shown to perform reasonably well, effectively capturing the specific interactions between the metal and its environment, which is essential for obtaining accurate results in the MD simulations.

### 5.3.3 Placement of Substrate or Reaction Intermediate

Once the initial coordinates of the *holo* state are obtained, either from an X-ray structure or MD simulations, one can proceed to place the substrate and intermediate of interest in the active site for subsequent QM, QM/MM, or MD simulations. A critical aspect of preparing for QM/MM calculations is the accurate placement of the substrate or reaction intermediate within the enzyme's active site. Substrates are typically docked into the active site using molecular docking. Achieving a reliable docking pose requires protocols that account for these metal properties, to ensure that the substrate's orientation and interactions with the active site are realistic. An alternative involves placing the substrate or intermediate based on truncated QM models of the enzyme's active site, which will provide an idea of the ideal arrangement of active site residues. Once the substrate is positioned in the active site, short MD simulations with positional restraints between the reactive atoms of the substrate and the cofactor are recommended. In this way, the substrate will adapt to the enzyme environment in a near-attack conformation. Finally, it is also possible to explore the binding of the substrate from the solvent to the enzyme's active site by spontaneous binding MD simulations, providing key details of the binding pathway [27]. All of these strategies can help refine the positioning of the substrate or intermediate, ensuring that QM and QM/MM calculations start from a geometrically and electronically plausible state.

Finally, it is also important to highlight the significance of "*fleeting*" chiral intermediates in enzyme-catalyzed reactions here. These transient, chiral molecules, arising from the symmetry breaking of prochiral precursors, are pivotal for fully understanding enzymatic mechanisms, activity, and selectivity and should therefore be properly considered when performing computational modeling studies [28].

In summary, the initial geometry used in MD and QM/MM simulations, encompassing both the geometric preorganization of the active site and the accurate positioning of the substrate or intermediate, is key to the successful application of computational pipelines to study metalloenzyme catalysis. This careful preparation allows researchers to unravel the complexities of metalloenzyme mechanisms, offering insights that are both predictive and explanatory guiding further experimental and theoretical investigations.

## 5.4 The Mechanistic Versatility of P450s Through the Lens of a Computational Microscope: Practical Examples

Cytochrome P450 are heme-dependent enzymes that utilize NAD(P)H to activate molecular oxygen and generate high-valent iron-oxo species (Compound I) [29].

This large family of enzymes is employed by Nature to catalyze important oxidation reactions, such as C—H hydroxylation, alkene epoxidation, fatty acid decarboxylation, and more, in a highly efficient and selective manner [30].

Lately, laboratory evolution and protein engineering techniques have been applied to expand the natural catalytic repertoire of P450 enzymes for synthetic applications, for example, in terms of substrate scope, catalytic efficiency improving electron transfer steps, and more recently [30], toward catalyzing unnatural (i.e. not naturally occurring) reactions [31, 32]. The latter includes abiological carbene and nitrene transfer reactions, but also unexpected oxidation reactions.

Typically, laboratory evolution of P450 enzymes exploits the intrinsic promiscuity they exhibit thanks to their usually large active sites and flexible protein scaffolds and the versatility of the heme cofactor and multiple oxidations (and spin) states accessible to the iron center.

P450s have largely evolved to access selective C—H bond hydroxylation in complex molecules where similar, nonequivalent, C—H positions are available, which represents a big challenge for chemical synthesis and small-molecule catalysis. Similarly, more recently, new abiological carbene and nitrene transfer reactions catalyzed by P450-engineered variants have been reported. Evolved enzyme variants are capable of catalyzing such reactions efficiently and offering selectivities hardly accessible for small-molecule catalysts. However, from a practical point of view, several rounds of protein engineering are required usually, involving the screening of large libraries of genes to introduce (several) mutations in the enzyme active site but also at distal positions, which also require tedious characterizations of the products (by MS-GC, nuclear magnetic resonance [NMR], etc.) being formed from each tested variant.

In this regard, computational modeling can provide extremely useful structural and mechanistic insights to guide protein engineering and laboratory evolution.

A practical computational strategy that our research group has been applying relies on a multi-scale consideration of the enzymatic reaction to account for different factors that contribute to enzyme catalysis, as described earlier. Combining different computational techniques, the objective is to understand how an enzyme can access new reactions or control the selectivity, and finally use these structural and mechanistic insights to rationally guide subsequent rounds of laboratory evolution and protein engineering, speeding up the process by limiting the number of sequences to be experimentally screened and tested [1, 2, 16, 16, 33–36].

Here, a case study focused on a newly characterized P450-catalyzed reaction reported by our research group in collaboration with the experimental group of Prof. Dr. Stephan Hammer will be discussed [37].

### 5.4.1 Practical Example: Anti-Markovnikov Alkene Oxidations Catalyzed by P450s

P450 enzymes can typically oxidize alkenes to form the corresponding epoxides in aerobic conditions, mediated by a high-valent iron-oxo catalytic species. This transformation is proposed to occur through a highly asynchronous TS, in which the oxygen group is transferred to the alkene moiety forming two new C—O bonds, or via a two-step mechanism in which a "fleeting," short-lived, radical intermediate is formed after the first C—O bond formation and prior the epoxide ring closure (see Figure 5.3a).

Recently, Hammer, Arnold et al. used laboratory evolution to engineer new P450-LA1 variants able to directly oxidizing terminal alkenes to the corresponding carbonyl (aldehyde), using styrene as a model substrate (Figure 5.2a,b) [38]. It was proposed that the reaction takes place through the formation of a carbocation covalent intermediate, generated after the C—O bond formation, followed by an intramolecular electron transfer and a 1,2-hydride migration leading to the carbonyl product (see Figure 5.3a). This mechanistic proposal was supported by experimental evidence (Figure 5.2c) that showed that: (i) no aldehyde is formed when the epoxide product is incubated with the enzyme, i.e. indicating that no isomerization occurs; and (ii) the use of deuterated styrene as substrate demonstrated that an intramolecular hydride migration was indeed occurring, and no H incorporation from other sources was taking place.

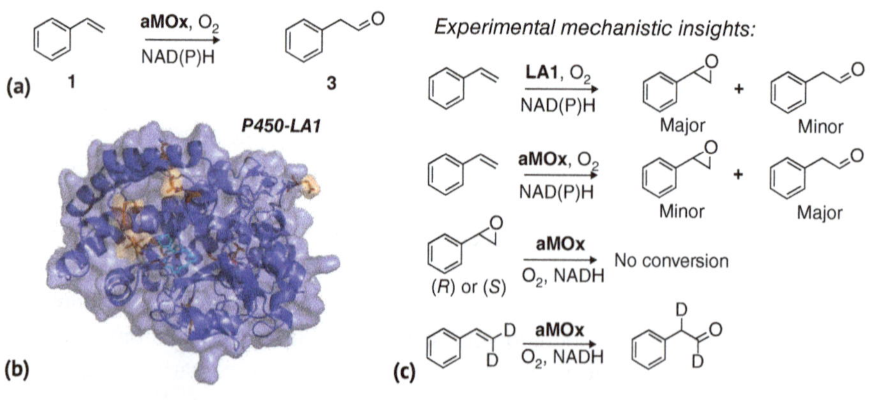


**Figure 5.2** (a) Anti-Markovnikov oxidation of alkenes to carbonyls by the laboratory-evolved aMOx enzyme considering styrene (**1**) as substrate model and aldehyde (**3**) as the final product; (b) Wild-type P450-LA1 (heme domain) used as parent enzyme for laboratory evolution. Mutated positions are highlighted in orange. (c) Experimental mechanistic insights supporting the direct alkene-to-aldehyde styrene oxidation by evolved aMOx variant.

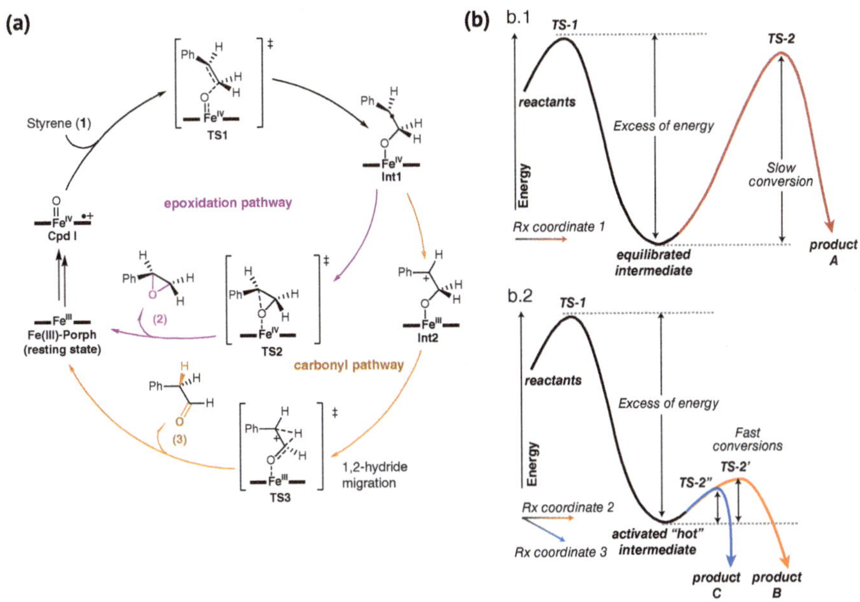


**Figure 5.3** (a) Proposed catalytic cycle for iron-oxo direct oxidation of alkene to carbonyl based on styrene (**1**) as substrate model and aldehyde (**3**) as final product. Competing epoxide (**2**) formation is also shown. (b) Dynamics of Energetically Activated "*Hot*" Intermediates. This illustration describes the dynamics of two intermediates emanating from an initial TS (**TS1**), both characterized by a significant excess of kinetic energy. (**B.1**) High-energy second TS, **TS2**, induces a slow conversion of the intermediate. This situation facilitates the thermal balancing of the intermediate, the dissipation of surplus energy to adjacent solvent molecules, and the rearrangement of intramolecular vibrational energy. The principles of transition state theory are effectively utilized to estimate rates of reaction, the stability of intermediates, and the selectivity of products. (**B.2**) A second TS with low energy, **TS2**′or **TS2**″, indicates a rapid transformation of the intermediate. Within these brief periods, the intermediate's reactivity may rival the processes of redistributing its excess energy and achieving thermal balance. As a result, the intermediate remains in an energetically "hot" state without thermal equilibrium. In this context, the reactivity is shaped by the dynamic properties of the "hot" intermediate. Controlled dynamically, the reaction may proceed to product **B** via **TS2**′ (a route outside the Intrinsic Reaction Coordinate, IRC, pathway) if a link exists between the intermediate's kinetic energy and the mechanisms facilitating its reaction, despite the presence of an energetically more favorable route to product **C** via **TS2**″ (the path of least energy according to the IRC). *Source*: Reference [37]/American Chemical Society/CC BY 4.0.

The aldehyde product was detected as a minor side product, in the reaction catalyzed by wild-type (wt) P450-LA1. By exploiting this promiscuous activity, laboratory evolution starting from this enzyme successfully provided a new variant, $P450_{aMOx}$ (aMOx enzyme), able to selectively form this carbonyl product after 10 rounds of protein engineering and screening [38].

In a first step, QM calculations, using density functional theory (DFT), using a computational truncated model were carried out (Figure 5.4). These model calculations aimed to evaluate the proposed catalytic cycle, and describe the intrinsic reactivity of styrene versus an Fe-porphyrin iron-oxo catalytic species in the absence of any protein scaffold. By using this truncated computational model, the role of the

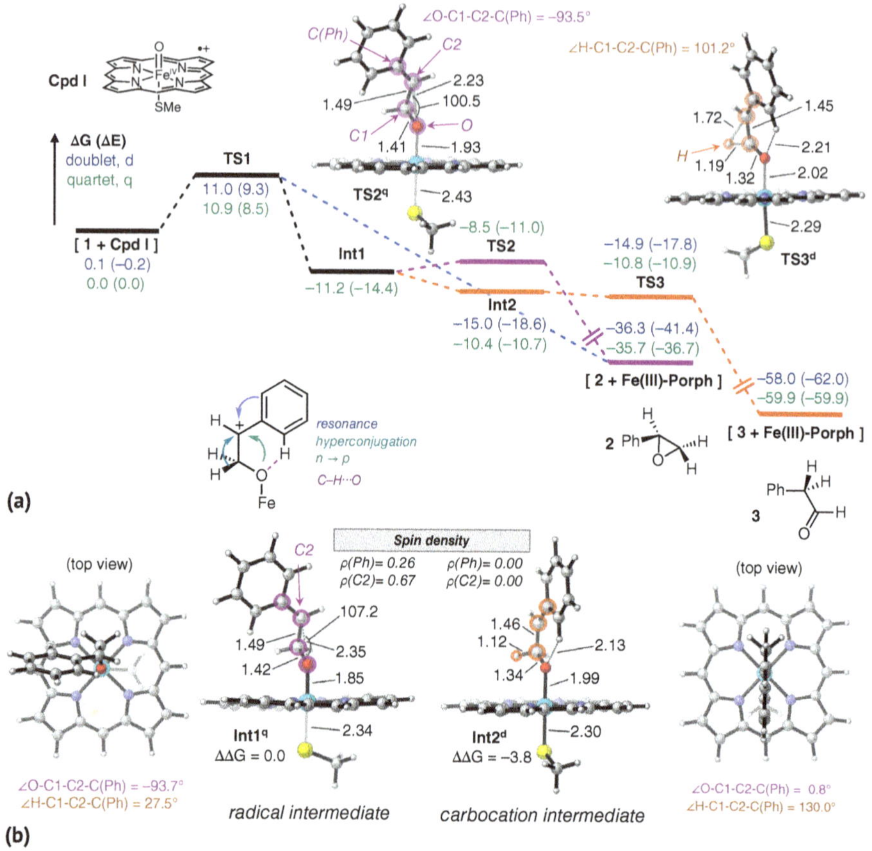


**Figure 5.4** Truncated model DFT calculations. (a) DFT-calculated competing reaction pathways for styrene (**1**) epoxidation (purple pathway) and carbonyl formation (orange pathway) catalyzed by the iron-oxo heme enzyme-free computational model. Relative Gibbs energies ($\Delta G$) and electronic energies ($\Delta E$, in parenthesis) are reported (in blue: doublet electronic state (d); in green: quartet electronic state (q)). DFT-optimized key TS geometries are shown. (b) DFT-optimized structures of key covalent radical and carbocation intermediates (**Int1** and **Int2**, respectively). Energies, distances, angles, and spin density values are given in kcal·mol$^{-1}$, angstroms (Å), degrees (°), and atomic units (a.u.), respectively. *Source*: Reference [37]/American Chemical Society/CC BY 4.0.

protein environment on controlling and promoting the unexpected chemical reactivity could be indirectly estimated by comparing this expected intrinsic behavior in the absence of the protein with the one observed for the enzyme-catalyzed reaction.

DFT model calculations described the expected mechanism for styrene epoxidation, in the line of previously reported similar truncated model calculations [39–42]. Calculations also described that the carbonyl-forming pathway is energetically accessible, and can indeed compete with the epoxidation pathway. As shown in Figure 5.4, in the quartet electronic state, the carbocation intermediate (**Int2**) can be formed from the radical intermediate (**Int1**) after the first C—O bond formation (**TS1**) and prior epoxide ring-closing (**TS2**). Calculations indicated that the formation of the carbocation intermediate is triggered by an energetically feasible conformational change, i.e. the rotation of the phenyl group (see Figure 5.4b). This new conformation of the intermediate allows the stabilization of the benzylic carbocation due to different stereoelectronic effects, including resonance with the aromatic ring, hyperconjugation with the former vinylic α-C—H bonds, a $n \rightarrow p$ interaction between the oxygen lone pairs and the empty carbocation $p$ orbital, and a C—H···O interaction that stabilizes the partial positive charge delocalized on the aromatic ring by resonance. Additionally, within this conformation the α-hydrogen is aligned with the empty benzylic carbocation $p$ orbital, favoring a fast and efficient 1,2-hydride migration.

The use of DFT calculations also allowed to consider different energetically accessible spin states for the iron-oxo catalytic species, which is important for an accurate description of the reaction mechanism, as further discussed in Chapter 2 (M. Swart).

Based on these model calculations, it was concluded that the carbonyl formation pathway is possible and can energetically compete with the epoxidation one. Considering the energy differences calculated, a product mixture would be predicted, which is not what it is observed for most of alkene oxidations by iron-oxo porphyrin complexes including P450s.

Due to the flat shape of the potential energy surface (PES) characterized around the **Int1**, **Int2**, **TS2**, and **TS3** region in both doublet and quartet electronic states, nonstatistical dynamic effects could also play an important role in determining the selectivity of the reaction (see Figure 5.3b) [43]. In order to study these effects, direct quasiclassical MD trajectories were carried out based on the computational truncated model and using DFT.

Quasiclassical trajectories (QCT) allow to explicitly consider the atomic motions during chemical reactions, and are able to describe the generation of nonequilibrated intermediates, i.e. activated ("hot") intermediates (Figure 5.3b). These short-lived intermediates are formed with an excess of kinetic energy that is not dissipated or redistributed between the different vibrational modes before further reacting, thus affecting to their chemical reactivity [44]. These activated intermediates can follow non-intrinsic reaction coordinate pathways (non-IRC), which can lead to products that do not correspond to those expected from the lowest-energy pathway (IRC pathways).

Starting from the optimized first TS1, reaction trajectories were propagated backward and forward (600 fs in each direction), following the direction of the vector corresponding to the imaginary normal mode (Figure 5.5).

QTC indicated that the epoxidation pathway is highly favored due to dynamic effects. Simulations showed that the direction of the atomic motions and the momentum of **Int1** after being formed from **TS1** are strongly coupled with the atomic motions that promote the low-energy barrier for epoxide ring closure (**TS2**). Consequently, the kinetic energy that the radical intermediate acquires when being formed from TS1 influences its lifetime and favors the formation of the second C—O bond and epoxide ring closure. This "*dynamic match*" between **TS1** and **TS2** reaction coordinates favors epoxidation over carbonyl formation, which requires a specific C1—C2 bond rotation to efficiently stabilize the carbocation intermediate and to allow the required 1,2-hydride migration (see Figure 5.5). These intrinsic

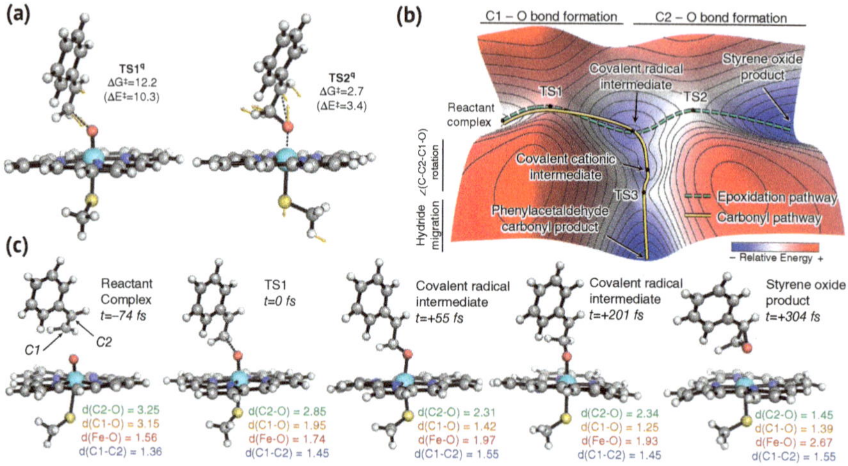


**Figure 5.5** Quasiclassical direct dynamic trajectories. (a) Truncated model DFT-optimized TS structures **TS1**-q (used as the starting points for direct QCT simulations) and **TS2**-q. Yellow arrows are the displacement vectors associated with the TS structure imaginary frequency. (b) Qualitative three-dimensional energy landscape where different possible pathways are highlighted: dynamically preferred epoxidation pathway highlighted as a green dashed line, the thermodynamically preferred carbonyl pathway as a yellow dashed line. (c) Selected snapshots along a quartet trajectory forming styrene oxide aldehyde product. Energies, distances, and time are given in kcal·mol$^{-1}$, Angstroms (Å), and femtoseconds (fs), respectively. *Source*: Reference [37]/American Chemical Society/CC BY 4.0.

dynamic preferences could explain the preferred alkene epoxidation exhibited by P450s and many small-molecule catalysts that use high-valent iron-oxo (and metal-oxo) species as oxidant.

Based on these results obtained from truncated model QM calculations, it was hypothesized that the laboratory-evolved enzyme is promoting the carbonyl-forming pathway by preventing the epoxidation to occur via the conformational control of the fleeting intermediates formed over the course of the reaction.

### 5.4.2 Molecular Dynamics Simulations for Modeling the Protein Environment

This hypothesis was first explored by performing extensive MD simulations. Starting from a homology model built for the holo aMOx evolved variant and P450-LA1 wild-type variant. MD simulations revealed a significant reshaping of the active site cavity due to mutations, which directly impacts on the catalytically relevant binding modes accessible for the styrene substrate (see Figure 5.6). These simulations described a more narrower substrate binding pocket, restricting the accessible conformations of the substrate styrene. Subsequent MD simulations with the substrate explicitly bound in both enzymes were carried out to characterize such reactive binding modes. Docking calculations followed by restrained-MD simulations were conducted to explore catalytically relevant

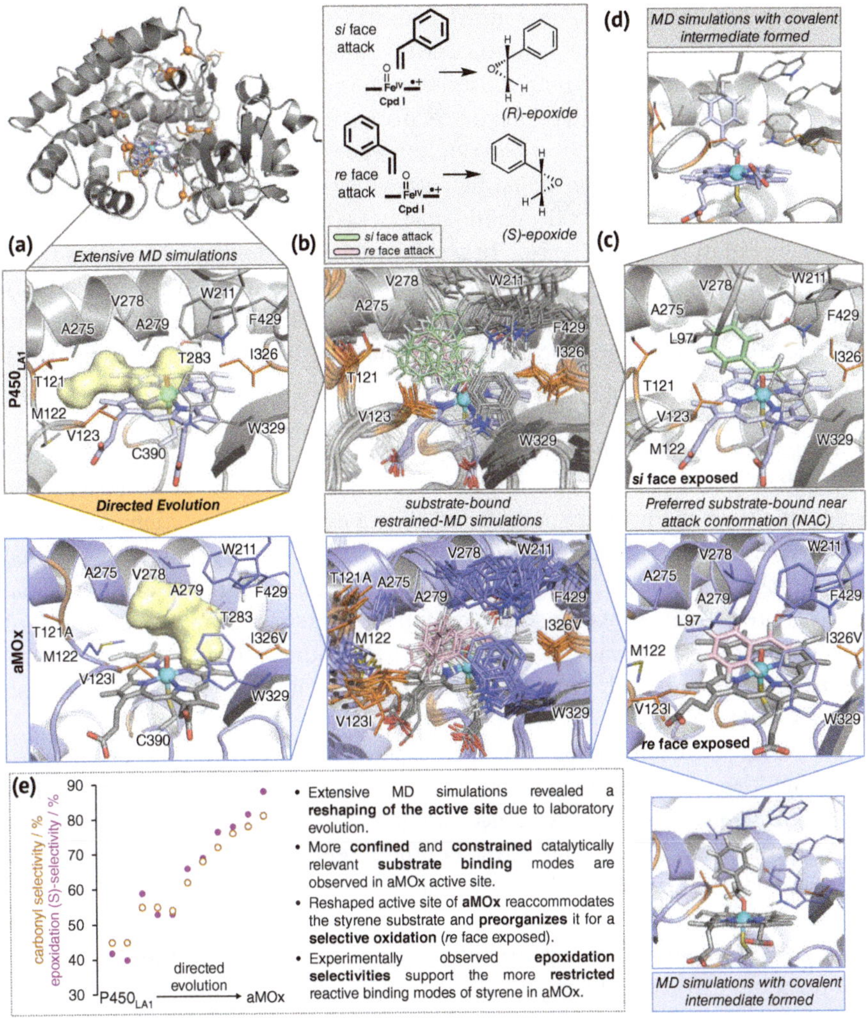


**Figure 5.6** Impact of the laboratory evolution on the enzyme active site. (a) The homology model generated for wild-type P450-LA1 (56% sequence identity with the template, Protein Data Bank: 6GII) and aMOx (12 mutations with respect to the wild-type). The homology model was refined by extensive molecular dynamics (MD) simulations (5.0 μs total simulation time). Positions mutated along directed evolution are highlighted in orange. The accessible volume in the active site is shown as a yellow surface. (b) Overlay of snapshots of reactive binding modes characterized from restrained-MD simulations of the bound styrene in wild-type P450-LA1 and aMOx active sites. Styrene substrate, shown in green, and the pink represents *si* face (*R*-epoxide) and *re* face (*S*-epoxide) binding poses, respectively. Mutated positions are highlighted in orange. (c) Preferential NACs of styrene in wildtype P450-LA1 and aMOx as found from restrained-MD simulations. (d) Binding modes explored by the covalent radical intermediate **Int1** when formed in wild-type P450-LA1 and aMOx active sites as characterized by MD simulations. (e) Enantioselectivity observed for styrene epoxidation as a function of directed evolution lineage. Epoxidation enantioselectivity changes during evolution in analogy to the aMOx selectivity. Wild-type P450-LA1 preferably generates the (*R*)-enantiomer (60:40 of *R*/*S*), aMOx leads to an enantioselectivity change reaching a ratio of 14:86 (*R*/*S*). *Source*: Reference [37]/American Chemical Society/CC BY 4.0.

binding poses, focusing on near-attack conformations (NACs) necessary for the oxidation reaction. To do so, restrained-MD simulations in which geometric constraints to maintain the styrene substrate in the active site during the whole MD trajectories were performed (a maximum distance between the center of mass of the alkene moiety and the iron-oxo catalytic species was defined).

These simulations demonstrated that P450-LA1 can orient styrene toward the iron-oxo species, exposing both enantiotopic faces (*re* and *si* face), with a preference for *si* face exposure (Figure 5.6b,c). In contrast, aMOx predominantly exposes the *re* face of the alkene. As a result, P450-LA1 wild-type is predicted to yield an (*R*)-selective epoxide formation, while aMOx variant is expected to produce (*S*)-selective epoxide due to the reshaped active site favoring selective oxidation (Figure 5.6b,c). These findings suggest a structural basis for the observed differences in substrate selectivity and reactivity between the two enzymes.

To experimentally validate the computational insights into the active site's restructuring and its influence on substrate binding and enzyme selectivity, the study analyzed the enantioselectivity of alkene epoxidation reactions across the directed evolution pathway (Figure 5.6e). The wild-type enzyme initially favored the formation of the (R)-epoxide enantiomer, aligning with the computational model's predictions. However, as the evolution progressed, a marked shift toward (S)-epoxide formation was observed, particularly in the aMOx variant, which produced this epoxide with notable (S)-enantioselectivity despite its lower overall epoxide yield.

Supporting the computational findings, experimental data from UV/Vis spectroscopy and kinetic analysis highlighted an optimized substrate binding in the aMOx variant, evidenced by a significant decrease in the dissociation constant for styrene and improved coupling efficiency. Notably, the absence of substantial hydrogen peroxide formation suggested an optimized catalytic pathway distinct from the less efficient peroxide shunt. The evolved aMOx enzyme demonstrated a substantial increase in catalytic efficiency, reaching levels comparable with those of natural enzymes and markedly higher than the P450-LA1 wild-type.

### 5.4.3 Insightful Revelations from QM/MM Calculations

The exploration of P450-LA1 and aMOx enzymes through QM/MM calculations has unveiled crucial aspects of their catalytic processes, particularly regarding the formation and stabilization of reaction intermediates. QM/MM calculations for both P450-LA1 and aMOx systems were carried out starting from representative snapshots obtained from MD simulations. In both cases, the competing epoxidation versus carbonyl formation pathways were calculated.

For P450-LA1 (Figure 5.7a–c), QM/MM calculations carried out pinpointed the formation of the first C1—O bond as the rate-limiting step in the reaction pathway, in line with truncated model DFT calculations. Notably, the QM/MM analysis highlighted the stabilization of a radical intermediate within the active site. This intermediate's stabilization is crucial, as it directs the pathway toward the formation of an epoxide product rather than alternative outcomes. The calculations revealed that this pathway involves either a low-energy TS or a barrierless progression, depending

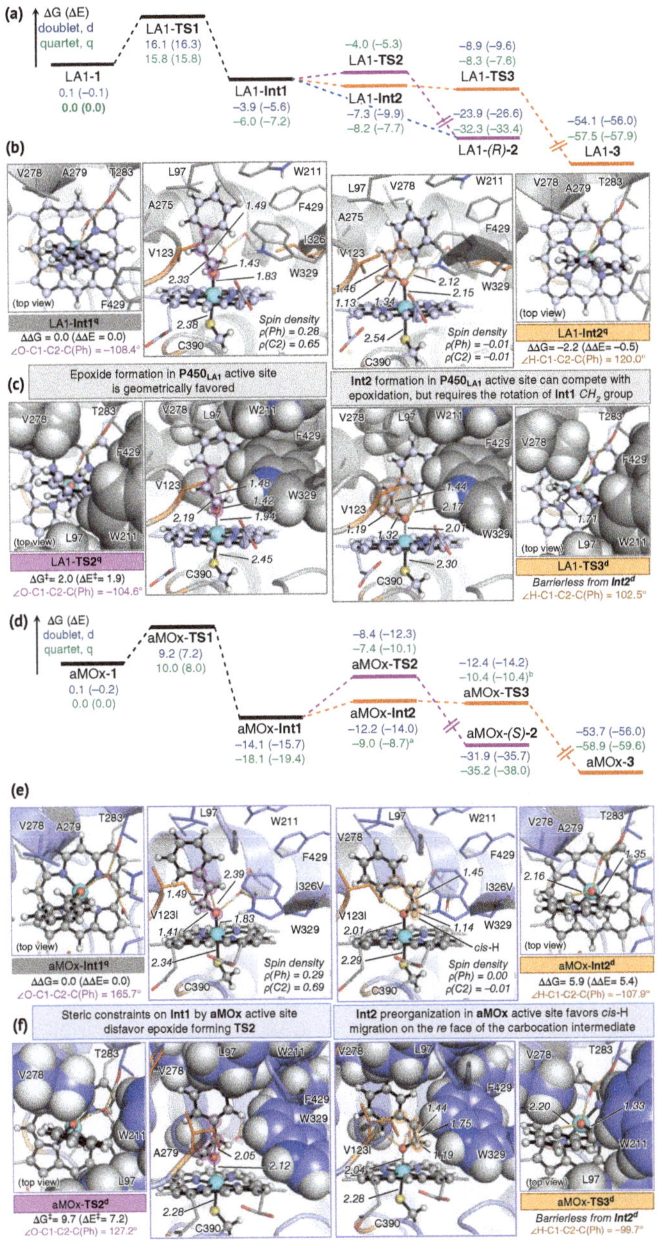

**Figure 5.7** QM/MM calculations on the P450-LA1- and aMOx-catalyzed reaction pathways. (a, d) QM/MM-calculated reaction mechanism for the (*S*)-selective epoxidation and carbonyl formation pathways of styrene catalyzed by wild-type P450-LA1 and aMOx variant, respectively. Relative Gibbs energies and electronic energies ($\Delta G$ and $\Delta E$, respectively) are given in blue for the doublet (d) electronic state and green for the quartet (q). Lowest-energy QM/MM-optimized structures for key (b, e) intermediates and (c, f) TS for each system are reported. Important steric constraints taking place in the active sites are described by space-filling representations. Energies, distances, angles, and spin density values are given in kcal· mol$^{-1}$, angstroms (Å), degrees (°), and a.u., respectively. *Source*: Reference [37]/American Chemical Society/CC BY 4.0.

on the electronic configuration of the system, emphasizing the role of the enzyme's structure in guiding the reaction dynamics.

In contrast, QM/MM calculations based on the aMOx enzyme, revealed a distinct preference for carbonyl product formation (Figure 5.7c–f). QM/MM calculated mechanism for aMOx described that carbonyl formation is energetically more favorable than epoxidation, which explains the chemoselectivity achieved by the evolved aMOx enzyme. This shift in chemoselectivity is attributed to the enzyme's ability to impose steric constraints on the formed radical intermediate, directing the reaction toward the carbonyl formation pathway.

The QM/MM calculations detailed how steric constraints within the active site limit the accessible conformations of the radical intermediate, allowing access to the reactive carbocation intermediate that effectively steers the reaction toward the carbonyl product.

This remarkable control over the reaction mechanism highlights the potential of enzyme engineering in tailoring catalysts for specific chemical transformations. The QM/MM results demonstrate how subtle changes in enzyme conformation can lead to significant differences in reaction outcomes, providing valuable insights for the design of novel enzymes with enhanced or altered functionalities.

### 5.4.4 The Role of Local Electric Fields in Enzymatic Reactions

Further enriching our understanding of enzymatic catalysis, computational modeling shed light on the role of local electric fields (LEFs) [45–49] generated within the active sites of P450-LA1 and aMOx. These LEFs, originating from the spatial arrangement of amino acid residues around the active site, influence the (de-) stabilization of charged intermediates, and were found to play a pivotal role in stabilizing the carbocation intermediate in aMOx evolved active site, thereby influencing the direction and efficiency of the reaction.

In particular, LEF generated at the center of P450-LA1 and aMOx active sites were estimated from representative snapshots selected from MD simulations (those used as starting points for QM/MM modeling).

Then, truncated model calculations were used to simulate the effect of the LEF generated in the aMOx active site. DFT model calculations showed that the generated LEF in aMOx active site largely favors the stabilization of the carbocation intermediate **Int2** as compared to the radical intermediate **Int1**, and that it has a suitable orientation and direction to favor the stabilization of such carbocation intermediate species.

This observation underscores the multifaceted strategies employed by enzymes to control reaction pathways, integrating both steric and electrostatic factors to achieve high selectivity and efficiency.

QM/MM calculations also revealed that the protein environment provides a significant degree of conformational control over reactive intermediates and the carbocation in particular, leading to a high level of stereoselectivity in the final hydride migration process during the carbonyl formation pathway. Specifically, the aMOx enzyme orients the cis-H atom of the styrene substrate for selective migration on the *re* face of the carbocation intermediate, through the lowest-energy TS. This selectivity is attributed to the specific orientation and rotation of the substrate within the enzyme's active site, allowing for a controlled and enantioselective migration process (Figure 5.7f).

To confirm these computational findings, deuterium-labeled *cis*- and *trans*-styrene substrates were synthesized and subjected to conversion by aMOx, with the reactions tailored to preserve the deuterium label. Mass spectrometry analysis of the reaction products demonstrated that aMOx catalyzes a *cis*-selective 1,2-hydride migration. Further experimental studies aimed at understanding the enantioselectivity of this migration confirmed that the *cis*-hydride selectively migrates on the *re* face of the prochiral substrate, resulting in predominantly (*R*)-configured products with a selectivity of greater than 90:10.

These findings underscore the unique capability of aMOx to catalyze enantioselective alkene to carbonyl oxidation through a catalytic enantioselective 1,2-hydride migration. Although this selectivity for the migration step was not pursued during evolution, it is a consequence of the geometric preorganization and steric constraints imposed by the evolved aMOx active site on the key carbocation intermediate.

Collectively, the integration of computational modeling with experimental observations underscores the effectiveness of directed evolution in enhancing enzyme selectivity and efficiency. This approach not only facilitated a deeper understanding of the structural and functional modifications imparted by directed evolution but also showcased the potential of combining these methodologies to refine enzyme functionalities for specific catalytic processes. Computational modeling can guide protein evolution by providing atomistic descriptions of key reactive intermediates that otherwise would be not possible to obtain. This information can be then used to select mutational spots to prepare smart libraries that could be efficiently screened experimentally.

In summary, this comprehensive study employs a blend of QM/MM calculations, MD simulations, and DFT analyses to dissect the complex mechanisms underpinning enzyme-catalyzed oxo-transfer reactions. By elucidating the nuanced interplay between steric constraints, electronic interactions, and electrostatic preorganization, the research sheds light on the molecular foundations of chemoselectivity in enzymatic reactions. It underscores the transformative impact of directed evolution in enhancing enzyme specificity and efficiency, paving the way for future explorations into enzyme design and catalysis.

## 5.5 Conclusions

This chapter aims to provide an overview of how computational modeling contributes to our understanding and utilization of metalloenzymes in catalysis. The discussions of computational strategies, from MD simulations to QM/MM hybrid approaches, have highlighted the capability of these methods to dissect the complex catalytic mechanisms of metalloenzymes, and to guide the engineering of these enzymes for novel functions.

The emphasis on metalloenzymes throughout the chapter reflects their significance in both biological systems and synthetic applications. By examining case studies, including the engineering of cytochrome P450 enzymes, we have demonstrated how computational tools can predict the effects of mutations and assist in the rational design of enzymes with desired catalytic properties. These examples underscore the utility of computational modeling in complementing experimental efforts to explore enzyme catalysis.

Looking forward, the integration of computational and experimental techniques appears promising for advancing our ability to design and optimize enzymes for specific tasks. The ongoing development of computational methods and the incorporation of advanced algorithms, including those from the field of artificial intelligence, are expected to enhance the accuracy and efficiency of these predictive models.

For designing purposes, one approach that our group is currently implementing is to use a combination of multi-scale computational modeling techniques to first understand how a targeted reaction, poorly catalyzed by a metalloenzyme (i.e. a *promiscuous* activity) can occur, and then use the gained mechanistic insights for aiding and guiding laboratory evolution efforts.

However, there are still challenges to overcome, particularly in accurately modeling the electronic states of metal centers and capturing transient states that are often critical for catalysis. Improvements in the simulation of environmental factors, such as solvent effects and substrate interactions, are also needed to increase the realism of our models.

In summary, computational modeling offers a valuable lens through which to examine the complex world of metalloenzyme catalysis. While significant progress has been made, the field continues to evolve, with ongoing research aimed at refining our tools and extending our understanding of these crucial biological catalysts. The continued collaboration between computational scientists and experimentalists will be key to unlocking the full potential of metalloenzymes in both nature and industry and to implementing new metalloenzyme-catalyzed reactions interesting from a synthetic perspective.

## Acknowledgments

This work was supported by the Spanish MICINN (Ministerio de Ciencia e Innovación) PID2022-141676NB-I00 and TED2021-130173B-C42 projects, and the Ramón y Cajal RYC2020-028628-I and RYC2020-029552-I fellowships; and the Generalitat de Catalunya AGAUR 2021-SGR-00623 and 2021-SGR-00487 projects. The authors would like to thank Dr. Carla Calvó-Tusell for providing original figures. The computational resources at the Barcelona Supercomputing Center – Red Española de Supercomputación (BSC-RES) are acknowledged (RES-QSB-2019-3-0009, RES-QSB-2020-2-0016, and RES-QH-2021-2-0013, BCV-2021-1-0015).

## References


1 Calvó-Tusell, C., Liu, Z., Chen, K. et al. Reversing the enantioselectivity of enzymatic carbene N—H insertion through mechanism-guided protein engineering**. *Angew. Chem. Int. Ed.* 2023;*62*(35):e202303879.

2 Garcia-Borràs, M., Kan, S.B.J., Lewis, R.D. et al. Origin and control of chemoselectivity in cytochrome c catalyzed carbene transfer into Si–H and N–H bonds. *J. Am. Chem. Soc.* 2021;*143*(18):7114–7123.

3 Reetz, M. Making enzymes suitable for organic chemistry by rational protein design. *ChemBioChem.* 2022;*23*(14):e202200049.

4 Reetz, M.T. Directed evolution of artificial metalloenzymes: a universal means to tune the selectivity of transition metal catalysts? *Acc. Chem. Res.* 2019;*52*(2): 336–344.

5 Arnold, F.H. Directed evolution: bringing new chemistry to life. *Angew. Chem. Int. Ed.* 2018;*57*(16):4143–4148.

6 Arnold, F.H. Innovation by evolution: bringing new chemistry to life (Nobel lecture). *Angew. Chem. Int. Ed.* 2019;*58*(41):14420–14426.

7 Chen, K. and Arnold, F.H. Engineering new catalytic activities in enzymes. *Nat. Catal.* 2020;*3*(3):203–213.

8 Boehr, D.D., Nussinov, R., and Wright, P.E. The role of dynamic conformational ensembles in biomolecular recognition. *Nat. Chem. Biol.* 2009;*5*(11):789–796.

9 Lewis, R.D., Garcia-Borràs, M., Chalkley, M.J. et al. Catalytic iron-carbene intermediate revealed in a cytochrome *c* carbene transferase. *Proc. Natl. Acad. Sci.* 2018;*115*(28):7308–7313.

10 Rutledge, H.L., Cook, B.D., Nguyen, H.P.M. et al. Structures of the nitrogenase complex prepared under catalytic turnover conditions. *Science* 2022; *377*(6608):865–869

11 Pochapsky, T.C. A dynamic understanding of cytochrome P450 structure and function through solution NMR. *Curr. Opin. Biotechnol.* 2021;*69*:35–42.

12 Romero-Rivera, A., Garcia-Borràs, M., and Osuna, S. Computational tools for the evaluation of laboratory-engineered biocatalysts. *Chem. Commun.* 2017; *53*(2):284–297.

13 Calvó-Tusell, C. Computational strategies for understanding the molecular basis of biochemical and biocatalytic processes. *TDX (Tesis Doctorals en Xarxa)* 2023.

14 Hénin, J., Lelièvre, T., Shirts, M.R. et al. Enhanced sampling methods for molecular dynamics simulations [Article v1.0]. *Living J. Comput. Mol. Sci.* 2022;*4*(1): 3.

15 Dodani, S.C., Kiss, G., Cahn, J.K.B. et al. Discovery of a regioselectivity switch in nitrating P450s guided by molecular dynamics simulations and Markov models. *Nat. Chem.* 2016;*8*(5):419–425.

16 Rui, J., Zhao, Q., Huls, A.J. et al. Directed evolution of nonheme iron enzymes to access a biological radical-relay C($sp^3$)—H azidation. *Science (1979)* 2022; *376*(6595):869–874.

**17** Caddell Haatveit, K., Garcia-Borràs, M., and Houk, K.N. Computational protocol to understand P450 mechanisms and design of efficient and selective biocatalysts. *Front. Chem.* 2019;*6*(JAN):663.

**18** Shaik, S., Cohen, S., Wang, Y. et al. P450 enzymes: their structure, reactivity, and selectivity—modeled by QM/MM calculations. *Chem. Rev.* 2010;*110*(2):949–1017.

**19** Narayan, A.R.H., Jiménez-Osés, G., Liu, P. et al. Enzymatic hydroxylation of an unactivated methylene C-H bond guided by molecular dynamics simulations. *Nat. Chem.* 2015;*7*(8):653–660.

**20** Senn, H.M. and Thiel, W. QM/MM methods for biomolecular systems. *Angew. Chem. Int. Ed.* 2009;*48*(7):1198–1229.

**21** Greife, P., Schönborn, M., Capone, M. et al. The electron–proton bottleneck of photosynthetic oxygen evolution. *Nature* 2023;*617*(7961):623–628.

**22** Jumper, J., Evans, R., Pritzel, A. et al. Highly accurate protein structure prediction with AlphaFold. *Nature.* 2021;*596*(7873):583–589.

**23** Baek, M., DiMaio, F., Anishchenko, I. et al. Accurate prediction of protein structures and interactions using a three-track neural network. *Science (1979)* 2021;*373*(6557), 871–876.

**24** Hekkelman, M.L., de Vries, I., Joosten, R.P. et al. AlphaFill: enriching AlphaFold models with ligands and cofactors. *Nat. Methods* 2023, *20*(2), 205–213.

**25** Krishna, R., Wang, J., Ahern, W. et al. Generalized biomolecular modeling and design with RoseTTAFold All-Atom. *Science (1979).* 2024;*384*:eadl2528.

**26** Li, P. and Merz, K.M. Metal ion modeling using classical mechanics. *Chem. Rev.* 2017;*117*(3):1564–1686.

**27** Calvó-Tusell, C., Maria-Solano, M.A., Osuna, S. et al. Time evolution of the millisecond allosteric activation of imidazole glycerol phosphate synthase. *J. Am. Chem. Soc.* 2022;*144*(16):7146–7159.

**28** Reetz, M.T. and Garcia-Borràs, M. The unexplored importance of fleeting chiral intermediates in enzyme-catalyzed reactions. *J. Am. Chem. Soc.* 2021;*143*(37): 14939–14950.

**29** Ortiz De Montellano, P.R. and de Montellano, P.R.O. *Cytochrome P450: Structure, Mechanism, and Biochemistry*, 4 ed. United States: Springer; 2015.

**30** O'Reilly, E., Köhler, V., and Flitsch, S.L. Cytochromes P450 as useful biocatalysts: addressing the limitations. *Chem. Commun.* 2011;*47*(9):2490–2501.

**31** Yang, Y. and Arnold, F.H. Navigating the unnatural reaction space: directed evolution of heme proteins for selective carbene and nitrene transfer. *Acc. Chem. Res.* 2021;*54*(5):1209–1225.

**32** Hammer, S.C., Knight, A.M., and Arnold, F.H. Design and evolution of enzymes for non-natural chemistry. *Curr. Opin. Green Sustain. Chem.* 2017;*7*:23–30.

**33** Huang, X., Garcia-Borràs, M., Miao, K. et al. A biocatalytic platform for synthesis of chiral α-trifluoromethylated organoborons. *ACS Cent. Sci.* 2019;*5*(2): 270.

**34** Liu, Z., Calvó-Tusell, C., Zhou, A.Z. et al. Dual-function enzyme catalysis for enantioselective carbon–nitrogen bond formation. *Nat. Chem.* 2021;*13*(12): 1166–1172.

**35** Gergel, S., Soler, J., Klein, A. et al. Engineered cytochrome P450 for direct arylalkene-to-ketone oxidation via highly reactive carbocation intermediates. *Nat. Catal.* 2023;*6*(7):606–617.

**36** Münch, J., Soler, J., Hünecke, N. et al. Computational-aided engineering of a selective unspecific peroxygenase toward enantiodivergent β-ionone hydroxylation. *ACS Catal.* 2023;*13*(13):8963–8972.

**37** Soler, J., Gergel, S., Klaus, C. et al. Enzymatic control over reactive intermediates enables direct oxidation of alkenes to carbonyls by a P450 iron-oxo species. *J. Am. Chem. Soc.* 2022;*144*(35):15954–15968.

**38** Hammer, S.C., Kubik, G., Watkins, E. et al. Anti-Markovnikov alkene oxidation by metal-oxo–mediated enzyme catalysis. *Science (1979)* 2017;*358*(6360):215–218.

**39** de Visser, S.P., Kumar, D., and Shaik, S. How do aldehyde side products occur during alkene epoxidation by cytochrome P450? Theory reveals a state-specific multi-state scenario where the high-spin component leads to all side products. *J. Inorg. Biochem.* 2004;*98*(7):1183–1193.

**40** Kumar, D., De Visser, S.P., and Shaik, S. Multistate reactivity in styrene epoxidation by compound I of cytochrome P450: mechanisms of products and side products formation. *Chem. Eur. J.* 2005;*11*(9):2825–2835.

**41** Meunier, B., de Visser, S.P., and Shaik, S. Mechanism of oxidation reactions catalyzed by cytochrome P450 enzymes. *Chem. Rev.* 2004;*104*(9):3947–3980.

**42** de Visser, S.P., Ogliaro, F., Harris, N. et al. Multi-state epoxidation of ethene by cytochrome P450: a quantum chemical study. *J. Am. Chem. Soc.*, 2001; *123*(13):3037–3047.

**43** Yang, Z., Jamieson, C.S., Xue, X.S. et al. Mechanisms and dynamics of reactions involving entropic intermediates. *Trends Chem.* 2019;*1*(1):22–34.

**44** Kurouchi, H. and Singleton, D.A. Labelling and determination of the energy in reactive intermediates in solution enabled by energy-dependent reaction selectivity. *Nat. Chem.* 2018;*10*(2):237–241.

**45** Shaik, S., Danovich, D., Joy, J. et al. Electric-field mediated chemistry: uncovering and exploiting the potential of (oriented) electric fields to exert chemical catalysis and reaction control. *J. Am. Chem. Soc.* 2020;*142*(29):12551–12562.

**46** Siddiqui, S.A., Stuyver, T., Shaik, S. et al. Designed local electric fields—promising tools for enzyme engineering. *JACS Au.* 2023;*3*(12):3259–3269.

**47** Shaik, S., Mandal, D., and Ramanan, R. Oriented electric fields as future smart reagents in chemistry. *Nat. Chem.* 2016;*8*(12):1091–1098.

**48** Stuyver, T., Danovich, D., Joy, J. et al. External electric field effects on chemical structure and reactivity. *Wiley Interdiscip. Rev. Comput. Mol. Sci.* 2020;*10*(2):1–22.

**49** Lai, W., Chen, H., Cho, K. et al. External electric field can control the catalytic cycle of cytochrome P450cam: a QM/MM study. *J. Phys. Chem. Lett.* 2010; *1*(14):2082–2087.

# 6

# Force Fields and Metals

*José-Emilio Sánchez-Aparicio, Lorena Roldán-Martín, Jean-Didier Maréchal, and Giuseppe Sciortino*

*Departament de Química, Universitat Autònoma de Barcelona, 08193 Cerdanyola del Vallés, Barcelona, Spain*

## 6.1 Introduction

Many molecular modeling studies can be initiated with a limited amount of structural information, including experimental structures (e.g. X-ray or nuclear magnetic resonance [NMR]). For certain processes, such as redox profiles or spectroscopic analysis, this could be sufficient to get simulations of good quality. Other molecular processes, though, cannot be addressed with such a limited set of geometries, especially those that require vast explorations of geometric ensembles, e.g. the identification of breathing motions of proteins, the characterization of diffusion pathways of substrates in metalloenzymes, the elucidation of the impact of the binding of a substrate to the structural and dynamical properties of the protein or even its aggregation profile with the membrane or other molecular partners. Studies implying large conformational sampling can only be achieved if the computational cost (and consequently the time) for the calculation of the energy of each geometry is sufficiently short. Now, this condition is not achievable under quantum mechanical (QM) frameworks and the solution stands on models based on classical physics.

The so-called classical molecular mechanics (MM) techniques are the common methodologies applied to calculate the energy of the system in a reasonable amount of time even for very large systems, such as proteins. MM is the generic name for a large family of methods and, based on the energetic calculations they provide, different approaches can be used to explore the energetic landscape, including deterministic approaches like molecular dynamics (MD) or stochastic ones like Monte Carlo or genetic algorithms. The former is by far the mainstream method when it comes to exploring molecular motions of biomolecules.

In this chapter, we first give the main concepts of MM methods and focus on the limitations, challenges, and advances when dealing with biometallic systems. Then, a series of illustrative cases are provided for each type of parameterization process.

*Computational Bioinorganics: From Description to Prediction*, First Edition. Edited by Jean-Didier Maréchal.

## 6.2 Overview on Molecular Mechanics and How to Deal with Metals

In MM methods, the system is modeled on the bases of classical physics. The system's potential energy is described as a function of the nuclear positions using a series of empirical parameters. The common parameters are available for well-defined functions for bonded and nonbonded interactions. Among bonded interactions, bonds, angles, and dihedrals are typically described in the most common force fields (such as AMBER, CHARMM, OPLS, etc.) using harmonic functions. Nonbonded interactions, like electrostatic and van der Waals (vdW), are generally modeled on the Coulombic framework using Lennard-Jones (LJ) potentials [1]. It can be highlighted that the accuracy and the general applicability of the techniques are limited by the number and type of functions that the specific force field accounts for, and a complete overview of the application of classical mechanics to model metal ions can be found in Ref. [2]. Advances in computational bioinorganics in the last decade are clearly related to the improvement of the viability and user friendliness to deal with metal-compatible force fields.

Despite their high potentiality and relatively low computational cost, most of the force fields lack standardized parameterization for metal ions, and the development of more accurate ones is a vivid field of research nowadays. The approaches that have been developed with the aim to overcome these limitations (Figure 6.1), include (i) representation of the metal ion as a simple 12-6 LJ interacting soft-sphere (with recent expansion) [3, 4]; (ii) bonded models for which the bond between the metal and their ligands are forced during the simulation [5–8]; (iii) nonbonded models using dummy atoms [9–15].

The LJ sphere model, describing the ion as a soft sphere with an integer charge, limits the metal interaction with its surroundings to simple van der Waals and

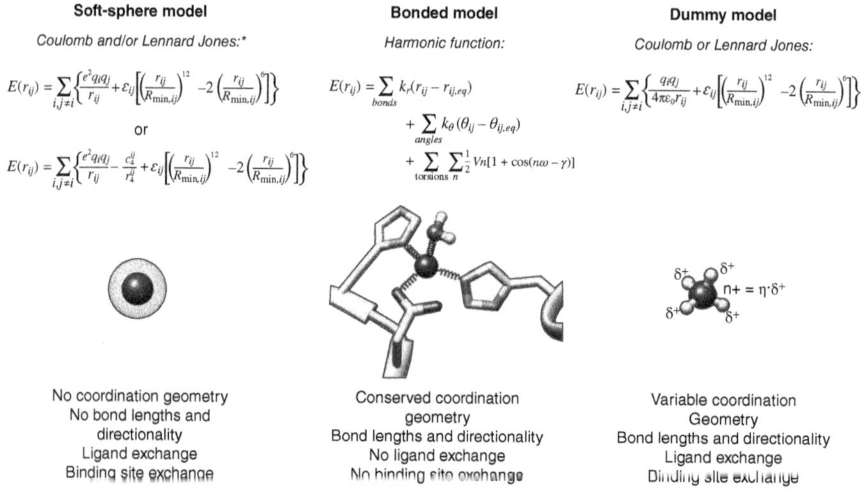


**Figure 6.1** Different approaches to dealing with metal ions in MM and MD frameworks.

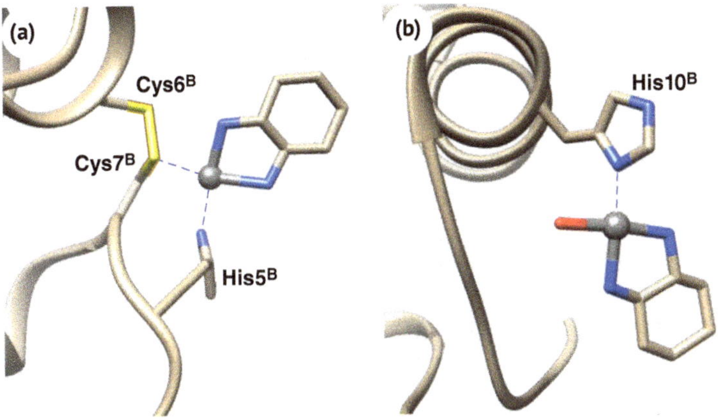


**Figure 6.2** Detail of two binding modes studied for the oxaliplatin-insulin complex. (a) The α-binding site with the coordination of Cys7$^B$ and His5$^B$. (b) The β-binding site with Pt$^{II}$ coordinated to His10$^B$. In both cases, coordination bonds are represented as dashed purple lines, with the oxaliplatin moiety and the donor amino acids highlighted in sticks. *Source*: Adapted from [23].

electrostatic interactions. Such approach was successfully applied to the main group metals but shows several limitations for transition metals, for which bond directionality, coordination geometry, and metal–donor affinity are fundamental. Moreover, the 12-6 LJ potential neglects metal-induced dipole interaction. Li and Merz recently proposed a 12-6-4 LJ-type nonbonded model to account for charge-induced dipole interactions for divalent [16] and highly charged metal ions (up to M4+) [17]. The additional empirical $C_4$ term was added to the standard 12-6 LJ potential and used to represent the ion-induced dipole interaction (Figure 6.2). This new parameter set is able to reproduce some experimental observables, such as hydration free energies (HFE), ion-oxygen distances (IOD), and coordination number (CN), in explicit water solvent with higher accuracy to respect the classical LJ potential, and has demonstrated good transferability on protein biomolecular systems [16, 17].

On the other hand, bonded models using specific parameters to describe bonds between metals and its specific donors, give a good description of the coordination geometry. In fact, bond distances, angle, dihedral, electrostatic, and van der Waals terms are specifically included in the force field. However, its application is limited to very few well-defined systems and the parameter portability is often unavailable [18]. Thankfully, in the last decade, solutions emerged to ease this task, and metal center parameters, if not available, could be derived for the specific system with several methods and tools, such as FUERZA [19] MCPB.py [20], or VFFDT [21]. However, one has to bear in mind that the approach obliges fix metal–ligand predefined bonds and is unable to account for coordination geometry interconversion or ligand exchange. The approach is, therefore, not applicable to all problems one could encounter in biometallic modeling since they are fundamental aspects of metal–protein interaction and in catalysis [7].

## 6.3 Bonded Model

Bonded models of metal–ligand bonds are, in a way, conceptually more restrictive than others because the dynamic effects of the first coordination sphere of the metal are absent in the simulations. The strength of standard coordination bonds is intermediate between ionic and covalent, which suggests that changes in the first coordination sphere of the metal can occur relatively frequently in solution; something these models do not allow. Still, when the system can or should be modeled with the fixed environment of the metal, those approaches are the most reliable and easy to set up. It is, therefore, not surprising that its application is widely spread in molecular modeling today. To show the applicability of the approach and its limitations, we selected three systems.

### 6.3.1 Cu(II) Binding to Polyhistidine Peptides

Bonded models were recently applied by Kozlowski et al. to determine the Cu(II) binding site on the polyhistidine domain of a peptide contained in the venom of *Atheris squamigera* (pHpG-1) [22]. The 24-residue peptide presents a repetitive sequence Glu-$(Asp)_2$-$(His)_9$Gly-Val-$(Gly)_{10}$-$NH_2$ showing a high abundancy of histidine that makes it a perfect candidate to bind a Cu(II) ion. Potentiometric, spectroscopic, and Electrospray Ionization Mass Spectrometry (ESI-MS) studies on the system pHpG-1/Cu(II) unveil eight different coordination modes as a function of pH. At low pH, a species with the coordination of one imidazolic or aminic nitrogen was identified whereas increasing the pH, up to three additional imidazolic nitrogens seems to bind the metal. Around the neutrality and a basic pH, the experimental evidence suggests the progressive substitution of the imidazolic nitrogens by amide donors up to form $[CuLH_{-3}]^{4-}$ complex. To shed light on the binding sites and structure upon binding, an MD study was carried out building 10 different models representing the most probable structures involving different side chains: the binding to the N-terminal domain as {Cu(Glu1)(Asp3)}–pHpG-1, {Cu(His7)(His9)}–pHpG-1, {Cu(Glu1)(Asp3)(His12)}–pHpG-1, and six models for the simultaneous coordination by three histidine residues {Cu$(His)_3$}–pHpG-1. The MD simulations of the solvated models were carried out in the NPT ensemble along 30 ns treating the coordination bonds by imposing empirical distance restraints. As a result, based on the stability of the models during the trajectories, the simulations confirm the high experimentally observed Cu(II)-pHpG-1 affinity. Another highlight regarding the {Cu$(His)_3$}–pHpG-1 models is a preferential binding motif with histidine residues separated by one amino acid and, in particular, the models involving His4, 6, and 8, or to His7, 9, and 11 seems to be favored. Finally, the models in which the Gly-rich domain has an α-helical shape appear more stable from a potential energy point of view with respect to models with a more disordered conformation. Despite the good results of the study using a bonded model, the exploration is quite limited in the case of exploring all possible metal binding sites. For the simple peptide we described above, if one considers only the side chain donors and the N-terminal end, the total models of models to consider with a tetracoordinate metal is 715! This would represent a massive tour de force in term of parameterization with a bonded model.

### 6.3.2 Interaction of Oxaliplatin with Insulin

The interaction of oxaliplatin (an anticancer drug) with insulin was recently studied through an integrated computational protocol that includes several steps of classical MD using bonded models for the metal environment [23].

Cisplatin, oxaliplatin, and other derivatives are mainly aimed at targeting DNA. However, those drugs also interact with other proteins resulting in off-target interactions that are the core of undesired effects, such as lower bioavailability, resistance etc [24–26]. Oxaliplatin, despite having much lower nephro- and ototoxicity when compared with first-generation anticancer drugs, has also shown resistance in certain cases, some of them being related to insulin interaction [27]. Therefore, understanding in detail such interactions is of major interest. However, no three-dimensional (3D) atomic structures of the oxaliplatin-insulin complex have been released and computation could, therefore, be the most interesting approach to deepen our knowledge.

Including docking and quantum mechanical calculations, the central part of the computational protocol was MD simulations to ascertain the stability and dynamics of the insulin-oxaliplatin adducts, and to obtain viable representative 3D structures. A bonded model was used to describe the metal moiety interaction with the amino acid donors. By definition, using a bonded model implies imposing what coordination bonds are established between the metal and the amino-acid donors. This information can be obtained either from experimental knowledge or a previous computational screening, or a combination of both. In the case under study, no crystallographic structures were available for the oxaliplatin-insulin complex. However, some structures containing insulin bound to other metal ions exist in the Protein Data Bank (PDB). In particular, the structure with the PDB code 1zni was considered a valid starting point because, besides its high resolution (1.498 Å), it contained several $Zn^{2+}$ ions bound to insulin, indicating a good sidechain and backbone pre-organization to allocate metals. At this point, it is interesting to highlight that in the absence of a valid preorganized structure, it could be necessary to add another previous step to the protocol, such as an MD simulation of an unbound structure to explore the conformational space of the protein.

Once the structure of the host was selected, docking of the oxaliplatin was carried out with the GOLD software [28] over the whole monomer of insulin. A protocol specialized in metal-containing ligands was applied [29, 30], leaving free to coordinate one or both equatorial vacancies of the oxaliplatin moiety and allowing side-chain flexibility to all the potentially coordinating amino acids. The docking experiment led to identify two binding modes (named $\alpha$ and $\beta$, Figure 6.2) and an additional binding mode $\gamma$, involving the same residues as in the binding mode $\alpha$ but considering the reduction of the disulfide bridge between Cys7A and Cys7B upon the binding of oxaliplatin. It should be noted that the exploration of the whole structure by docking was possible due to the relatively small size of insulin (30 residues). When analyzing larger systems, a previous screening of the most probable metal-binding sites could be necessary, for example, by means of BioMetAll [31].

After the identification of these three binding modes, appropriate parameterization for metal ions was performed so to run MD simulation. This is without a doubt the most specific point of the integration of metal into force field experiments. Pt bonding parameters were obtained using the Seminario's method computing the geometry and harmonic frequencies with Gaussian09 [32] at a Density Functional Theory (DFT) level with the B3LYP functional combined with the Stuttgart–Dresden (SDD) pseudopotential and its associated double-ξ basis plus a set of *f* polarization functions [33] for Pt, whereas the 6-31G(d,p) basis set was used for H, C, N, and S atoms. Point charges were derived using the Restrained Electrostatic Potential (RESP) model [34] and the force field building was carried out by means of MCPB.py [20], a software mentioned earlier and that has represented a major advance in metal parameters for standard force fields.

Apart from the assessment of the dynamical behavior of the system through the analysis of the MD simulations, one of the objectives of this study was to obtain reliable 3D structures of the oxaliplatin-insulin complex. Consequently, clustering was carried out on each of the three MD trajectories. Ten clusters were generated in each case using the MSMBuilder implementation of the K-medoid clustering algorithm [35], taking the Root Mean Square Deviation (RMSD) on the alpha carbons as a metric to measure the similarity between frames. The representative structure of the most populated cluster was selected for each MD. Finally, a QM/MM geometry optimization was performed on each representative structure using the Garleek framework [36] describing the Pt atom and its first coordination sphere at the QM level using Gaussian09 [32], and the rest of the system at the MM level using Tinker [37].

One of the crucial steps of this study is the possibility to identify stable geometric space of the metallodrug-protein adducts and the bonded model has been fundamental to obtained such information.

### 6.3.3 Artificial Metallohydratase

As a last study case we selected corresponds to the work performed by Alonso et al. [38] on the study of copper-mediated hydration by an artificial metalloenzyme (ArM). ArMs are systems generated by the insertion of synthetic metallic cofactors, generally homogeneous catalysts, into biological receptors, mostly proteins. Reaching high yields with these *de novo* enzymes is still quite challenging but, many of these systems have shown interesting enantioselective patterns.

Under this framework, copper metallohydratases have been developed by Roelfes et al. and tend to be formed by on side a Cu(II) baring cofactor (typically phenanthroline or bipyridine) the transcription factor *Lactococcal multidrug resistance Regulator* (LmrR) (Figure 6.3) [39, 40].

The mechanism of copper-mediated hydration performed by these ArMs consists of two consecutive steps: first a nucleophilic attack by the water molecule, which specifies the enantioselectivity, followed by a *syn* or *anti* electrophilic protonation step [41]. The catalytic mechanism of the copper was studied in an integrative manner, combining QM and MM tools, such as protein–ligand docking and MD simulations with bonded parameters. For such purpose, the catalysis was analyzed and compared both in an aqueous system and inside the LmrR protein. In this case,

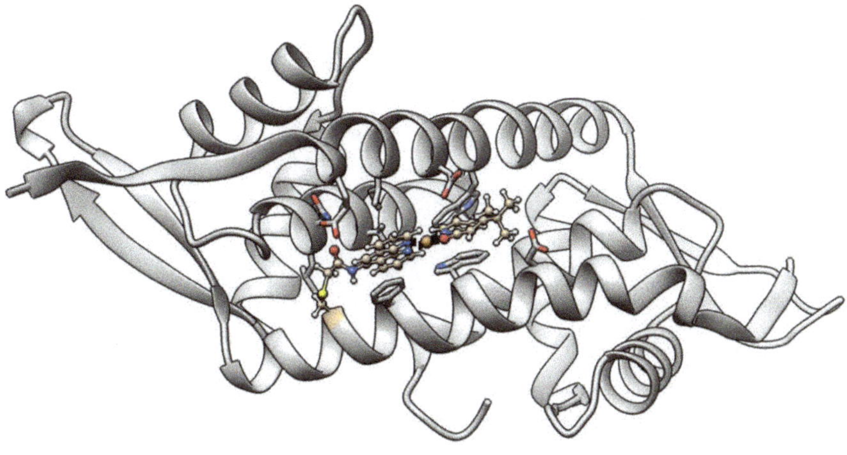

**Figure 6.3** Image of an LmrR-based copper hydratase. At the center, the Cu(II) ion (orange), with phenanthroline as a ligand (light brown). In grey, the LmrR, the biomolecular scaffold.

MM-based approaches were used for both types of systems using the bonded model for the metal center. The force field parameters for the copper complexes were calculated with the Seminario's method, and the charges with the RESP method in antechamber. Moreover, the Generalized Amber Force Field (GAFF) was used for the remaining atoms; the TIP3P model for the water molecules, and AMBER for the protein, when present.

This was crucial for the case of the isolated phen-Cu(II) complex in solution. Indeed, one of the main issues to solve was the mechanism by which the protonated water transferred the proton to the Cα of the substrate, the very last step of the reaction. Therefore, MD simulations were applied to discern the conformational preferences of the water molecules around the complex, demonstrating that H-bond water chains between the proton donor and the acceptor atoms were sporadic. This observation, together with ab initio MD (AIMD) analysis, discarded the proton-shuttle mechanism through a water chain in favor of the Cα protonation due to an acid-base mechanism of the solvent.

For the embedding of the phen-Cu(II) catalyst inside the dimeric LmrR protein, the binding site of the complex was identified through protein–ligand docking, and a posteriori MD simulations were carried out. Once the MD trajectories were achieved, two lectures could be performed. On the one hand, several essential interactions between the copper complex and the protein were identified, as they kept stable during the whole MD simulation. First, Phe93 and Phe93' residues stabilize the phen-Cu(II) complex in the interdimeric region, allowing the proper rate of conversion of the hydration reaction. Second, a hydrogen bond with Asn19. Third, the interaction of Asp100' with the metal center of the complex. It is remarkable the role of the latter interaction, as the Asp100' residue not only stabilizes the complex in the active site but also participates in the activation of the water nucleophile. On the other hand, precatalytic configurations were identified if two constraints were accomplished. The first constraint is a distance shorter than 5 Å between the Cβ and the surrounding carboxylate oxygens of negatively charged residues. The second is the orientation of

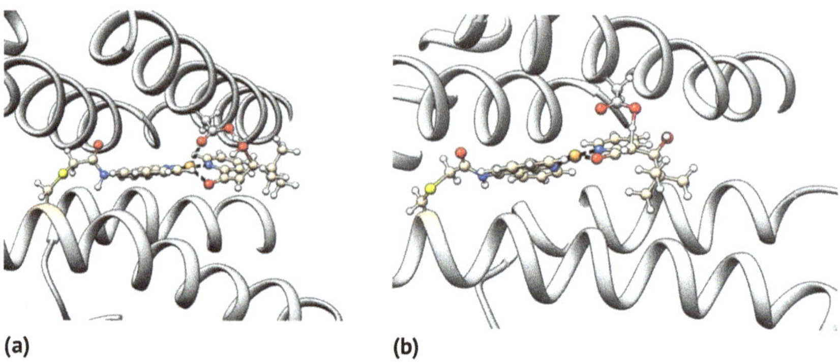

**Figure 6.4** Transition-state geometries calculated with QM cluster models for the hydration reaction. Two different attacks happen in this image, (a) nucleophilic and (b) electrophilic, for D100-Cu and D100-w-Cu configurations.

the surrounding water molecules with respect to the double-bond substrate, which should allow the formation of an H-bond. With such analysis, the residues participating in the hydration mechanism can be spotted, as well as an estimation of the enantioselectivity of the system according to the side of exposition of the double bond to the catalytic water. In the present case, the authors identified two different precatalytic configurations: one with Asp100' directly coordinating the copper and another one with an indirect interaction, bridged by a water molecule (Figure 6.4). Once the possible precatalytic frames were chosen, QM calculations were performed, demonstrating that the conformation with the in-between water is the most favorable catalytic mechanism. Here again, MD calculations performed with the bonded model of the metal were fundamental not only to analyze the general motions of the protein but to decode several key steps in the catalytic process.

As it has been demonstrated through the present study case, the integration of force field approaches with QM can provide with a deep study of a system. While QM can properly describe the different steps of a reaction, MD simulations based on a bonded model can follow the arrangements of the atoms over a period of time, allowing the identification of essential conformations, such as the distribution of water molecules around the system or pre-catalytic forms if a biological scaffold is involved. Altogether, the combination of all these computational tools made possible the elucidation of the mechanism of copper-mediated hydration of an alkene, even if the nucleophile involved is a water molecule of the solvent.

The sum of the case studies selected in this section provides a vision of the limitations and strengths of the bonded model for metal. First, one should be aware of the main assumption of the model: the coordination bonds between the metal and its donors are assumed to exist and cannot be broken during the simulation, which limits the conclusions that can be drawn from this framework. Also, the description of these bonds is needed previously to run the simulation, which can be challenging to obtain depending on the experimental information available. However, taking into account, these considerations, bonded models give a good description of the coordination geometry and allow to make large conformational exploration, something determinant for key molecular events including pre-organized geometries for catalysis.

## 6.4 Nonbonded Models

### 6.4.1 The Dummy Atom

The nonbonded dummy model, originally proposed by Aqvist and Warshel [12] consists of describing the metal ion as a hard sphere with negligible dimension and dummy particles placed in the coordination vacancies and bonded to the central atom. The total charge of the metal ($n^+$) is distributed between the dummies that result with partial positive charge ($\delta^+$), which can interact by electrostatic contacts with donors present in the surroundings. At the state of the art, several models have been proposed for $Zn^{2+}$, $Ca^{2+}$, $Mg^{2+}$, $Fe^{2+}$, $Ni^{2+}$, $Co^{2+}$, $Mn^{2+}$, $Cu^{2+}$, $Al^{3+}$, $Fe^{3+}$, and $Cr^{3+}$ in tetrahedral, octahedral, or pentagonal bipyramid geometries [9–15]. The method, despite some limitations in bond directionality and metal–donor relative affinities, represents a promising approach for studying metal ions–protein interactions. In a recent study, Strodel et al. successfully validated the transferability of the approach to biological systems, studying the metal binding to the amyloid-β peptide [14]. $Cu^{2+}$ and $Zn^{2+}$ interaction with $A\beta_{1-16}$ was studied by building two starting models with different coordination modes. Both metals were treated by placing the dummy atoms around the ions in an octahedral geometry. In both ions, the charge of the central metal was set to −1, for Zn the dummies' partial charges were set to +0.5 while for Cu the equatorial dummies were set to +0.725 and the axial ones to +0.05, a distribution that account with Jahn–Teller distortion. The starting models involved H6, E11, H13, and H14 (model α) for both ions and an additional model involving A2, H6, H13, and H14 (model β) was built for $Cu^{2+}$. The NMR structure (PDB 1ZE9) was used as initial coordinates for the model α, while the model for β, a theoretical geometry was taken from Ref. [2] (Figure 6.5). MD simulations of the TIP4P [3] solvated models were performed with GROMACS [44] at 300 K on the NPT framework along 100 ns using the OPLS-AA/L force field [45, 46].

The results highlight a metal coordination geometry retained during the trajectories with RMSD with deviations about 0.5 Å, despite the high flexibility of the whole system and large conformational changes reporting RMSD values around 2.6 Å. Along the MD simulation, the model α shows, for both Cu and Zn, a coordination

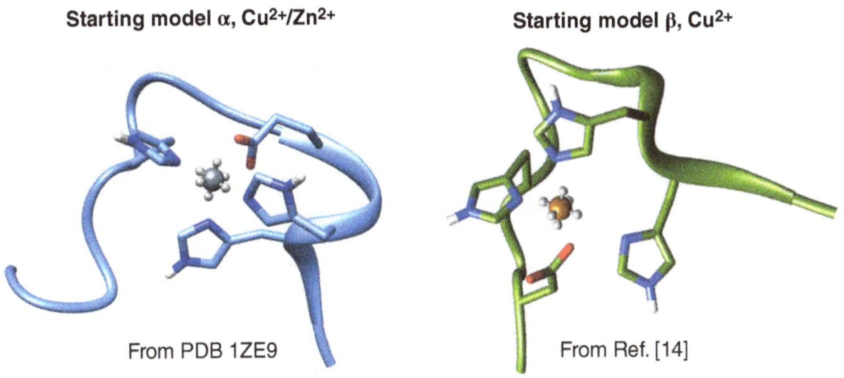


**Figure 6.5** Different models adapted from the study by Strodel et al.

set N-His6, N-His13, and N-His14 defining the equatorial plane and the bidentate $O_2C$-Glu11 chelating the metal, with a water molecule completing the coordination in a distorted elongated octahedral arrangement. The Cu-model β evolves in an elongated octahedron with the equatorial plane defined by N-His6, N-His13, N-His14, and the monodentate Asp11, with two water axial molecules defining the octahedral. From an energetic point of view, the Cu-metallated peptide, model α, appears about 30 kcal $mol^{-1}$ more stable with respect to the Zn homolog due to stronger Coulomb interactions between the metal- and the donor-containing residues. The simulations were also repeated starting from a disordered model obtaining comparable coordination geometries, highlighting the potentiality of the method for metal-binding site prediction.

In principle, all these approximations can be applied to the ion-binding site prediction problem. However, using a pure MD protocol could be very time-consuming for a global exploration of the potential energy surface and the system could fall and be trapped in local minima. For a complete exploration of the chemical space by these methods, parallel simulations could be used generating different ion–protein relative orientations as starting points, but the protocol could be very tedious and highly time-consuming.

## 6.5 The 12-6(-4) Lennard–Jones Potentials

While standard 12-6 model has shown to be a good model for certain metals like aluminum, it misses some key physico-chemical contents. Indeed, one of the limitations of this widely used nonbonded model is to neglect of metal-induced dipole interaction, which leads to an underestimation of the experimentally obtained free energies, especially on systems with highly charged metal ions. Several force fields have been developed for a long time to specifically deal with polarization effects of metals like the SIBFA, YETI, or LFMM approaches. These approaches, despite resulting from a major scientific tour de force and having shown interesting successes, have been developed outside the most common packages for MM simulations, which limited the spreading of their used by the community.

In recent years though, Li and Merz proposed the 12-6-4 LJ-type model, which accounts for the addition of an empirical $C_4$ term and offers a very practical and efficient way to work new metal parameterization in standard force fields like AMBER [16]. After the groundwork of their concept, they subsequently parameterized this nonbonded model for 24 different metal ions with three widely used explicit water models (TIP3P, SPC/E, and $TIP4P_{EW}$), and validated these parameters in a modeling exercise on a protein biomolecular system [17]. In this third and last case study, we aim to show the improvements that the $C_4$ term provides to the 12-6 LJ potential, and discuss the limitations and advantages of the nonbonded model throughout the practical example of the latter referenced work.

To assess the improvement of the 12-6-4 LJ-type model in its reproducing experimental properties, it was compared with the traditional 12-6 LJ nonbonded model in the AMBER force field against the hydration of several ions [47]. Two experimental properties were considered: HFE and IOD. The overall results for the 12-6

LJ model showed a marked underestimation of the experimental values that increase dramatically along with the formal charge of the ion. Interestingly, it was not possible to reproduce simultaneously the HFE and IOD experimental values: parameters aimed at reproducing the HFE experimental values lead to increase the error in IOD values and vice versa, when using parameters to reproduce the IOD experimental values, the error in the HFE values would increase. However, with the 12-6-4 LJ-type model, both HFE and IOD values could be reproduced simultaneously with one set of parameters. Furthermore, the 12-6-4 LJ-type model significantly reduces the error even when compared with the best 12-6 LJ parameter set for each individual property, being the experimental HFE values reproduced within $\pm 2$ kcal mol$^{-1}$ for M(IV) ions, and the IOD values within of $\pm 0.01$ Å.

Once the performance of the 12-6-4 LJ-type model was demonstrated for the ion in a solvated environment, it would be interesting to see if the approach can be easily transferred to a protein biomolecular system, as it has been shown for the 12-6 LJ model [48, 49]. To address this challenge, the work of reference proposed to validate the 12-6-4 LJ-type model on a superoxide reductase found in *Nanoarchaeum equitans*. The chain C of the PDB code 4bv1, with a $Fe^{3+}$ ion bound to His10/His35/His41/Cys97/His100 and a water molecule (Figure 6.6), was solvated with a TIP3P water model and used as starting structure for MD simulations of 10 ns of length, which were conducted for both 12-6 LJ parameter sets (HFE and IOD) and the 12-6-4 parameter set. In this latter case, the $C_4$ terms between the $Fe^{3+}$ ion and atom types other than water oxygen were evaluated using Eq. (6.1). Although this procedure could seem in principle much simpler than the one used for the preparation of the bonded MD simulations shown in the previous case studies, it must be highlighted that

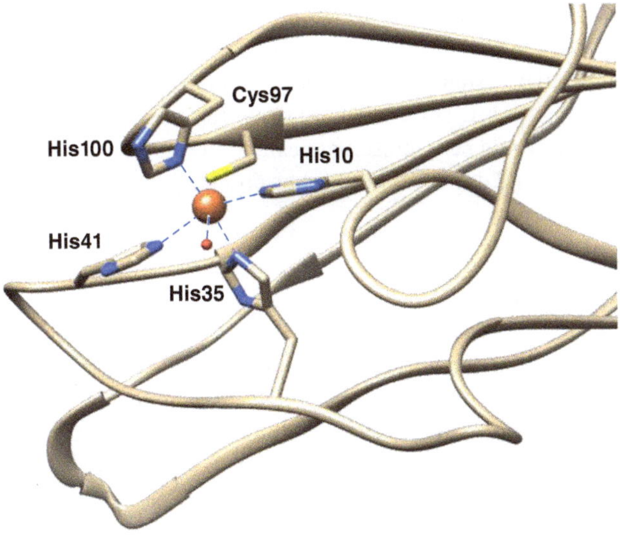


**Figure 6.6** Detail of the metal-binding site for the chain C of the PDB code 4bv1. The ferric ion is represented as a brown sphere, the water molecule as a smaller red sphere, and the five coordinating amino acids are highlighted in sticks. The coordination bonds are represented with blue dashed lines. *Source*: Adapted from [17].

polarizability values ($\alpha_0$) for the corresponding ion and all the atom types used in the system must be either calculated or obtained from the literature. In this case, the polarizability of the $Fe^{3+}$ ion was calculated at the B3LYP/6-311++G(2d,2p) level of theory using Gaussian09 [32], whereas the polarizability of the water oxygen and the rest of atom types were taken from the bibliography [50, 51].

$$C_4\,(\text{atom type}) = \frac{C_4(\text{H}_2\text{O})}{\alpha_0(\text{H}_2\text{O})} \times \alpha_0\,(\text{atom type}) \tag{6.1}$$

In contrast to bonded models, here the ion atom has the possibility to form or break coordination bonds. Therefore, the evaluation of the MD trajectories focused on whether the crystallographic binding mode was maintained during the course of the simulations. For the 12-6 LJ with HFE parameter set, the metal moves out from the binding pocket, with a smaller coordination number (four) than the crystallographic structure. However, for the 12-6 LJ with IOD parameter set and the 12-6-4 LJ-type model, the metal-binding site is stable during the whole simulation, being the RMSD of the metal site of ~0.5 Å from the initial structure. This test, therefore, revealed good transferability to a protein biomolecular system of the parameters determined for the hydrated ions. More tests or longer simulations would be necessary to evaluate whether the model is capable to reproduce phenomena like chemical bond dissociation produced, for example, by a conformational change in the protein.

The 12-6-4 LJ-type model, in summary, addresses some of the limitations of the traditional nonbonded models. In particular, a clear improvement has been demonstrated when reproducing HFE and IOD experimental values. As one of the strengths of the nonbonded models is their low computational cost, it must be highlighted that this improvement was achieved with only a slight increase due to the incorporation of the $C_4$ term. However, other limitations of the nonbonded models remain, such as bond directionality or coordination geometry. The parameters obtained for hydrated ions have been successfully transferred for the case of a $Fe^{3+}$ ion into a protein biomolecular system, showing that the crystallographic binding mode is maintained during an MD simulation. To show the full potential of the model in protein systems, it would be interesting to extend the tests to systems where chemical bond formation and dissociation could be observed.

## 6.6 Conclusions

When it is necessary to consider the dynamical behavior in the study of bioinorganic systems, the use of MM becomes unavoidable among the current state-of-the-art plethora of methods and in this field, MD has become mainstream. Whereas all force fields incorporate the parameterization of natural amino acids natively, this is rarely the case for the inorganic part of the system. We have introduced in this chapter the most common approaches to deal with metal ions in an MM framework. The truth is that none of them is a gold standard in the field, and the choice strongly depends on the nature of the problem and the available experimental knowledge. In this sense, we have illustrated scope of each model and limitations through a series of practical examples.

Undoubtedly, one of the current limitations of the use of metal ions in MM continues to be the low usability of the models. Following in the vein of recent years, you can envision the development of new and easier-to-use approaches to managing metal ion parameterization soon. Their native integration with the most common MD packages could lower the usability barrier that nowadays discourages some researchers from addressing bioinorganic systems. Besides user-friendliness, accuracy can also be improved. Nonbonded models must address their limitations with bond directionality, coordination geometry, and metal–donor affinity, whereas bonded models are inherently constrained to specific problems. Perhaps a cutting-edge technology could emerge that enables full-quantum mechanics calculations. In the meantime, those improvements should strike a good balance between accuracy and the computational resources employed. It is expected that the emergence of machine learning-based force field could bring some novelty to our abilities to model metal in classical models of biometallic molecules.

## References


1 Ponder, J.W. and Case, D.A. Force fields for protein simulations. In *Advances in Protein Chemistry*. San Diego: Academic Press, 2003, Vol. 66, pp. 27–85.

2 Li, P. and Merz, K.M. Metal ion modeling using classical mechanics. *Chem. Rev.* 2017;*117*:1564–1686.

3 Li, P., Roberts, B.P., Chakravorty, D.K., et al. Rational design of particle mesh Ewald compatible Lennard-Jones parameters for +2 metal cations in explicit solvent. *J. Chem. Theory Comput.* 2013;*9*:2733–2748.

4 Torras, J. and Alemán, C. Determination of new Cu+, Cu2+, and Zn2+ Lennard-Jones ion parameters in acetonitrile. *J. Phys. Chem. B.* 2013;*117*:10513–10522.

5 Hancock, R.D. Molecular mechanics calculations as a tool in coordination chemistry. In *Progress in Inorganic Chemistry*. New York: John Wiley & Sons, Ltd, 1989, pp. 187–291.

6 Hancock, R.D. Molecular mechanics calculations and metal ion recognition. *Acc. Chem. Res.* 1990;*23*: 253–257.

7 Lin, F. and Wang, R. Systematic derivation of AMBER force field parameters applicable to zinc-containing systems. *J. Chem. Theory Comput.* 2010;*6*(1): 1852–1870.

8 Peters, M.B., Yang, Y., Wang, B., et al. Structural survey of zinc-containing proteins and development of the zinc AMBER force field (ZAFF). *J. Chem. Theory Comput.* 2010;*6*:2935–2947.

9 Pang, Y.-P. Novel zinc protein molecular dynamics simulations: steps toward antiangiogenesis for cancer treatment. *Mol. Model. Annu.* 1999;*5*:196–202.

10 Oelschlaeger, P., Klahn, M., Beard, W.A., et al. Magnesium-cationic dummy atom molecules enhance representation of DNA polymerase β in molecular dynamics

simulations: improved accuracy in studies of structural features and mutational effects. *J. Mol. Biol.* 2007;*366*:687–701.

**11** Saxena, A. and Sept, D. Multisite ion models that improve coordination and free energy calculations in molecular dynamics simulations. *J. Chem. Theory Comput.* 2013;*9*:3538–3542.

**12** Aqvist, J. and Warshel, A. Calculations of free energy profiles for the staphylococcal nuclease catalyzed reaction. *Biochemistry* 1989;*28*:4680–4689.

**13** Duarte, F., Bauer, P., Barrozo, A., et al. Force field independent metal parameters using a nonbonded dummy model. *J. Phys. Chem. B.* 2014;*118*:4351–4362.

**14** Liao, Q., Kamerlin, S.C.L., and Strodel, B. Development and application of a nonbonded Cu2+ model that includes the Jahn–Teller effect. *J. Phys. Chem. Lett.* 2015;*6*:2657–2662.

**15** Liao, Q., Pabis, A., Strodel, B., et al. Extending the nonbonded cationic dummy model to account for ion-induced dipole interactions. *J. Phys. Chem. Lett.* 2017;*8*:5408–5414.

**16** Li, P. and Merz, K.M. Taking into account the ion-induced dipole interaction in the nonbonded model of ions. *J. Chem. Theory Comput.* 2014;*10*:289–297.

**17** Li, P., Song, L.F., and Merz, K.M. Parameterization of highly charged metal ions using the 12-6-4 LJ-type nonbonded model in explicit water. *J. Phys. Chem. B.* 2015;*119*:883–895.

**18** Hu, L. and Ryde, U. Comparison of methods to obtain force-field parameters for metal sites. *J. Chem. Theory Comput.* 2011;*7*:2452–2463.

**19** Seminario, J.M. Calculation of intramolecular force fields from second-derivative tensors. *Int. J. Quantum Chem.* 1996;*60*:1271–1277.

**20** Li, P. and Merz, K.M. MCPB.Py: a python based metal center parameter builder. *J. Chem. Inf. Model.* 2016;*56*:599–604.

**21** Zheng, S., Tang, Q., He, J., et al. VFFDT: a new software for preparing AMBER force field parameters for metal-containing molecular systems. *J. Chem. Inf. Model.* 2016;*56*:811–818.

**22** Pontecchiani, F., Simonovsky, E., Wieczorek, R., et al. The unusual binding mechanism of Cu(Ii) ions to the poly-histidyl domain of a peptide found in the venom of an African viper. *Dalt. Trans.* 2014;*43*:16680–16689.

**23** Sciortino, G., Sánchez-Aparicio, J.-E., Rodríguez-Guerra Pedregal, J., et al. Computational insight into the interaction of oxaliplatin with insulin. *Metallomics* 2019;*11*:765–773.

**24** Medici, S., Peana, M., Nurchi, V.M., et al. Noble metals in medicine: latest advances. *Coord. Chem. Rev.* 2015;*284*:329–350.

**25** Jones, C.J. and Thornback, J.R. *Medicinal Applications of Coordination Chemistry.* Cambridge: The Royal Society of Chemistry, 2007.

**26** Hagrman, D., Goodisman, J., Dabrowiak, J.C., et al. Kinetic study on the reaction of cisplatin with metallothionein. *Drug Metab. Dispos.* 2003;*31*:916.

**27** Chen, J., Huang, X.-F., Qiao, L., et al. Insulin caused drug resistance to oxaliplatin in colon cancer cell line HT29. *J. Gastrointest. Oncol. J Gastrointest. Oncol.* 2011;*2*:27–33.

**28** Jones, G., Willett, P., Glen, R.C., et al. Development and validation of a genetic algorithm for flexible docking. *J. Mol. Biol.* 1997;*267*:727–748.

**29** Sciortino, G., Garribba, E., and Maréchal, J.-D. Validation and applications of protein–ligand docking approaches improved for metalloligands with multiple vacant sites. *Inorg. Chem.* 2019;*58*:294–306.

**30** Sciortino, G., Rodríguez-Guerra Pedregal, J., Lledós, A., et al. Prediction of the interaction of metallic moieties with proteins: an update for protein-ligand docking techniques. *J. Comput. Chem.* 2018;*39*:42–51.

**31** Sánchez-Aparicio, J.-E., Tiessler-Sala, L., Velasco-Carneros, L., et al. BioMetAll: identifying metal-binding sites in proteins from backbone preorganization. *J. Chem. Inf. Model.* 2021;*61*(1):311–323.

**32** Frisch, M.J., Trucks, G.W., Schlegel, H.B., et al. *Gaussian 09, Revision D.01*. Gaussian Inc., Wallingford, CT, 2009.

**33** Ehlers, A.W., Böhme, M., Dapprich, S., et al. A set of F-polarization functions for pseudo-potential basis sets of the transition metals ScCu, YAg and LaAu. *Chem. Phys. Lett.* 1993;*208*:111–114.

**34** Bayly, C.I., Cieplak, P., Cornell, W.D., et al. A well-behaved electrostatic potential based method using charge restraints for deriving atomic charges: the RESP model. *J. Phys. Chem.* 1993;*97*:10269–10280.

**35** Harrigan, M.P., Sultan, M.M., Hernández, C.X., et al. MSMBuilder: statistical models for biomolecular dynamics. *Biophys. J.* 2017;*112*:10–15.

**36** Pedregal, J.R.-G., Funes-Ardoiz, I., Sciortino, G., et al. GARLEEK: adding an extra flavor to ONIOM. *J. Comput. Chem.* 2019;*40*:381–386.

**37** Pappu, R.V., Hart, R.K., Ponder, J.W. Analysis and application of potential energy smoothing and search methods for global optimization. *J. Phys. Chem. B.* 1998;*102*:9725–9742.

**38** Alonso-Cotchico, L., Rodríguez-Guerra Pedregal, J., Lledós, A., et al. The effect of cofactor binding on the conformational plasticity of the biological receptors in artificial metalloenzymes: the case study of LmrR. *Front. Chem.* 2019;*7*:211.

**39** Rosati, F. and Roelfes, G. Artificial metalloenzymes. *ChemCatChem* 2010;*2*:916–927.

**40** Drienovská, I., Alonso-Cotchico, L., Vidossich, P., et al. Design of an enantioselective artificial metallo-hydratase enzyme containing an unnatural metal-binding amino acid. *Chem. Sci.* 2017;*8*:7228–7235.

**41** Bos, J., García-Herraiz, A., Roelfes, G. An enantioselective artificial metallo-hydratase. *Chem. Sci.* 2013;*4*:3578–3582.

**42** Alí-Torres, J., Maréchal, J.-D., Rodríguez-Santiago, L., et al. Three dimensional models of $Cu^{2+}$-Aβ(1–16) complexes from computational approaches. *J. Am. Chem. Soc.* 2011;*133*:15008–15014.

**43** Jorgensen, W.L., Chandrasekhar, J., Madura, J.D., et al. Comparison of simple potential functions for simulating liquid water. *J. Chem. Phys.* 1983;*79*:926–935.

**44** Hess, B., Kutzner, C., van der Spoel, D., and Lindahl, E. GROMACS 4: algorithms for highly efficient, load-balanced, and scalable molecular simulation. *J. Chem. Theory Comput.* 2008;*4*:435–447.

**45** Jorgensen, W.L., Maxwell, D.S., and Tirado-Rives, J. Development and testing of the OPLS all-atom force field on conformational energetics and properties of organic liquids. *J. Am. Chem. Soc.* 1996;*118*:11225–11236.

**46** Kaminski, G.A., Friesner, R.A., Tirado-Rives, J., et al. Evaluation and reparametrization of the OPLS-AA force field for proteins via comparison with accurate quantum chemical calculations on peptides. *J. Phys. Chem. B.* 2001;*105*:6474–6487.

**47** Case, D.A., Cheatham III, T.E., Darden, T., et al., Simmerling, C., Wang, B. The amber biomolecular simulation programs. *J. Comput. Chem.* 2005;*26*:1668–1688.

**48** Merz, K.M. Carbon dioxide binding to human carbonic anhydrase II. *J. Am. Chem. Soc.* 1991;*113*:406–411.

**49** Stote, R.H. and Karplus, M. Zinc binding in proteins and solution: a simple but accurate nonbonded representation. *Proteins Struct. Funct. Bioinform.* 1995;*23*:12–31.

**50** Eisenberg, D. and Kauzmann, W. *The Structure and Properties of Water*. Oxford: Oxford University Press, 2005.

**51** Miller, K.J. Additivity methods in molecular polarizability. *J. Am. Chem. Soc.* 1990;*112*:8533–8542.

7

# Modeling Thermally Activated Processes

*Pietro Vidossich[1] and Alessandra Magistrato[2]*

[1]*Laboratory of Molecular Modeling and Drug Discovery, Istituto Italiano di Tecnologia (IIT), Genoa, Italy*
[2]*National Research Council-Institute of Material Foundry (CNR-IOM), Scuola Internazionale Superiore di Studi Avanzati (SISSA), Trieste, Italy*

## 7.1 Introduction

The function of biomolecules depends on their conformational dynamics [1]. Signaling proteins undergo conformational changes in response to some stimulus, allowing them to transmit a signal [2]. Enzymes have to recognize their respective substrates and reach a catalytically competent conformation [3]. The DNA double helix has to unwind to allow for its transcription and replication [4]. The type of conformational changes involved in these processes spans a broad range of time and length scales, which are governed by the underlying free energy landscape [5]. The currently accepted model pictures the free energy landscape as a rugged surface, in which wells correspond to stable states, and is organized in a hierarchy of tiers depending on the depth of the wells and the magnitude of the barriers separating them [6]. The conformational changes may involve the rearrangement of domains and secondary structural elements, or local reorganization of groups of atoms, such as amino acid side chains. These concepts apply to any biomolecule, including metalloproteins. In this chapter, we are interested in processes, which are specific to bioinorganic systems, such as those taking place at the interface between metals and the bioorganic material. Specifically, we discuss modeling techniques, which allow for describing transformations at the metal center, namely, the exchange of ligands and chemical reactions.

The reasons for modeling such processes are twofold. On one side, we seek structural information. The structures corresponding to the stable states between which the transition occurs may be known, for instance from X-ray, nuclear magnetic resonance, cryo-electron microscopy studies, guessed based on available information, or predicted [7]. Modeling is thus used to reconstruct the intermediate structures not available experimentally, through which the system proceeds evolving from one state to the other. This knowledge is of interest because it allows pinpointing the molecular determinants that allow (or hamper) the process under study. The other key aspect

*Computational Bioinorganics: From Description to Prediction*, First Edition. Edited by Jean-Didier Maréchal.
 Published 2026 by John Wiley & Sons Ltd.

underlying a bioinorganic transformation is the energetics of the process. If this is known experimentally, comparison between the experimental and computational estimates allows to assess the accuracy of modeling. If it is not known, the computational estimates may be used to predict the behavior of the system. Together, structural and energetic information may be used to rationalize the functioning of the system and compare it to related systems, for instance, contrasting the enzymatic efficiency of different enzyme mutants or with respect to different substrates.

The computational techniques used to model transitions should thus provide both the transition pathway and the associated free energy profile. In this chapter, we describe some of the approaches that have been developed for this purpose. The focus will be on established and popular methods that allow realistic models to be treated. Indeed, it is recognized that to capture the fine details of biological processes, the full complexity of the system should be approached. Successful modeling of the binding of a drug or of the native reaction of a metalloenzyme requires models that explicitly account for the environment around the active site, including the protein matrix, solvent, and counterions. As the number of particles in the model increases, sampling of configuration space may be performed by simulation methods, such as molecular dynamics (MD) or Monte Carlo simulations [8]. Furthermore, modeling ligand exchange or reactions at the metal center requires interatomic potentials that allow for the formation and breaking of chemical bonds. Quantum mechanical (QM) methods and density functional theory, in particular, are suitable choices [9]. However, the high computational cost often limits their applicability, and specialized molecular mechanical (MM) potentials have been developed to treat these events (note that QM and MM interatomic potentials, and their coupling (QM/MM), are discussed in other chapters of this book). The computational resources now available allow for the modeling of large systems. However, the timescale accessible to simulations remains limited. Thus, enhanced sampling methods, some of which are described in this chapter, were developed to address this issue.

## 7.2 Modeling Transitions

Transitions between stable states – two different conformations of a molecule, the bound and unbound states of a ligand, reactants and products of a chemical reaction – are infrequent events compared to the timescale of molecular vibrations. Statistical mechanics explains this phenomenon in terms of the low probability, $P(\boldsymbol{r})$, of occurrence of intermediate configurations along the transition [10]. Indeed, Boltzmann statistics (Eq. 7.1), which apply to systems in thermal equilibrium, disfavors high-energy configurations:

$$P(\boldsymbol{r}) = \frac{e^{-\beta U(\boldsymbol{r})}}{\int d\boldsymbol{r} e^{-\beta U(\boldsymbol{r})}} \tag{7.1}$$

where $\boldsymbol{r}$ is the set of Cartesian coordinates describing the position of each atom in the system, $U(\boldsymbol{r})$ is the potential energy, and $\beta\text{–}1 = k_B T$, with $T$ the temperature and

$k_B$ Boltzmann's constant. Furthermore, it is rare that enough thermal energy accumulates in the appropriate direction to push the system through the transition. Because of this, MD simulations rarely visit high-energy configurations, and direct simulation of transitions starting from one stable state is thus not an efficient approach, if not impossible for large systems within a reasonable computer time. We also note that a proper description of the dynamics of the transition would require multiple observations to accumulate sufficient statistics to identify the most likely transition pathway and to provide a meaningful representation of the events taking place during the transition. Different strategies must be devised. The statistical nature of molecular behavior suggests a way to proceed. To move from one state to another, the most likely pathway is the one characterized by the lowest free energy increase [11]. In order to search for this pathway, it is convenient to reduce the dimensionality of the problem, focusing on a subset of variables that are expected to be of relevance during the transition. In the following, we discuss how to choose these variables and reconstruct the associated free energy change using MD simulations.

### 7.2.1 Theoretical Framework

Assume that the transition proceeds along a pathway described by some variable $s = s(\mathbf{r})$, the reaction coordinate. The probability that the system reaches a configuration $\mathbf{r}$ characterized by the value $s$ of $s(\mathbf{r})$ is

$$P(s) = \int d\mathbf{r}\delta\left(s - s(\mathbf{r})\right)P(\mathbf{r}) \tag{7.2}$$

where $\delta$ is Dirac's delta function. $P(s)$ is related to the free-energy $F(s)$ by

$$F(s) = -\frac{1}{\beta}\ln P(s) \tag{7.3}$$

$F(s)$ is the potential of mean force acting on the system at $s(\mathbf{r}) = s$. For an ergodic system, $P(s)$ may be estimated from simulations as the frequency of observing the system in a configuration corresponding to the value $s$ of $s(\mathbf{r})$

$$P(s) = \lim_{t\to\infty}\frac{1}{t}\int_0^t dt'\delta(s - s(\mathbf{r}(t'))) \tag{7.4}$$

Given the transition pathway and the associated free energy, transition state theory (TST) allows to connect the microscopic behavior revealed by simulations with the reaction rate constant, a macroscopic observable [12]. Within the limits of applicability of TST, the rate constant $k^{\text{TST}}$ of the process is related to the free energy change $F(s)$ along the pathway by the relation

$$k^{\text{TST}} = \frac{k_B T}{h} e^{-\frac{\Delta F^{\#}}{k_B T}} \tag{7.5}$$

where $h$ is Plank's constant and $\Delta F^{\#}$ is the difference between the free energies of the transition state, i.e. the state characterized by the highest free energy along the pathway, and that of the reactant state. The transition state plays a key role in theories of transitions. It comprises the set of configurations that separate, in both dynamical and statistical senses, the end points of the transition. An MD simulation started at the transition state should have an equal probability of falling in either the reactants or products basins. Furthermore, it is the least probable state along any transition pathway connecting reactants to products. Of all the possible pathways connecting reactants and products, the most likely is the minimum free energy pathway [11]. Transition state theory has proved highly successful, and developments to cope with its limitations have been proposed [13]. What is lost in this description is the dynamical information of the transition, i.e. the sequence of events taking place during the transition, which are only described in a statistical sense. Special techniques, such as transition path sampling [14], have been devised to recover this information.

### 7.2.2 Collective Variables

As the number of atoms in the model system increases, exploring the whole set of coordinates in search of reactive pathways becomes impractical. To reduce the dimensionality of the search space, a few variables suitable for describing the transition must be identified (called collective variables, CVs). These variables are continuous functions of the atomic coordinates $\boldsymbol{r}$ or a subset of them

$$s = s(\boldsymbol{r}) \tag{7.6}$$

The $s$ should allow to distinguish the end states of the transition and to describe the progress of the transition. Furthermore, motion along these variables is slower compared to other variables. Identifying these variables is a major challenge, especially when the transition involves the collective motion of a large number of atoms. For the kind of transitions that are characteristic of bioinorganic systems, namely, ligand exchange and catalysis, in which a reduced number of atoms reorganize during the process, chemical intuition may suggest the use of few geometrical parameters. Types of variables that have been used to model ligand binding include the distance between the interacting atoms. For a ligand exchange process, two distances may be used: those between the metal and the leaving ligand and that between the metal and the incoming ligand. Since the computational cost of the simulation increases with the number of variables, the trick is to use as few variables as needed. In a chemical reaction such as an $SN_2$ displacement, we may combine the distances between the site of substitution and the attacking and leaving atoms in one variable: the difference between them. More complex variables may be required. Elegant examples include those used to describe hydrogen dissociation from acids [15], ring conformations [16], and variables that describe position and distance along a predefined path [17]. The coordination number ($CN$) is a useful CV for describing interactions and bonds between two sets of atoms, particularly useful in the context of bioinorganic and organometallic applications

$$\mathrm{CN}_{AB} = \sum_{i}^{N_A}\sum_{j}^{N_B} f_{ij} = \sum_{i}^{N_A}\sum_{j}^{N_B} \frac{1 - \left(\frac{r_{ij}}{d_0}\right)^p}{1 - \left(\frac{r_{ij}}{d_0}\right)^q} \tag{7.7}$$

where $r_{ij}$ is the distance between atoms $i$ and $j$, $d_0$ is a reference distance, $p$ and $q$ are parameters [18]. This variable counts the number of atoms in set B interacting with atoms in set A. The function $f$ is a step function, which counts ~1 when $j$ is close to $i$ $(r_{ij} < d_0)$ and ~0 otherwise. Running the sum over all atoms in the two sets tells how many atoms $j$ are bound to atoms $i$.

When it is not possible to identify the relevant $s$ based on intuition and chemical knowledge, statistical and deep learning techniques may conveniently be used to compress a large number of descriptors into a single (or few) CVs [19, 20]. Alternatively, a trial-and-error strategy may be used, in which the CV is improved based on the outcome of auxiliary simulations and the aid of machine learning approaches [21–23]. Libraries of collective variables, which can be interfaced with popular MD codes, have been developed, facilitating their dissemination [24–26].

**Example**

Consider the aqua-ion of Zn(II). Experimental evidence supports an octahedral complex, $Zn(H_2O)_6^{2+}$ (Figure 7.1a) [27]. Zinc, however, is known to present variable coordination patterns in proteins, with tetra-coordination being the most represented in a survey of crystal structures [28]. We may thus be interested in determining the free energy cost of varying the coordination of Zn in water solution. To describe this process, a model comprising one Zn(II) ions and 200 water molecules, treated under periodic boundary conditions and described using the BLYP functional supplemented with a dispersion correction, is considered [29–31]. A CV capable of distinguishing, for instance, penta from hexa coordinated Zn is necessary. It is not possible to simply pull one water molecule from the Zn coordination sphere, by acting on the distance between a selected water molecule and Zn. This is because any other water molecule would be free to diffuse and replace the pulled one. The coordination number (CN) was designed to address this issue. In this case, set A in Eq. (7.7) comprises only the Zn atom, while set B includes all oxygen atoms of water molecules. To define CN, three parameters have to be given. In order to decide the value of $d_0$, we inspect the radial distribution function (rdf) of oxygen atoms around Zn from an equilibrium *ab initio* MD simulation [32]. The rdf in Figure 7.1b shows a first peak centered at 2.18 Å, which accounts for six water molecules, and a second peak at 4.08 Å. Parameter $d_0$ should thus allow to distinguish bound water molecules (those accounting for the first peak of the rdf) from water molecules in the second shell (the second peak). A value of $d_0 = 2.8$ Å appears appropriate. The parameters $p$ and $q$ determine the steepness of the step function $f$. A steep function would allow for a clear-cut distinction between bound and unbound molecules, and provide CN values close to whole numbers. However, if the CN is to be used in combination with an enhanced sampling method, it would allow to act on the position of water molecules only in the varying region of the step function. With values $p = 8$ and $q = 16$, the step function takes the form in Figure 7.1b. The so-defined CN allows manipulating the coordination at Zn. For instance, using steered MD simulation, the penta-coordinated species may be generated (Figure 7.1d). The time series of CN during some exploratory simulations is shown in Figure 7.1c.

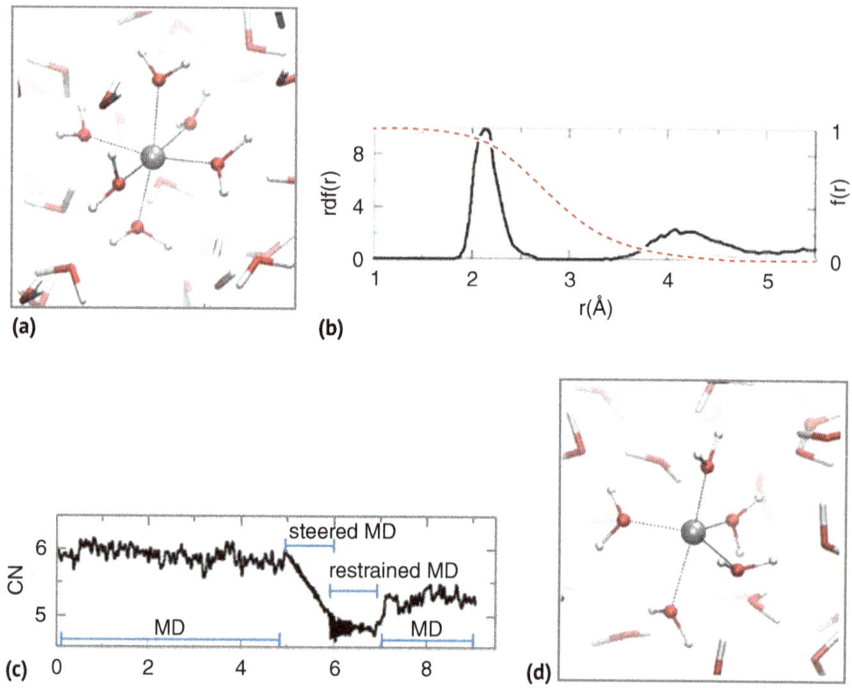


**Figure 7.1** (a, b) Setting the parameters defining a CN between a metal ion (Zn(II) – gray sphere) and ligands (water oxygen atoms – red spheres) requires the analysis of equilibrium simulations, particularly the radial distribution function (rdf) characterizing the hydration of Zn(II). (c, d) The capability of CN to control the configuration of the system may be tested via steered or restrained MD simulations.

### 7.2.3 Biased Methods

Estimating $P(s)$, and thus $F(s)$, from an MD simulation using Eq. (7.4) for the whole range of $s$ characterizing the transition is not possible if the system is nonergodic on the timescale of the simulation. Biased methods improve the sampling of $s$ values corresponding to high-energy configurations modifying the potential energy of the system

$$U(\mathbf{r}) \rightarrow U(\mathbf{r}) + V(s(\mathbf{r})) \tag{7.8}$$

where $V(s(\mathbf{r}))$ is chosen to reduce and possibly remove the barriers in $F(s)$ [33]. Performing the simulation using the modified potential improves the sampling of regions of $s$ which would otherwise result inaccessible using $U(\mathbf{r})$. The simulation returns $P_V(s)$, the probability distribution of $s$ in the biased ensemble, which is related to $P(s)$, the probability distribution in the unbiased ensemble, by

$$P(s) \propto P_V(s) e^{\beta V(s)} \tag{7.9}$$

This reweighting scheme allows the estimation of thermal averages of any configurational property from biased simulations [33].

#### 7.2.3.1 Umbrella Sampling

If the bias potential $V(s)$ compensates $F(s)$, a uniform distribution will result [34]. However, bias potentials are often used to restrict sampling to a region of interest [35]. To this scope, $V(s)$ may take the form of a harmonic potential, which restraints the value of $s(\mathbf{r})$ around a reference value $s_0$

$$V(s(\mathbf{r})) = k\left(s(\mathbf{r}) - s_0\right)^2 \tag{7.10}$$

However, this might not be sufficient to explore the whole range of interest in $s$. Multiple simulations are thus needed, running with $V(s)$ centered at different $s_0$ values, each covering a different interval (window) of $s$. Simulations may be performed in parallel, since each simulation is independent of the others. The data will then be combined to provide an estimate $F(s)$ using a weighted average of the distributions of the individual windows $P_i(s)$

$$P(s) = \sum_i w_i(s) P_i(s) \tag{7.11}$$

The weights $w_i$ may be determined to minimize the statistical error of $P(s)$, leading to the popular weighted histogram analysis method (WHAM) [36]. An alternative approach is umbrella integration, which averages the mean force and integrates it [37]. To set up an umbrella sampling simulation, decisions have to be taken on the number of windows, their reference values and the force constants, which may vary from window to window. The value of the force constant should be strong enough to maintain the system around $s_0$ in each window. However, large $k$ narrow the distribution $P_i(s)$, and thus more windows are required to cover the range of interest, since the analysis methods, such as WHAM, used to combine the data from the individual simulations, require the distributions $P_i(s)$ to overlap. If an estimate of the barrier is available, suggestions on how to choose k and the spacing between windows have been proposed [38]. Umbrella sampling is particularly useful when a single CV is to be sampled. Other methods, such as metadynamics, are computationally more efficient when the problem requires the use of two or more CVs.

#### 7.2.3.2 Metadynamics

Metadynamics accelerates transitions between states by pushing the system away from already visited configurations [39]. The potential energy surface is progressively modified during the course of the simulation, by adding at time intervals $t_V$ small repulsive potentials centered on the current values of the CVs. At simulation time $t$, the history-dependent biasing potential is

$$V(s,t) = \sum_i^{t/t_V} H e^{-\sum_j^{N} CV \frac{(s_j - s_j(it_V))^2}{2W_j^2}} \tag{7.12}$$

where the sum runs over the deposited repulsive potentials, which take the form of Gaussian functions of height $H$ and width $W$. In Eq. (7.12), $V(s,t)$ depends on $N_{CV}$ collective variables. At the beginning of the simulations, the system, prepared in state A, will sample the local free energy minimum corresponding to state A. As the simulation proceeds and the biasing potential builds up, the system will commence to explore regions higher in free energy and eventually fall into a neighboring minimum, say B, as enough energy is provided to cross the barrier separating states A and B. Now the bias builds for state B, until further transitions may take place. Thus, contrary to umbrella sampling, in which the whole range of interest of the CVs has to be covered, metadynamics will first sample transition corresponding to the lowest free energy pathway, and only later, as the simulation continues, explore higher energy pathways, if they exist. These features allow metadynamics to explore CVs space of higher dimensionality at a lower computational cost compared to umbrella sampling, and, doing so, reveals the most favorable reaction pathway [40, 41].

The bias potential depends on three parameters, namely the height $H$ and width $W$ of the Gaussian functions, and the deposition interval $t_V$. To choose them, a compromise has to be adopted between the rate of acceleration (favored by large values of $H$ and $W$, and small $t_V$) and the desire to least perturb the system (favored by small $H$ and $W$, and long $t_V$) [42]. The height $H$ should be set smaller than the thermal energy $k_BT$, to avoid excessive sudden changes in the potential energy. The choice of $W$ and $t_V$ may be guided by analyzing the fluctuations of the CVs in a preliminary equilibrium simulation. The width $W$ should be set at a fraction of the standard deviation observed in the time series of the CVs. The deposition interval $t_V$ should be large enough to allow the system to relax in between successive additions. Given the higher cost of *ab initio* (and hybrid QM/MM) simulations compared to force field-based simulations, the former often adopts a faster deposition rate.

Besides accelerating the exploration of configurational space, metadynamics also provides quantitative information on the energetics of the transition between states. Indeed, the long-time limit of the bias potential provides an estimate of the free energy profile along the CVs

$$\lim_{t\to\infty} V(s,t) = -F(s) + \text{constant} \tag{7.13}$$

However, since repulsive Gaussians continue to be added to the bias potential, $F(s)$ continues fluctuating. Well-tempered metadynamics addresses this issue progressively reducing the height of the repulsive Gaussian at already visited $s$ values, according to a parameter $\Delta T$ [43]:

$$H = He^{-\frac{V(s,t)}{k_B\Delta T}} \tag{7.14}$$

With this formulation, the bias potential converges to a limiting value, which depends on $\Delta T$

$$\lim_{t\to\infty} V(s,t) = -\frac{\Delta T}{T + \Delta T} F(s) + \text{constant} \tag{7.15}$$

By this means, fluctuations of the CVs are enhanced, which in turn facilitates the observation of transitions [44]. A proper choice of $\Delta T$, in units of temperature, allows limiting the free energy range explored during the simulation, such that only the region of interest is sampled. The value of $\Delta T$ is dictated by the size of the free energy barrier to overcome, which may be estimated from preliminary simulations or available from literature data [42]. Being a nonequilibrium technique, the probability distributions of variables other than the CVs are distorted from the canonical distributions. Reweighting schemes have been proposed to address this issue in the context of well-tempered metadynamics [45, 46].

## 7.3 Illustrative Studies

Many successful examples exist of the use of enhanced sampling simulations applied to the study of biological systems [47–49]. Here, we report a few selected cases, in which the techniques described in this chapter were adopted. In describing these studies, emphasis will be placed on the choice of the CVs used to investigate the transition mechanisms and reconstruct the associated free energy profiles, referring interested reader to the original publications for the details of the simulations.

### 7.3.1 Metal Coordination

The coordination properties of metals are of wide interest in bioinorganics. Here, we consider ligand exchange at the metal center, a process involved in a broad range of applications, from drug design to biocatalysis [50]. As a case study, we first describe how to investigate the hydration of a Zn(II) complex [51]. Clioquinol (5-chloro-7-iodo-hydroxyquinoline, CQ, Figure 7.2) is a drug-like metal-coordinating molecule, which has been evaluated as an agent for metal ion chelating therapy in Alzheimer's disease [52]. Upon deprotonation, CQ binds Zn(II) in a 2:1 stoichiometry, leading to the neutral $Zn(CQ)_2$ complex. However, given the flexible Zn coordination behavior, the actual coordination sphere of Zn in solution is not known and should be addressed, as it is of relevance for modeling studies of clioquinol's mechanism of action. To do this computationally, extended model systems should be considered to avoid artifacts arising from competing interactions in reduced models. When a minimal model was chosen to investigate the hydration of the $Zn(CQ)_2$ complex, a tendency to favor low coordination numbers was observed because water molecules prefer to form H-bonds with CQ's oxygen atoms rather than binding to Zn. When the model is extended adding about 250 explicit water molecules, the hydrogen (H)-bonding properties of ligand atoms are saturated, and metal coordination can be properly investigated. Such large systems require proper sampling of configuration space, which may be achieved via *ab initio* MD simulations based on

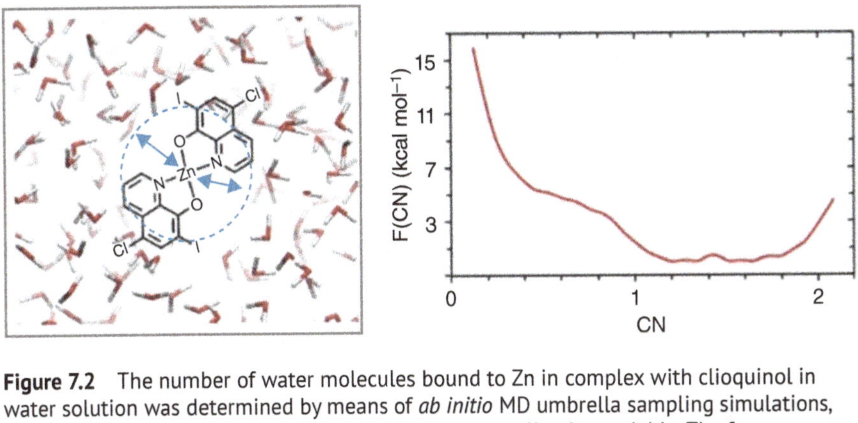


**Figure 7.2** The number of water molecules bound to Zn in complex with clioquinol in water solution was determined by means of *ab initio* MD umbrella sampling simulations, using the *CN* between Zn and water oxygen atoms as collective variable. The free energy profile (right panel) indicates that one or two water molecules have similar probabilities of binding to Zn.

density functional theory (DFT). Furthermore, in order to probe different coordination environments, a suitable order parameter to describe the number of water molecules coordinated to Zn has to be identified. The issue here is that water molecules are all equivalent and we thus cannot focus on only one of them, otherwise, while the selected water would, for instance, unbind from Zn, any other could replace it and the value of the order parameter would thus not correspond to the configuration of the system. The CN (Eq. (7.7)) between Zn and any water molecule has been designed to take into account the equivalence between atoms. This variable can conveniently be used to control the coordination at Zn, allowing the exploration configurations away from stable structures. A set of umbrella sampling simulations were used to enhance the sampling of such configurations, each applying a harmonic potential centered at a distinct CN value, covering the range between 0 (corresponding to the $Zn(CQ)_2$ complex) to 2 (corresponding to the $Zn(CQ)_2(H_2O)_2$ complex). The CN distributions from these independent simulations were combined using the umbrella integration method, allowing to reconstruct the free energy profile, which features a broad minimum in the range $1.2 < CN < 1.8$, indicating that one or two water molecules are bound to Zn in water solution. Species $Zn(CQ)_2(H_2O)$ and $Zn(CQ)_2(H_2O)_2$ are equally likely, while $Zn(CQ)_2$ does not correspond to a minimum of the free energy profile.

Similar approaches may be used to model ligand exchange processes, as, for instance, those involved in metal transport mediated by proteins. The copper transporter 1 (CTR1) is the only known mammalian Cu(I) importer, which regulates the concentration of Cu(I) ions in cells [53]. CTR1 is an integral membrane protein, composed of three identical monomers, each organized in a three-helix bundle fold. After recruiting Cu(II) ions from the extracellular matrix and reducing it to Cu(I) via its unstructured N-terminal domain [54], CTR1 delivers a Cu(I) ion to the selectivity filter placed at the central interface between the monomers. Each monomer contributes a MX3M motif (X stands for any amino acid) to the selectivity filter, which thus features two stacked triads of methionine (Met) residues, one facing the extracellular side (Site 1) and the other the inner cytosol-exposed cavity of the transporter (Site 2) (Figure 7.3a). Hybrid QM/MM simulations, based on DFT, were undertaken to investigate the Cu(I) transport mechanism, focusing on how the metal ion crosses the selectivity filter [55]. Based on the experimental structure of the transmembrane domain, three models of the protein embedded in a lipid bilayer were built, differing in the Cu(I) loading at the selectivity filter: only Site 1 occupied, only Site 2, and both sites occupied by Cu(I) ions.

Conformational analysis of the methionine residues in Sites 1 and 2 during classical and QM/MM MD simulations of the three models revealed that the side chain orientations with respect to the axis of the selectivity filter are interlinked and, furthermore, dependent on ion loading. Metadynamics was next employed to model the flow of a single Cu(I) ion. Translocation from Site 1 to Site 2 was achieved using two CVs, namely the distance of Cu(I) from Site 1 and the *CN* between Cu(I) and the sulfur atoms of Site 2. The reconstructed free energy surface shows two intermediate states, corresponding to the progressive exchange of the Met ligands from those of Site 1 to those of Site 2 (Figure 7.3b). The free energy barriers associated with these ligand-exchange events do not exceed ~6 kcal $mol^{-1}$.

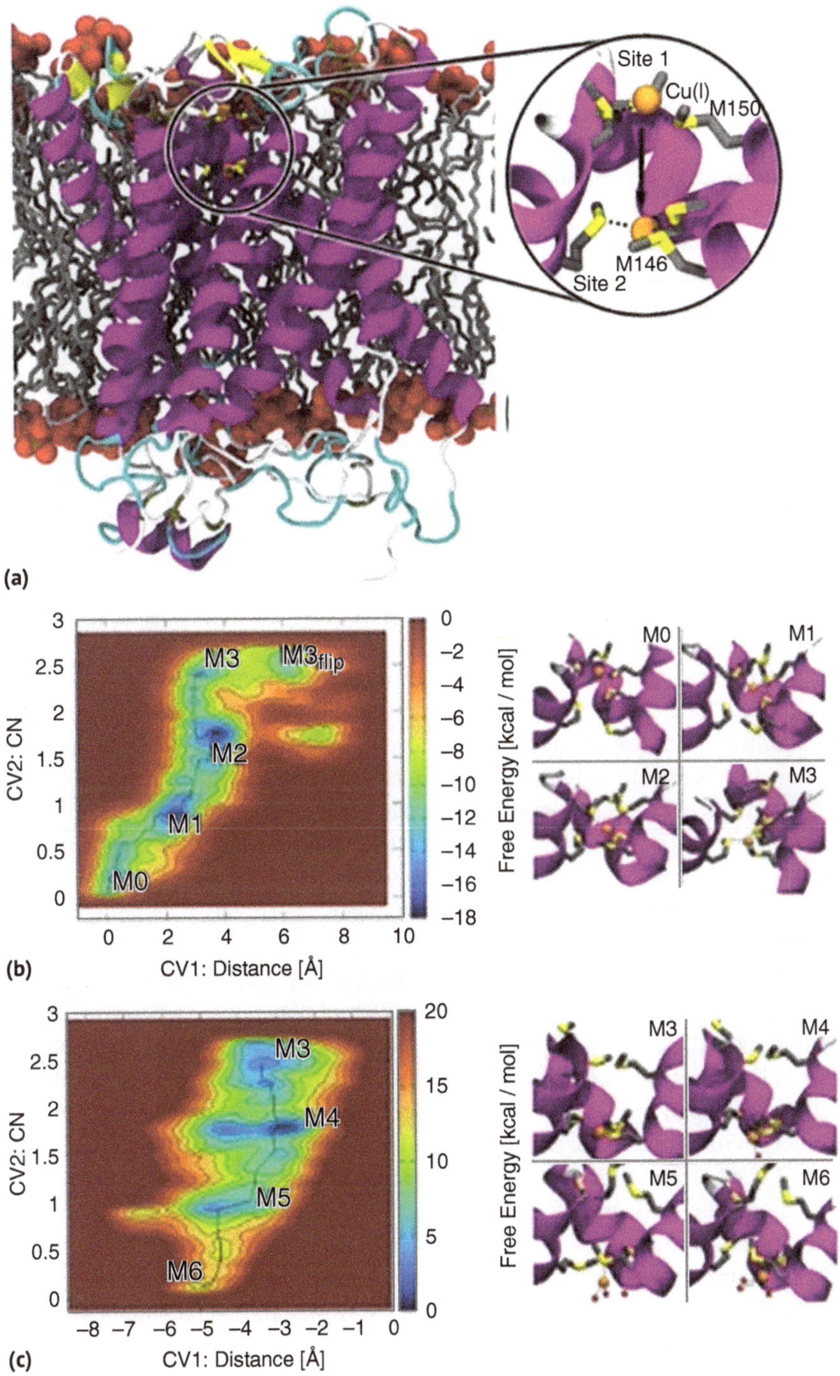


**Figure 7.3** (a) Model system of the CTR1 transporter embedded in a lipid bilayer, with inset showing the selectivity filter; (b) free energy surface of Cu(I) translocation from Site 1 to Site 2 and representative structures along the translocation path; (c) free energy surface of Cu(I) dissociation from Site 2 to the inner solvent-exposed cavity and representative structures along the dissociation path. *Source*: Janoš et al. 2022/Cambridge University Press/ CC BY 4.0.

Importantly, the most stable configuration of the Met side chains of Site 2 with bound Cu(I) allows for nearby water molecules to approach the ion. This occurrence is a requirement for the subsequent dissociation of Cu(I) from Site 2, which involves the exchange of the Met ligands for water molecules. This step was modeled using similar CVs as for the translocation step. The path followed is again stepwise, with the progressive replacement of Met residues by water molecules at the metal center (Figure 7.3c). The last of these exchanges was found to be rate limiting, with a barrier of 12.5 kcal $mol^{-1}$. Conversely, modeling dissociation from Site 2, when also Site 1 is loaded, revealed a considerably higher activation energy (>20 kcal $mol^{-1}$). This occurrence was rationalized in terms of the reduced plasticity of the Met side chains imposed by the presence of a second Cu(I) at Site 1, and the adoption by the Site 2 Mets of orientations, which hamper water access to the metal ion. Based on these findings, the authors put forward that the coupling of ion occupancy, ligand plasticity and hydration at the selectivity filter enable a gated mechanism, according to which Cu(I) uptake is hindered at high extracellular copper concentrations, which would result in double occupancy of the selectivity filter.

### 7.3.2 Enzymatic Reactions

Metalloenzymes use metal ions, directly bound to the protein or part of a prosthetic group, to facilitate chemical reactions. Here, we describe two modeling studies: the first addressing a redox reaction, and the second focuses on hydrolysis. In the redox reaction, iron is used to store oxidation equivalents, used to reduce a reactive oxygen species. In the hydrolysis reaction, two magnesium ions are used to assemble the reactants in a proper geometry and increase the acidity of the water nucleophile. In one case, one set of CVs was able to reveal two competitive pathways. In the other, different mechanistic possibilities were investigated separately using different collective variables.

The heme prosthetic group is highly versatile [56]. Depending on the protein matrix in which it is buried, it can display different reactivities to perform a broad range of activities, including transport, sensing and oxidation. Heme catalases degrade hydrogen peroxide ($H_2O_2$) into water and oxygen, protecting the cell from oxidative damage [57]. The reaction proceeds in two steps, in which the iron-porphyrin is initially oxidized by $H_2O_2$ to a high valent iron intermediate (compound I, Cpd I), and then reduced by a second molecule of $H_2O_2$. The first step is common to all peroxidases (for a computational study of compound I formation in horseradish peroxidase, see e.g. Refs. [58, 59]), while the second is highly efficient only in catalases. The mechanism of Cpd I reduction by $H_2O_2$ in *Helicobacter pylori* catalase (HPC) was studied by *ab initio* QM/MM MD and metadynamics simulations [60]. Initially, a model of the Michaelis complex was built based on an early proposal (Figure 7.4) [61]. It was observed that a hydrogen atom was spontaneously transferred to the oxoferryl group, quenching the porphyrin radical and leading to the species $\text{Por–Fe}^{IV} = \text{OH} \cdots {}^{\bullet}\text{O}_2\text{H}$ (Cpd II-like $\cdots {}^{\bullet}\text{O}_2\text{H}$). This intermediate resulted stable on the time scale of the simulation. To complete the reaction, a further H has to reach the hydroxoferryl, and this step required activation. This was achieved via

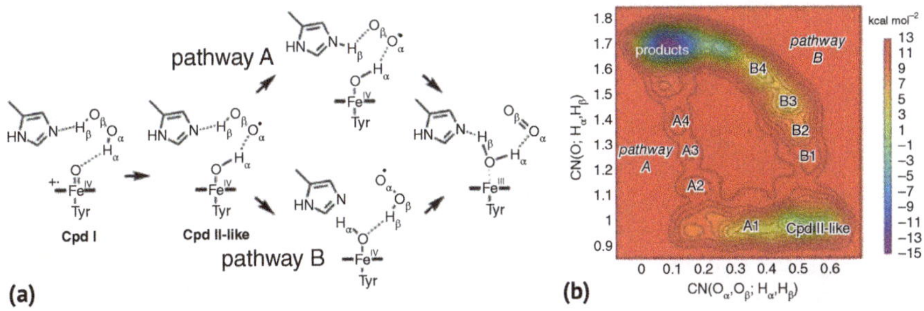


**Figure 7.4** (a) The mechanism of Cpd I reduction by $H_2O_2$—the second step of the catalase reaction—in HPC catalase was modeled by means of *ab initio* QM/MM metadynamics simulations, using two *CN* as collective variables; (b) the free energy surface shows that the reaction may proceed via two competing pathways. *Source*: Adapted from Alfonso-Prieto et al. (2009)/American Chemical Society.

metadynamics, using two collective variables to bias the breaking and forming of bonds. The first CV was the CN (Eq. (7.7)) between the oxygens and the hydrogens of $H_2O_2$, CN $(O_\alpha, O_\beta; H_\alpha, H_\beta)$. The second CV was the CN between the hydrogens of $H_2O_2$, and the oxyferryl oxygen, CN $(O; H_\alpha, H_\beta)$. These variables distinguish reactant from the product states, and do not assume a priori any pathway. Metadynamics, acting on these two variables, promoted the reaction allowing the system to follow the most convenient route. In HPC catalase, the reduction of Cpd I may take place along two competing pathways (Figure 7.4a). Along pathway A, a proton is transferred from $^\bullet O_2H$ to the distal His, which then changes H-bonding partner, switching to the oxoferryl oxygen and delivering the proton. The proton transfer is accompanied by the transfer of an electron from $O_2^{\bullet-}$ to iron, generating a water molecule and restoring the Fe(III) resting state. Along pathway B, $^\bullet O_2H$ flips orientation, breaking the H-bond with the distal His and forming a new one with the hydroxoferryl, to which it delivers a hydrogen atom, leading to the product state. The study was extended, using the same methodology, to compare catalases from different clades, catalase mutants and to unravel the origin of the low catalytic activity of peroxidases [60, 62].

As a further example of the use of biased simulation methods to study enzymatic reactions mechanisms, we describe the cleavage of phosphodiester bonds by flap endonuclease 1 (FEN1), a structure-specific nucleic acid hydrolyzing enzyme, which removes 5′-single-stranded protrusions resulting during DNA replication and repair [63]. Structural determination of the enzyme complexed with DNA in the reactant and product states provided valuable insight into the mechanism used to recognize the structural feature—the 5′ single stranded flap—to be processed. The hydrolytic reaction is mediated by two divalent magnesium ions, which were, however, substituted by Sm(III) in the experimental structure. The first issue addressed in the computational study was thus to reconstruct the catalytically competent Mg(II)-bound state from the available structural information. Classical and hybrid QM/MM MD simulations were performed to equilibrate the reactant state. The simulation resulted in an active site configuration consistent with the expected two-metal-ions mechanism used by other nucleic acids processing enzymes [64]. In particular, a water molecule was observed to bind $Mg_B$ and properly orient for an in-line attack on the phosphorous atom, as expected for an $SN_2$-type reaction (Figure 7.5a), according to which the water molecule attacks the phosphate phosphorus (step I, Figure 7.5b), forming a phosphorene intermediate, which, after protonation of the leaving group, dissociates into products (step II, Figure 7.5b). Step I requires activation of the water molecule via a general acid-base mechanism, the details of which are often system-specific and debated in the literature [65, 66]. Inspection of the active site and consideration of previous mechanistic proposals for the similar reaction performed by ribonuclease H (RNase H), suggested that a phosphate oxygen from either nucleotides T1 or T2 could abstract a proton from the nucleophilic water (Figure 7.5a). Nucleophilic attack and activation could be concerted events. This possibility was explored for each of the two possible deprotonation pathways via

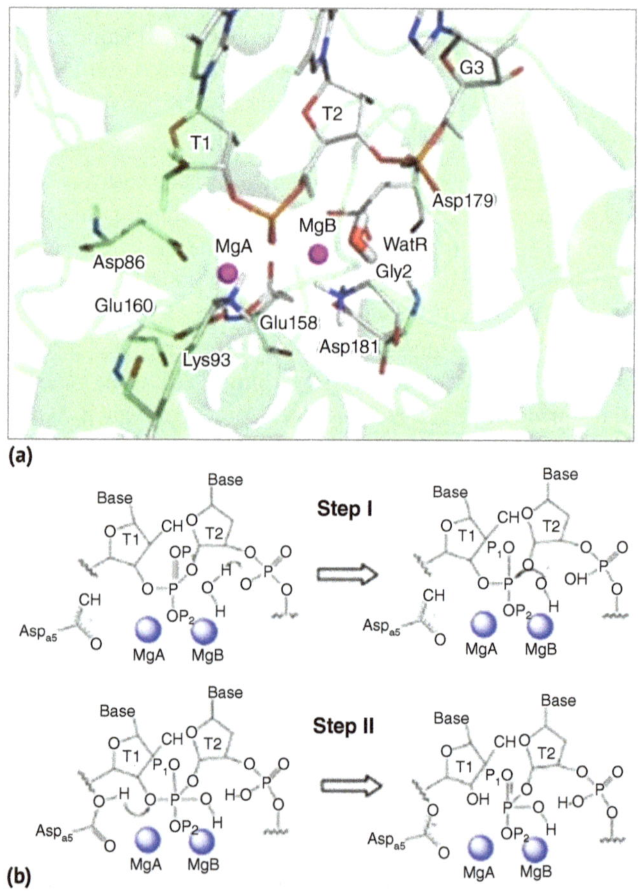


**Figure 7.5** (a) Michaelis complex of hFEN1 bound to DNA and catalytic Mg(II) ions. (b) Stepwise mechanism of water activation and nucleophilic attack leading to the phosphorene intermediate (Step I), followed by the protonation of the leaving group, leading to ester bond hydrolysis and dissociation into products (Step II). *Source*: Reproduced from Sgrignani and Magistrato (2015)/American Chemical Society.

metadynamics simulations using two CVs, one acting on the nucleophile's approach and the other on proton exchange to a given phosphate oxygen. Comparison of the free energy surfaces allowed ruling out the involvement of the phosphate oxygen from T1 as the general base. The authors also investigated the possibility that water activation and attack proceed stepwise. This route was explored using umbrella sampling simulations, each based on a single CV. First the energetics of proton transfer from the nucleophilic water to the phosphate oxygen of T2 were reconstructed, followed by the nucleophilic attack of the resulting hydroxide on the phosphorus atom. The stepwise mechanism turned to be energetically favored compared to the concerted one. In addition, the authors also considered the possibility of a hydroxide being present in the active site, either directly bound to $Mg_B$ or acting as a base for the activation of the nucleophilic water.

For the reaction to proceed further (step II), protonation of the leaving oxygen of the phosphorane is required. The authors considered different possible sources for the proton, including a protonated active-site aspartate and an acidic lysine. The latter required proton hopping to a phosphate oxygen of T1, followed by the transfer to O3′, and resulted more energetically demanding than the direct proton transfer from the aspartate to the leaving group. As may be appreciated, in this study, alternative mechanistic hypothesis were compared through simulations, allowing highlighting competitive reaction pathways and discarding energetically unfavorable ones. Substantial sampling was necessary, for which the authors relied on a tuned semiempirical method to lower the computational cost without sacrificing accuracy [67].

## 7.4 Conclusion

Atomistic simulations constitute a valuable tool to investigate biological systems at length and time scales difficult to access experimentally. The structural insight provided by the modeling of conformational and chemical transformations advances the understanding of these processes, unveiling the molecular determinants facilitating the transitions, which may eventually be converted into designing principles for biotechnological applications. The problem faced in the modeling of activated processes, besides the accuracy offered by the currently available potential energy functions, comes from the limited simulation time compared to the timescale of biological phenomena. In this context, advanced simulation techniques play a key role, allowing for bridging the gap with experiments. Among the many techniques that have been proposed, we may distinguish between those that bias the sampling of selected collective variables and those that enhance the sampling of all degrees of freedom. Biased methods (such as umbrella sampling, metadynamics, adaptive bias force, steered MD) are particularly useful when we can recognize the collective variables involved in the process, while the latter class (including accelerated MD, parallel tempering, replica exchange) does not require such knowledge in advance. However, to rationalize the outcome of these simulations, analysis of the trajectories to unravel the key features of the transitions is necessary. Identifying these few CVs in the high-dimensional space of atomic coordinates is certainly a challenge, but highly rewarding, as it allows to reduce the complex atomic motion to more familiar and intuitive concepts. It is thus not surprising that much research effort is currently placed in developing methods to identify, possibly automatically, the relevant collective motions underlying complex transformations. For processes unique to bioinorganics, namely, metal–ligand binding and metal-mediated reactions, knowledge of the end states, together with chemical intuition, may guide the initial choice of CVs, as we have shown in this chapter.

## References


1 Roux, B. *Molecular Machines*. World Scientific Publishing Co Pte Ltd; 2011.
2 Rosenbaum, D.M., Rasmussen, S.G.F., and Kobilka, B.K. The structure and function of G-protein-coupled receptors. *Nature* 2009;*459*(7245):356–363.
3 Saleh, T. and Kalodimos, C.G. Enzymes at work are enzymes in motion. *Science* 2017;*355*(6322):247–248.
4 Alberts, B. *Molecular Biology of the Cell/Bruce Alberts ... et al.* 4ed. New York: Garland; 2002.
5 Frauenfelder, H., Sligar, S.G., and Wolynes, P.G. The energy landscapes and motions of proteins. *Science* 1991;*254*(5038):1598–1603.
6 Henzler-Wildman, K. and Kern, D. Dynamic personalities of proteins. *Nature* 2007;*450*(7172):964–972.
7 Jumper, J., Evans, R., Pritzel, A. et al. Highly accurate protein structure prediction with AlphaFold. *Nature* 2021;*596*(7873):583–589.
8 Frenkel, D. and Smit, B. *Understanding Molecular Simulation*. London: Academic Press, Inc.; 2001.
9 Cramer, C.J. and Truhlar, D.G. Density functional theory for transition metals and transition metal chemistry. *Phys. Chem. Chem. Phys.* 2009;*11*(46):10757–10816.
10 Allen, M.P. and Tildesley, D.J. *Computer Simulation of Liquids*. Oxford: Oxford University Press; 2017.
11 Vanden-Eijnden, E. Transition-path theory and path-finding algorithms for the study of rare events. *Annu. Rev. Phys. Chem.* 2010;*61*:391–420.
12 Laidler, K.J. and King, M.C. Development of transition-state theory. *J. Phys. Chem.* 1983;*87*(15):2657–2664.
13 Truhlar, D.G. Transition state theory for enzyme kinetics. *Arch. Biochem. Biophys.* 2015;*582*:10–17.
14 Bolhuis, P.G., Chandler, D., Dellago, C. et al. Transition path sampling: throwing ropes over rough mountain passes, in the dark. *Annu. Rev. Phys. Chem.* 2002; *53*:291–318.
15 Park, J.M., Laio, A., Iannuzzi, M. et al. Dissociation mechanism of acetic acid in water. *J. Am. Chem. Soc.* 2006;*128*(35):11318–11319.
16 Biarnés, X., Ardèvol, A., Planas, A. et al. The conformational free energy landscape of $\beta$-d-glucopyranose. Implications for substrate preactivation in $\beta$-glucoside hydrolases. *J. Am. Chem. Soc.* 2007;*129*(35):10686–10693.
17 Branduardi, D., Gervasio, F.L., and Parrinello, M. From A to B in free energy space. *J. Chem. Phys.* 2007;*126*(5):054103.

**18** Iannuzzi, M., Laio, A., and Parrinello, M. Efficient exploration of reactive potential energy surfaces using Car-Parrinello molecular dynamics. *Phys. Rev. Lett.* 2003;*90*(23):238302.

**19** Bonati, L., Rizzi, V., and Parrinello, M. Data-driven collective variables for enhanced sampling. *J. Phys. Chem. Lett.* 2020;*11*(8):2998–3004.

**20** Piccini, G., Mendels, D., and Parrinello, M. Metadynamics with discriminants: a tool for understanding chemistry. *J. Chem. Theory Comput.* 2018;*14*(10): 5040–5044.

**21** Ray, D., Trizio, E., and Parrinello, M. Deep learning collective variables from transition path ensemble. *J. Chem. Phys.* 2023;*158*(20):204102.

**22** Peters, B. and Trout, B.L. Obtaining reaction coordinates by likelihood maximization. *J. Chem. Phys.* 2006;*125*(5):054108.

**23** Bonati, L., Piccini, G., and Parrinello, M. Deep learning the slow modes for rare events sampling. *Proc. Natl. Acad. Sci.* 2021;*118*(44):e2113533118.

**24** Fiorin, G., Klein, M.L., and Hénin, J. Using collective variables to drive molecular dynamics simulations. *Mol. Phys.* 2013;*111*(22–23):3345–3362.

**25** Tribello, G.A., Bonomi, M., Branduardi, D. et al. PLUMED 2: new feathers for an old bird. *Comput. Phys. Commun.* 2014;*185*(2):604–613.

**26** Bonati, L., Trizio, E., Rizzi, A. et al. A unified framework for machine learning collective variables for enhanced sampling simulations: mlcolvar. *J. Chem. Phys.* 2023;*159*(1):014801.

**27** Marcus, Y. Ionic radii in aqueous solutions. *Chem. Rev.* 1988;*88*(8):1475–1498.

**28** Laitaoja, M., Valjakka, J., and Jänis, J. Zinc coordination spheres in protein structures. *Inorg. Chem.* 2013;*52*(19):10983–10991.

**29** Lee, C., Yang, W., and Parr, R.G. Development of the Colle-Salvetti correlation-energy formula into a functional of the electron density. *Phys. Rev. B* 1988;*37*(2):785–789.

**30** Becke, A.D. Density-functional exchange-energy approximation with correct asymptotic behavior. *Phys. Rev. A* 1988;*38*(6):3098–3100.

**31** Grimme, S., Antony, J., Ehrlich, S. et al. A consistent and accurate ab initio parametrization of density functional dispersion correction (DFT-D) for the 94 elements H-Pu. *J. Chem. Phys.* 2010;*132*(15):154104.

**32** Marx, D. and Hutter, J. *Ab Initio Molecular Dynamics: Basic Theory and Advanced Methods.* Cambridge: Cambridge University Press; 2009.

**33** Chipot, C. and Pohorille, A. *Free Energy Calculations: Theory and Applications in Chemistry and Biology.* 1st ed. Berlin: Springer, 2007.

**34** Torrie, G.M. and Valleau, J.P. Nonphysical sampling distributions in Monte Carlo free-energy estimation: umbrella sampling. *J. Comput. Phys.* 1977;*23*(2):187–199.

**35** Kästner, J. Umbrella sampling. *WIREs Comput. Mol. Sci.* 2011;*1*(6):932–942.

**36** Kumar, S., Rosenberg, J.M., Bouzida, D. et al. The weighted histogram analysis method for free-energy calculations on biomolecules. I. The method. *J. Comput. Chem.* 1992;*13*:1011–102.

**37** Kästner, J. and Thiel, W. Bridging the gap between thermodynamic integration and umbrella sampling provides a novel analysis method: "Umbrella integration." *J. Chem. Phys.* 2005;*123*(14):144104.

38 Kästner, J. and Thiel, W. Analysis of the statistical error in umbrella sampling simulations by umbrella integration. *J. Chem. Phys.* 2006;*124*(23):144104.
39 Laio, A. and Parrinello, M. Escaping free-energy minima. *Proc. Natl. Acad. Sci.* 2002;*99*(20):12562–12566.
40 Barducci, A., Bonomi, M., and Parrinello, M. Metadynamics. *WIREs Comput. Mol. Sci.* 2011;*1*(5):826–843.
41 Bussi, G. and Laio, A. Using metadynamics to explore complex free-energy landscapes. *Nat. Rev. Phys.* 2020;*2*(4):200–212.
42 Bussi, G. and Branduardi, D. Free-energy calculations with metadynamics: theory and practice. In *Reviews in Computational Chemistry Volume 28*. Wiley Blackwell; 2015, pp. 1–49.
43 Barducci, A., Bussi, G., and Parrinello, M. Well-tempered metadynamics: a smoothly converging and tunable free-energy method. *Phys. Rev. Lett.* 2008;*100*(2): 020603.
44 Valsson, O., Tiwary, P., and Parrinello, M. Enhancing important fluctuations: rare events and metadynamics from a conceptual viewpoint. *Annu. Rev. Phys. Chem.* 2016;*67*:159–184.
45 Bonomi, M., Barducci, A., and Parrinello, M. Reconstructing the equilibrium Boltzmann distribution from well-tempered metadynamics. *J. Comput. Chem.* 2009;*30*(11):1615–1621.
46 Tiwary, P. and Parrinello, M. A time-independent free energy estimator for metadynamics. *J. Phys. Chem. B* 2015;*119*(3):736–742.
47 Leone, V., Marinelli, F., Carloni, P. et al. Targeting biomolecular flexibility with metadynamics. *Curr. Opin. Struct. Biol.* 2010;*20*(2):148–154.
48 Sinko, W., Lindert, S., and McCammon, J.A. Accounting for receptor flexibility and enhanced sampling methods in computer-aided drug design. *Chem. Boil. Drug Des.* 2013;*81*(1):41–49.
49 Bernardi, R.C., Melo, M.C.R., and Schulten, K. Enhanced sampling techniques in molecular dynamics simulations of biological systems. *Biochim. et Biophys Acta* 2015;*1850*(5):872–877.
50 Riccardi, L., Genna, V., and De Vivo, M. Metal–ligand interactions in drug design. *Nat. Rev. Chem.* 2018;*2*(7):100–112.
51 Rodríguez-Santiago, L., Alí-Torres, J., Vidossich, P. et al. Coordination properties of a metal chelator clioquinol to Zn2+ studied by static DFT and ab initio molecular dynamics. *Phys. Chem. Chem. Phys.* 2015;*17*(20):13582–13589.
52 Ritchie, C.W., Bush, A.I., Mackinnon, A. et al. Metal-protein attenuation with iodochlorhydroxyquin (clioquinol) targeting Abeta amyloid deposition and toxicity in Alzheimer disease: a pilot phase 2 clinical trial. *Arch. Neurol* 2003;*60*(12):1685–1691.
53 Magistrato, A., Pavlin, M., Qasem, Z. et al. Copper trafficking in eukaryotic systems: current knowledge from experimental and computational efforts. *Curr. Opin. Struct. Biol.* 2019;*58*:26–33.
54 Aupič, J., Lapenta, F., Janoš, P. et al. Intrinsically disordered ectodomain modulates ion permeation through a metal transporter. *Proc. Natl. Acad. Sci.* 2022;*119*(48):e2214602119.

**55** Janoš, P., Aupič, J., Ruthstein, S. et al. The conformational plasticity of the selectivity filter methionines controls the in-cell Cu(I) uptake through the CTR1 transporter. *QRB Discovery* 2022;*3*:e3.

**56** Poulos, T.L. Heme enzyme structure and function. *Chem. Rev.* 2014;*114*(7): 3919–3962.

**57** Dunford, H.B. *Peroxidases and Catalases: Biochemistry, Biophysics, Biotechnology, and Physiology*. Hoboken, NJ: Wiley;2010.

**58** Derat, E., Shaik, S., Rovira, C. et al. The effect of a water molecule on the mechanism of formation of compound 0 in horseradish peroxidase. *J. Am. Chem. Soc.* 2007;*129*(20):6346–6347.

**59** Vidossich, P., Fiorin, G., Alfonso-Prieto, M. et al. On the role of water in peroxidase catalysis: a theoretical investigation of HRP compound I formation. *J. Phys. Chem. B* 2010;*114*(15):5161–5169.

**60** Alfonso-Prieto, M., Biarnés, X., Vidossich, P. et al. The molecular mechanism of the catalase reaction. *J. Am. Chem. Soc.* 2009;*131*(33):11751–11761.

**61** Fita, I. and Rossmann, M.G. The active center of catalase. *J. Mol. Biol.* 1985;*185*(1):21–37.

**62** Campomanes, P., Rothlisberger, U., Alfonso-Prieto, M. et al. The molecular mechanism of the catalase-like activity in horseradish peroxidase. *J. Am. Chem. Soc.* 2015;*137*(34):11170–11178.

**63** Sgrignani, J. and Magistrato, A. QM/MM MD simulations on the enzymatic pathway of the human flap endonuclease (hFEN1) elucidating common cleavage pathways to RNase H enzymes. *ACS Catal.* 2015;*5*(6):3864–3875.

**64** Palermo, G., Cavalli, A., Klein, M.L. et al. Catalytic metal ions and enzymatic processing of DNA and RNA. *Acc. Chem. Res.* 2015;*48*(2):220–228.

**65** De Vivo, M., Dal Peraro, M., and Klein, M.L. Phosphodiester cleavage in ribonuclease H occurs via an associative two-metal-aided catalytic mechanism. *J. Am. Chem. Soc.* 2008;*130*(33):10955–10962.

**66** Rosta, E., Nowotny, M., Yang, W. et al. Catalytic mechanism of RNA backbone cleavage by ribonuclease H from quantum mechanics/molecular mechanics simulations. *J. Am. Chem. Soc.* 2011;*133*(23):8934–8941.

**67** Lopez, X. and York, D.M. Parameterization of semiempirical methods to treat nucleophilic attacks to biological phosphates: AM1/d parameters for phosphorus. *Theor. Chem. Acc.* 2003;*109*(3):149–159.

# 8

# Metal-induced Folding of Oligopeptides

*L. Roldán-Martín[1], L. Rodríguez-Santiago[1], J. Alí-Torres[2], J. D. Maréchal[1], and M. Sodupe[1]*

[1] *Departament de Química, Universitat Autònoma Barcelona, Bellaterra, Spain*
[2] *Departamento de Química, Universidad Nacional de Colombia – Sede Bogotá, Bogotá, Colombia*

## 8.1 Introduction

It has been estimated that about one-third of all proteins and enzymes enclose metal ions for their proper functioning in several biological processes, such as catalysis, signal transduction, or storage. Furthermore, dysfunctional processes, such as protein misfolding or aggregation in neurodegenerative diseases, have also been linked to the presence of metal ions, such as Fe(III), Cu(II), or Zn(II) [1]. In the last two decades, many efforts have been devoted to understanding the role of metal ions in the structure and folding of metallopeptides.

Structural biology, biomimetic chemistry, and molecular modeling constantly enrich our understanding of metal–protein synergies. Many physicochemical variables must be considered, including coordination rules, redox properties, spin states, preorganization of the biological receptor, and desolvation [2]. Despite intensive efforts, the complete picture of their relative contributions must be included. Like nonmetallic peptides, many variables could contribute to the metallopeptides' folding process, and which dominates in the metal–protein recognition depends on the system and the binding stage [3].

Current opinions agree that solvation and preorganization are fundamental at the initial stages of the interaction. Still, subtle electronic effects define the optimum geometry of the metal–protein complex. To date, though, there is no definitive computational framework, and computational bioinorganics studies on metal–protein models are crucial to give deeper insights.

Among relevant models to understand metal-mediated foldings are synthetic and natural metallopeptides with sequences lower than 100 amino acids, including non-natural ones [4]. Those small biomolecular constructs are almost omnipresent at the interface between chemistry and biology. Some examples are those involved in the early stages of Life's molecular evolution [5, 6], possible templates for *de novo*

*Computational Bioinorganics: From Description to Prediction*, First Edition. Edited by Jean-Didier Maréchal.

enzymes design [7, 8], biosensors [9], and, finally, some are landmarks of neurological diseases, such as Alzheimer's and Parkinson's [10–13].

From a molecular modeling perspective, small metallopeptides represent a unique framework for various methods. Quantum-mechanical (QM)-based approaches allow studying their fine electronic properties with realistic dimensional models (all or almost all systems can be studied at the QM level). Molecular mechanics approaches allow for exploring the entire conformational space intensively. Multiscale approaches can provide a complete framework to study metal-induced folding processes [14, 15].

Despite being idealistic systems in appearance, the study of metallopeptides by computational means, particularly their folding, does not lack challenges. The more remarkable ones are (i) the identification of conformational patterns in a vast conformational landscape (especially for peptides with a sequence upper than 10–14 amino acids); (ii) the high number of possible first coordination sphere environments compatible with metal binding, and how to energetically discriminate between them; and (iii) the relationship between both magnitudes. Metal-induced metallopeptide folding is an excellent case study for metal–protein synergies and an essential milestone in the progress of computational bioinorganics.

This chapter provides the crucial concepts for studying metal-induced peptide folding and presents how we could address them through different computational strategies. Illustrative cases are presented by increasing sequence length and amino acid variability. The procedures we report constitute a general framework that the readers could use for their purposes and a dissertation on the molecular insights in those systems.

## 8.2 Metal–Ligand Interactions with Short Peptides ($n = 2–8$)

The properties of metal cation complexes of amino acids have been widely studied at a quantum chemical level, mainly for alkali, alkaline-earth, and closed-shell transition metal cations. However, the possible interaction sites for metal cations in peptides and proteins involve chelation with basic sites like amide linkages and amino acid side chains. The advances in computational tools have allowed the theoretical study of increasingly larger peptide systems. Most of the studies on the interaction of metal cations with small peptides have been carried out in the gas phase, which, in turn, allows a direct comparison between calculations and experimental results obtained using infrared multiphoton dissociation spectroscopy.

### 8.2.1 $Gly_n$ and $Ala_n$ Metal Complexes

The obvious choice for a bottom-up approach from the amino acids to peptides is the study of oligopeptides of the simplest amino acids, $Gly_n$ and $Ala_n$. Several groups have addressed the interaction of alkali [16–31], alkaline-earth [16, 17, 24], or transition metal [16, 32–36] cations with these peptides. Glycine and alanine have no functionalized side chains, and the interaction with the metal cation is restricted to the amide oxygens of the peptide bonds, the amino nitrogen at the N-terminal, and the carboxyl oxygen atoms at the C-terminal.

While the structures of bare metallated amino acids could be determined by constructing the initial complexes based on chemical intuition, this strategy becomes impossible as the peptide chain grows. Thus, most studies combine a conformational search, usually based on molecular mechanics force fields, with the quantum chemical refinement of the most stable structures that emerged from the conformational analysis. For the smallest oligopeptides ($n$ = 2–5) several works [3, 14, 19–21] have used Monte Carlo sampling (for example, through the Monte Carlo multiple minima conformational searching technique [37]) combined with different versions of the AMBER force field. Initial structures of different isomers of the complex are considered, for example, those corresponding to the neutral or the zwitterionic forms of the peptide moiety. The final structures are then ordered according to the AMBER energy, and all within a given energy range are selected for quantum chemical minimization. A relatively large energy window must be considered since force field structures can differ significantly from those obtained from *ab initio* calculations. Other studies have used simulated annealing molecular dynamics (MD) to find reasonable candidate structures [22, 23]. Again, the lowest energy structures are reoptimized through quantum chemical methods. The conformational landscape of larger systems [18, 19, 26] ($n$>5) has been explored using replica exchange MD (REMD) simulations with the AMOEBA force field [38]. The most stable structures obtained in the REMD simulations were further refined through electronic structure calculations. Regarding the quantum chemical methods, most of the studies use the B3LYP density functional both for the geometries and energies, although in some cases, energetics are refined through MP2 calculations.

#### 8.2.1.1 Alkali and Alkali-earth Complexes

Results in gas phase for the smallest alkali metalated complexes show that the non-zwitterionic form (usually referred to as charge solvated, CS) of the peptide is always more stable than the zwitterionic one (referred to as salt bridge, SB, structure). Both CS and SB structures of $Na^+$-$Gly_2$ are shown in Figure 8.1a). For $M^+$-$Gly_n$ complexes, the preferred coordination maximizes the interaction of the metal cation with the available basic sites of the peptide; thus, resulting in the coordination of all the peptidic CO groups up to $M^+$-$Gly_6$ (see Figure 8.1b) [20–23, 26–28]. Larger peptides lead to more complex folding and internal peptide structuration through $\gamma$ or $\beta$ turns; therefore, in the case of sodiated complexes, a structural transition between $Na^+$-$Gly_6$ and $Na^+$-$Gly_7$ is observed [26]. Similar trends are found for the smallest metallated $Ala_n$ compounds up to $n = 3$ with all the CO groups coordinated to the metal cation [20]. However, $Na^+$ complexes of larger Ala-based peptides ($n = 4, 5$) present structures with lower degrees of metal chelation and the presence of intramolecular hydrogen bonds [20, 24]. Actually, Martens et al. [18] determined the turning point from globular to helix between $n = 8$ and $n = 9$ for $Na^+$-$Ala_n$ systems. Thus, for systems larger than $Na^+$-$Ala_8$, the most stable structure corresponds to the metal cation interacting at the C-terminus of a $\alpha$-helix (Figure 8.1c).

Increasing the metal charge from the alkali to alkaline-earth metal cations results in a similar preference for the neutral structure in front of the zwitterionic one in

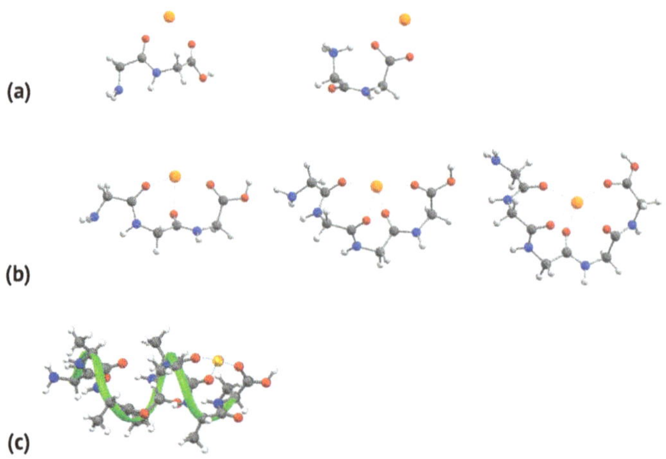

**Figure 8.1** (a) Zwitterionic (salt bridge, SB) non-zwitterionic (charge solvated, CS) complexes of $Na^+$-$Gly_2$. (b) CS forms of $Na^+$-Glyn ($n = 3-5$). *Source*: Adapted from [22]. (c) CS structure of $Na^+$-$Ala_8$.

the gas phase, and only in the case of $Ba^{2+}$ interacting with the smallest $Ala_n$ peptide ($n = 2$) a zwitterionic structure is found [17]. For larger peptides and other metal cations, the neutral structure of the peptide is the preferred one, and complexes show more significant coordination degrees than those of alkali metals [17, 24]. Regarding the coordination site, alkali-earth metals also bind to the available CO groups of the peptide [16, 17, 24]. This binding pattern is characteristic of alkali and alkaline-earth-containing proteins, such as calcium transporter proteins [39].

Solution phase studies on peptide folding around the metal cations are scarce. For example, Benzakour et al. have shown that $Li^+$ and $Na^+$ interact with the neutral form of $Gly_2$ also in solution, although the coordination mode changes from the one observed in the gas phase [27].

#### 8.2.1.2 Transition Metal Complexes

For transition metal cations, the nature of the metal–ligand interaction is more complex than for alkali metal cations because, in addition to the electrostatic terms, other effects, such as metal–ligand repulsion, *s*–*d* promotion and hybridization or dative interactions can significantly influence the mode of binding. Moreover, complexes containing open-shell transition metals can present several low-lying electronic states that arise from different metal *d* occupations. Therefore, the multiplicity of the complexes must be considered, and the relative stability of the different electronic states depend on the ligand-field splitting of the metal valence orbitals.

In contrast to what is found for alkali metals, results for transition metal mono-cations of metallated $Gly_n$ systems show remarkable differences depending on the nature of the metal cation. For example, $Cu^+$ complexes prefer linear di-coordinated structures due to metal *sd* hybridization. The binding site is formed by the terminal carbonyl oxygen and the amino group in the case of $Gly_2$ and by two CO amide groups for larger peptides (see Figure 8.2a) [35, 36]. On the contrary, $Ag^+$, $Co^+$, and $Ni^+$ form tri- and tetra-coordinated structures with the metal interacting

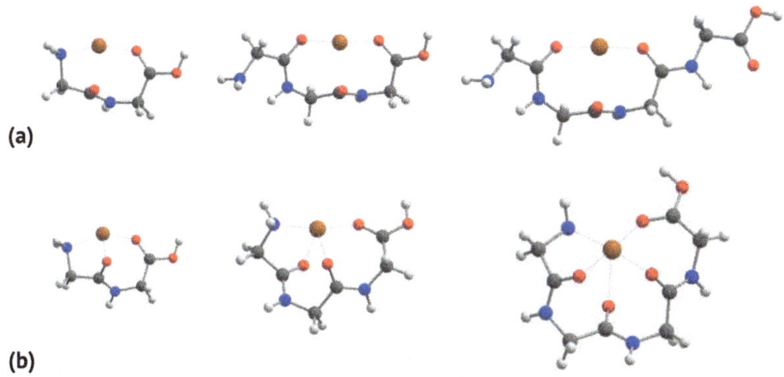

**Figure 8.2** Most stable structures of (a) $Cu^{+}$-$Gly_n$ ($n$ = 2–4) and (b) $Cu^{2+}$-$Gly_n$ ($n$ = 2–4). *Source*: Adapted from [35].

with the amide CO groups except for the smallest $Gly_2$ system, where an additional interaction with the N-terminal atom is established to maximize the metal–ligand interactions [33–36].

From a biological point of view, transition metal di-cations, such as $Co^{2+}$, $Ni^{2+}$, and $Cu^{2+}$, are of particular interest because many fundamental biological processes involve their interaction with proteins or peptides. Studies on the complexation of these cations by short $Gly_n$ oligomers ($n$ = 2–4) show that the metal interacts with all the CO groups of the peptide and the terminal amino group, thus reaching a penta-coordinated binding site for $M^{2+}$-$Gly_4$ systems (see Figure 8.2b) [32, 34, 35]. Interestingly, it has been shown that the interaction of these, and other late transition metal cations, with short $Ala_n$ peptides ($n = 2, 3$) shifts the amide/iminol tautomerism (analogous to the keto/enol tautomerism of ketones) toward the iminol form. In the iminol binding pattern, the metal cation coordinates one ($Ala_2$) or two ($Ala_3$) negatively charged deprotonated amide nitrogen atoms [16, 32]. The other coordinating atoms are the terminal carbonyl oxygen and the amino group. Figure 8.3a shows $Ni^{2+}$ binding patterns for the iminol form in $Ni^{2+}$-$Ala_3$. Thus, short nonfunctionalized peptides with few accessible CO groups for binding tend to favor the iminol binding form. This coordination pattern is very similar to relevant binding sites observed in biological environments for $Ni^{2+}$ and $Cu^{2+}$, for example, the ATCUN binding pattern (Figure 8.3b, as described subsequently). The difference is that in biological media, coordination to the nitrogen atom is accompanied by a transfer of the proton to the environment.

### 8.2.2 Metal–Peptide Complexes Involving Residues Other than Gly and Ala

Amino acid residues with functionalized side chains also play a critical role in the interaction of metal ions with peptides and proteins. A large number of studies have considered the interaction of transition metals as well as alkali and alkaline-earth metal cations with functionalized peptides, mainly His [32, 40–43] and Phe [29, 42, 44–46] containing di- and tripeptides. It should be noted that these tripeptides are long enough to wrap the metal cation with four or five binding contacts.

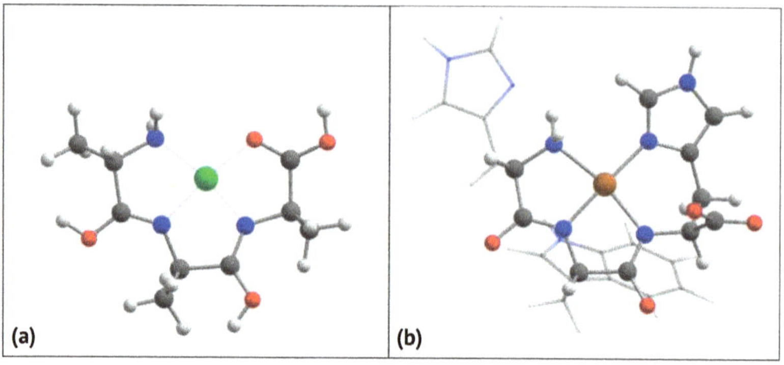


**Figure 8.3** (a) $Ni^{2+}$-$Ala_3$ iminol binding to the deprotonated amide nitrogen and the terminal groups. *Source*: Adapted from [32]. (b) ATCUN binding in $Cu^{2+}$-(HisTrpHis-2H$^+$).

Other amino acids such as Arg [47] and Pro [48, 49] have also been considered. In general, for alkali metal cation systems, a molecular mechanics conformational analysis using different search techniques with MMFFs or AMBER force fields is carried out to identify the low-lying energy structures. Then, structures within a given energy window are evaluated at a quantum chemical level, usually using density functional theory (DFT) methods. For transition metal cations, conformational searches are less exhaustive, and sometimes the initial structures are derived from those obtained with $Gly_n$ or $Ala_n$ metallated peptides.

#### 8.2.2.1 Alkali and Alkaline Earth Complexes

The coordination of alkali and alkaline-earth metal cations to Phe-containing di- and tripeptides shows similar binding patterns. That is, the most stable structure corresponds to the neutral peptide coordinating the metal through its CO groups as in the case of nonfunctionalized peptides, with the additional chelation of the aromatic ring in a cation-$\pi$ interaction [29, 42, 44–46].

Results for His-containing dipeptides, HisGly or GlyHis, show a sequence-dependent binding [42]. $Ba^{2+}$ and $Ca^{2+}$ prefer binding to the zwitterionic form of GlyHis (through the carbonyl oxygens), but prefer binding to the neutral structure of HisGly (through two carbonyl oxygens, the N-terminal, and the imidazole group).

#### 8.2.2.2 Transition Metal Complexes

Transition metal complexes, particularly $Cu^{2+}$ or $Ni^{2+}$ [32, 42] ones, also induce the proton transfer from the NH to the O atom of the amide bond, thus, resulting in the interaction of the metal with deprotonated nitrogen (iminol). For the smallest Phe or His-containing dipeptides, the coordination shell is completed by interacting with the terminal amino and CO groups and the aromatic ring (in a cation-$\pi$ manner in the case of Phe) [42]. For larger tri- and tetrapeptides with the aromatic amino acid in the first position, the metal cation also interacts with the iminol form of the peptide through two deprotonated nitrogen atoms in a square planar binding site. In this case, the aromatic side chain does not interact directly with the metal cation but with the hydrogen atom of the OH iminol group. These results led the authors to conclude that the interaction with the iminol form is favored by side chains with basic sites or by deprotonating amide nitrogens, as observed in the condensed phase [32]. As noted earlier, this type of interaction is similar to the ATCUN binding pattern (Figure 8.3b) [32, 41]. However, in the ATCUN motif, a His residue is located in the third position of the sequence. Thus, this motif is a peptidic sequence of type Aa1—Aa2—His (Aa = any amino acid), which was first identified in human serum albumin but subsequently found in other natural proteins. This particular sequence allows the interaction of the aromatic side chain of His with the metal cation, as shown in Figure 8.3b. In biological media, this interaction is accompanied by deprotonation of amide nitrogens, in contrast to gas-phase complexes where the proton is transferred to the amide oxygen. The nature of the other two amino acids also determines the properties of the complex. Thus, in a recent study of Aa1—Aa2—His (where Aa1 and Aa2 can be either His or Trp) interacting with $Cu^{2+}$, it was observed that the binding properties and redox behavior of the resulting copper complexes drastically depend on the position of the additional His residue [50].

Finally, the gas-phase study on the interaction of different transition metal di-cations with the deprotonated (PheGlyGly—$H^+$) peptide concludes that the deprotonation takes place at the C-terminal group, and the most stable structure always corresponds to the metal cation interacting with three CO groups and the aromatic ring [45].

## 8.3 Metal-oligopeptides (*n* >8)

The computational study of the folding of a peptide around a transition metal cation becomes more and more challenging upon increasing the length of the peptide since many possible metal coordination and peptide configurations may arise. In this sense, computational modeling allows the construction of reliable structures for the metal binding site based on the properties obtained by experiments, such as electron paramagnetic resonance (EPR), IR, and other techniques depending on the system. Usually, the available experimental information is used to reduce the number of possible coordinating ligands, and high-level QM calculations provide a good description of the geometry and electronic properties of the metal site, which are essential for the description of the complete metal–peptide system.

A clear example is the study of copper interaction with amyloid-β (Aβ), a 40–42 residue-long peptide, highly relevant in Alzheimer's disease since it aggregates, leading to brain amyloid plaques, one of the hallmarks of this disease. Redox-active metal cations, particularly $Cu^{2+}$ have been shown to be involved in the oxidative stress present in the brain of Alzheimer's disease patients. Thus, during the last two decades, many experimental and theoretical works have been devoted to studying $Cu^{2+}$ coordination to the peptide with conflicting results in some cases [51, 52]. Different coordination spheres have been proposed based mainly on spectroscopic results. The most solid evidence supports a 3N1O coordination shell of $Cu^{2+}$, but different sites are involved depending on the pH. The N-terminal region of the peptide has numerous possible metal coordination sites, among which there are three histidines (His6, His13, and His14). Thus, several computational studies have considered the interaction of metal cations with model ligands of these possible coordination sites to determine the most plausible coordination shell at each pH by computation of the energetics of each environment at a quantum chemical level [53–56]. The reduced metal cation, $Cu^+$, has also been studied due to its relevance in the redox processes leading to oxidative stress [56–58]. Calculations are usually carried out using DFT and adding the solvent effect with a continuum model. In some cases, *ab initio* MD simulations have been used to select structures of each coordination environment, which are then optimized at the DFT level [56, 58]. These studies have allowed us to propose structures for the metal–peptide binding site, and even $Cu^{2+}$-A$\beta$/$Cu^+$-A$\beta$ reduction mechanisms, according to the experimental information available at each moment.

### 8.3.1 Construction of Three-dimensional Models. Application to $Cu^{2+}$-A$\beta_{1-16}$

Beyond the coordination shell around the metal cation, experimental elucidation of the three-dimensional (3D) structures of metal–peptide complexes is challenging. When experimental results are unavailable (or inconclusive), computational modeling can obtain the information. Constructing 3D metal–peptide complexes requires correctly describing the coordination sphere (as mentioned earlier) and the peptide conformation upon metal coordination. The study of the metal coordination sphere in the case of transition metals requires QM methods, as described in the previous section. On the other hand, the elucidation of the conformation adopted by the peptide requires very large simulation times, in the order of microseconds to milliseconds on a timescale for the smallest peptides or proteins [59] that cannot be covered at that level of calculation. The alternative is using an integrative strategy that combines QM calculation of the metal site with homology modeling simulations, as shown in Figure 8.4.

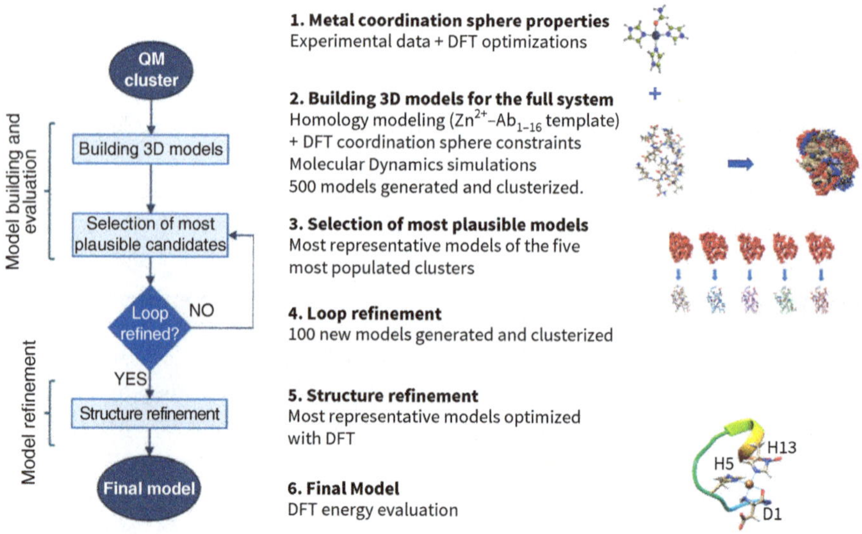


**Figure 8.4** Protocol used for the construction of 3D models of metal-oligopeptides.

In this strategy, the metal sphere is obtained by means of QM methods, and the peptide is added by using as a reference a similar metal–peptide complex by comparative modeling (as described subsequently). The information on the peptide conformation is taken from one or more references with a good sequence alignment and known structure (templates). These templates need to have similar coordination properties as the target metal–peptide system and similar size to have a good initial structure for the modeling process. Once the coordination sphere and templates are chosen, the folding process consists of building models exploring the conformational space by taking the information of the coordination sphere from experimental information or QM calculations and the peptide configuration from the templates.

One of these template-based modeling strategy with structural constraints is that implemented in Modeler [60]. This program performs comparative protein structure modeling by satisfaction of spatial restraints. Comparative modeling involves four main stages: fold assignment, target-template alignments, model building, and model evaluation [61].

The fold assignment and target-template alignments consist of identifying one or more structures similar to the target sequences (templates). This task is performed by searching reported structures in protein databases (PDB) and comparing them with the target sequence. This procedure provides a score value indicating the grade of similarity among the target sequence and the possible templates. Usually, a structure is considered a good template when the percentage of sequence identity is over 30% [62]. With all the possible structures, the template or templates are chosen considering several factors. Sometimes, the selection considers only the most significant percentage of target-template sequence identity. In other cases, additional aspects, such as experimental conditions and resolution, active site conservation, interaction with ligands of interest, and solvents, among other factors, are considered. From this filter, more than one proper template may result, and so, in some cases, the use of more than one template improves the model-building process [63].

The third stage in the process is to build a full-size model. In the case of metal–peptide complexes, a coordination sphere obtained by QM calculations is taken as a starting point and used as a restraint in the building process of the entire peptide. The construction of the peptide part is made by simulating annealing procedures assuming the distances and angles from the QM complexes as constraints for the coordination sphere, and the geometrical parameters for the peptide are taken from the protein template(s). The spatial restraints are combined with the CHARMM force field to obtain the most plausible conformations evaluated through an objective function [64]. This objective function depends on the Cartesian coordinates of the atoms that form the modeled complexes, and with this information, the restraints of the systems are established. The restraints consider geometrical parameters, such as distances, angles, and dihedrals. In addition, it includes the steric effect by considering the solvent accessibility and atom densities. The models are obtained by optimizing this objective function in Cartesian space by the use of the variable target function method [65], which uses methods of conjugate gradients and MD with simulated annealing [66].

The construction process uses information from a similar metal–peptide system whose structure has been solved experimentally. This information is used to build a good initial structure, and from this initial model, the building process consists of the construction and evaluation of a sufficient number of models that widely cover the peptide conformational space. The number of models to build depends on the system's size, the peptidic part's flexibility, and the user's criteria.

In peptides, loop modeling is also essential to the building process. In the case of metallopeptides, a specific loop refinement process is carried out after the first stage of model construction by fixing all the systems but the loops. For short loops, it is possible to explore sufficiently the conformational space. There are several methods to do this, such as those that consider structural information of similar regions from databases [67], those that reoptimize the evaluation function by doing a conformational search only in the loop regions [68], or those that combine both approximations [69].

Due to the high number of models resulting from the modeling process, it is necessary to apply a filter to select the best candidates. This selection implies energetic and geometrical criteria. Cluster analysis is usually done by grouping the models according to their conformation similarities for the geometrical evaluation. This process generates several clusters with different numbers of models. To select the best candidates from each cluster, the most representative ones, those whose geometries are close to the centroid of their respective cluster, are selected. With all the representative models from the most populated clusters, an energetic and visual evaluation is performed to choose the best candidate. There are several energetic functions to evaluate models, including the discrete optimized protein energy) score [70]. This method uses probability density functions to evaluate and compare the models with crystallographic data. However, an energetic evaluation with a high-level computational method is recommended for a better selection of the candidates [14].

We have applied this protocol (Figure 8.4) to construct 3D models of $Cu^{2+}$-$A\beta_{1-16}$ metal–peptide complexes. These complexes are relevant in Alzheimer's disease since metal cations have been shown to interfere with amyloid aggregation, one of the hallmarks of Alzheimer's disease. Furthermore, copper is a redox-active metal cation that is involved in an increase of oxidative stress by leading to the formation of reactive oxygen species. Thus, we computationally built 3D models for several $Cu^{2+}$-$A\beta_{1-16}$ complexes enclosing the different coordination spheres proposed experimentally at different pH and evaluated their redox properties. For this case, the coordination sphere information was obtained from CW-EPR spectroscopy experiments on $Cu^{2+}$ $A\beta_{1-16}$, which revealed a square planar coordination and two main species at physiological pH: one referred to as Component I at low pH (<7), involving the binding of the terminal amino group, His6 and His13 (or His14) and the CO from Asp1 and another one, named Component II, at higher pH (>7), for which two coordination spheres have been proposed: one at pH = 8 coordination involving three histidines and the CO of Ala2 (IIa) and another one at pH = 9 involving one His, the $NH_2$-terminus, the deprotonated amide nitrogen of Ala2 and the carbonyl oxygen (IIc) [52] (Scheme 8.1).

Model systems, including the experimentally proposed metal coordination spheres, were optimized with DFT QM techniques, whereas the peptide part was built by taking into account the structural information of the Zn- $A\beta_{1-16}$, complex reported in the PDB (PDB code 1ZE9) [71]. Homology modeling (HM) calculations

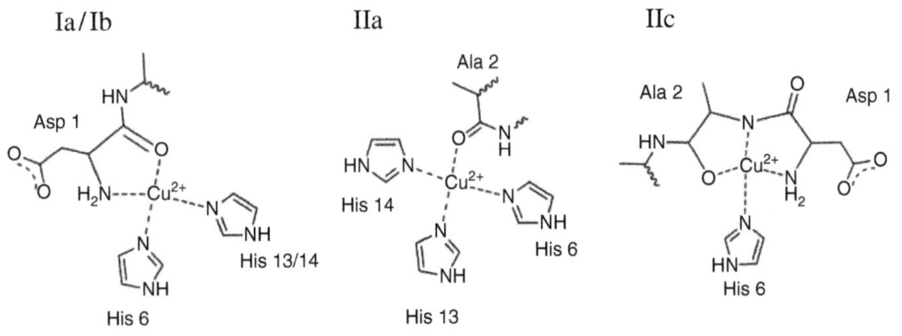


**Scheme 8.1** Coordination environments of $Cu^{2+}$ in $Cu^{2+}$ $A\beta_{1-16}$ complexes proposed from CW-EPR spectroscopy. In the case of complex **I**, two different structures, **Ia** and **Ib**, are possible depending on the second coordinating His residue (His13 or His14, respectively).

were then carried out, including geometry restraints (distances and angles) obtained from DFT calculations on small copper complexes representing the metal binding site. 500 models were built from the metal complex and the template. These models were grouped in clusters, with five more populated ones containing 175, 68, 31, 16, and 14 models, respectively. From each cluster, the most representative ones were selected and evaluated energetically and geometrically. The most plausible model was then subjected to a loop refinement. From this process, 100 models were built, and the clustering and evaluation process was also applied to this new group of models. At the end of the process, a plausible candidate of the metal peptide in that configuration was obtained. The process was applied again with all the possible coordination spheres, and the resulting models from each coordination were evaluated at a higher level of theory (DFT) to compare the different metal–peptide complexes [72]. Representative models for each coordination are shown in Figure 8.5.

QM optimizations indicated that all $Cu^{2+}$-$A\beta_{1-16}$ structures exhibit a distorted square planar metal coordination and that copper reduction induces the $CO_{backbone}$ decoordination leading to a tricoordinated T-shape species (Figure 8.5). Furthermore, results highlighted the importance of the nature of the ligands (neutral vs. charged) on the standard reduction potential, the computed value for the complex with deprotonated amide (−0.4 V for IIc) being significantly lower than those computed for models with neutral ligands (0.2–0.3 V for Ia and IIa). Overall, building 3D models for $Cu^{2+}$-$A\beta_{1-16}$ systems allowed us to unravel the role of the metal site and peptide on the whole metallopeptide stability and determine the trends in redox properties according to their coordination sphere.

### 8.3.2 Impact of $Cu^{2+}$ and $Al^{3+}$ Binding on the $A\beta_{1-42}$ Conformational Landscape

*In vitro* experimental studies have shown that metal ions significantly interfere in amyloid aggregation and fibril formation. However, their role in forming aggregates and at which molecular stage they act is unclear. In this context, MDs simulations can provide hints on metal cations' impact on the amyloid conformational landscape. However, detecting even simple structural patterns is difficult due to

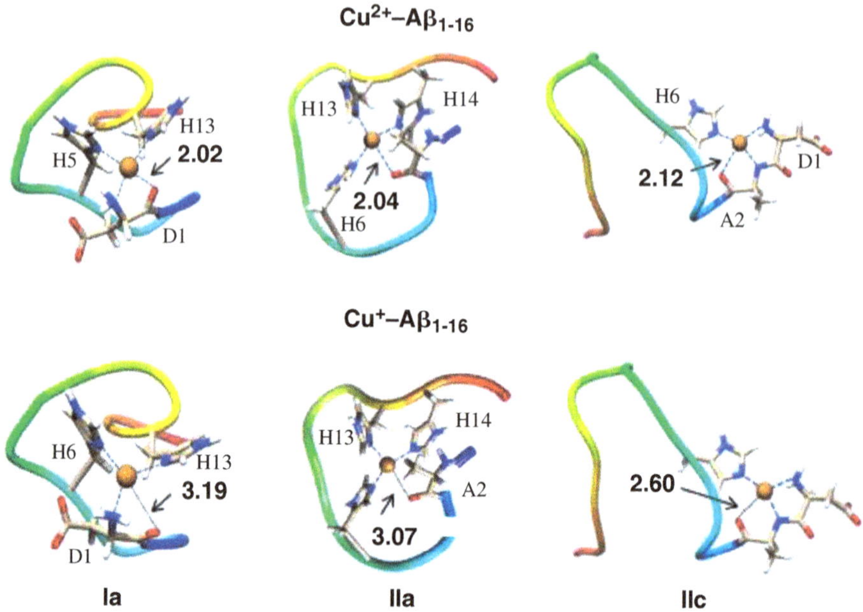


**Figure 8.5** Representative models for Ia, IIa, and IIc metal coordination modes. Distances measured in unit Å. *Source*: Reference [73]/AIP Publishing/CC BY 3.0.

the high conformational plasticity of metallopeptides and the fact that standard MD simulations can easily get trapped into local minima. The study of these systems requires enhanced sampling techniques, such as the Gaussian accelerated MD (GaMD) approach [74], to identify the distinct low-energy states of the biomolecule. This approach has recently been applied by us to analyze the impact of $Cu^{2+}$ and $Al^{3+}$ binding in the conformational landscape of $A\beta_{1-42}$ [75] 3D models of metal-bound complexes were built starting from the most stable models reported in our previous studies of $Cu^{2+}A\beta_{1-16}$ [72] and $Al^{3+}$-$A\beta_{1-40}$ [15] and extending them up to the 42 residues of Aβ. The systems were solvated with a cubic box of preequilibrated TIP3P water molecules, and NPT GaMD MD simulations up to 4 μs were carried out with the AMBER ff14SB force field [76]. Energy profiles were computed using the generalized born implicit model for 800 frames extracted from each 4 μs GaMD trajectory, stripping water molecules, and after performing a short minimization.

Next, we summarize the results obtained from NPT GaMD MD simulations up to 4 μs for one representative model of $Cu^{2+}$-$A\beta_{1-42}$ , $Al^{3+}$-$A\beta_{1-42}$ , and the free peptide. The resulting energy profiles obtained from GaMD simulations, the radius of gyration, the timeline analysis, and Frequency Contact Maps for the free peptide and for one representative aluminum and copper complex are shown in Figure 8.6.

Regarding the free peptide, GaMD simulations show it exhibits significant conformational shifts in which up to three short α-helix regions fold and unfold along the simulation or reorient between them. Accordingly, three regions with a

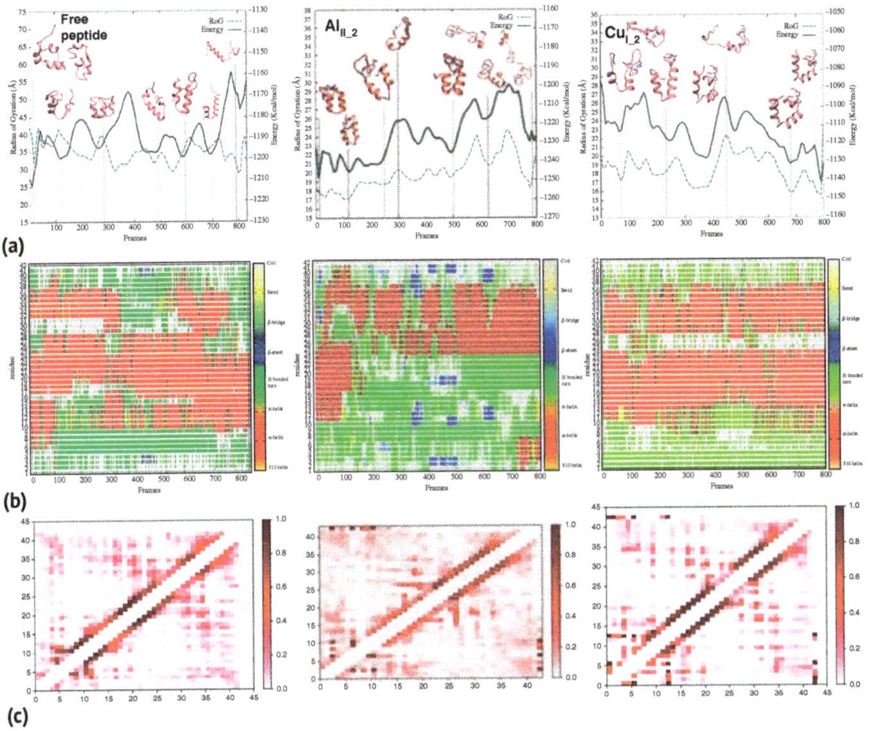


**Figure 8.6** (a) Implicit solvent energies and radius of gyration, (b) timeline analysis, and (c) frequency contact map.

recurrent α-helix structure are identified in the timeline analysis. Moreover, in short, both in time extension and residue implication, β-sheet structures are observed. The free peptide is flexible, adopting several configurations and continually reorganizing along the trajectory. Consequently, the frequency contact map exhibits a large number of transient contacts.

Metal coordination increases the compactness of the metallopeptides, as shown by the RoG values, which are 35.9 in the free peptide and around 20 in metal-coordinated complexes. Furthermore, metal binding induces the formation of low-energy U-shaped structures with two α-helical regions: Gln15-Gly25 and Ala30-Val40. These structures are stabilized by H-bond contacts, either between Asp23 and Ser26, or Glu22 and Lys28, as well as by inter-helical hydrophobic contacts. Those contacts have also been previously reported in monomeric Aβ forms [77–80] and remind the ones typically found in reported structures of Aβ fibrils [81–83]. Major differences between the two metals are observed. For Cu(II), metal binding increases the α-helix content, and the two formed helices exhibit a pronounced tendency to organize in a U-shape. In contrast, for most Al(III) complexes, metal binding induces a decrease in the helical content, in agreement with experimental studies [84]. In addition, more extended structures are observed, which favors the appearance of transitory β-sheet regions, thereby suggesting more aggregation-prone structures in the presence of $Al^{3+}$. The implication of these trends on the aggregation process is unclear and would require further studies since metal cations can also play a significant role in the following steps of aggregation, either by stabilizing intermediate oligomeric species or the final fibrils through interactions with other residues.

## 8.4 Conclusions

Metal ions are known to play a significant structural role in the biological processes of metallopeptides, and thus, information at the atomic level is essential to understanding their structure and function. In this context, computational simulations are particularly informative. However, despite metallopeptides being relatively simple systems, their computational study is challenging due to their high conformational plasticity and several metal binding sites that could induce different foldings. This chapter provides how different computational strategies can address metal-induced peptide folding according to the peptide features. For short peptides (up to $n = 8$), the presence of a few basic sites available for coordination leads to a reduced number of possible coordination modes, and the different conformational complexes can easily be addressed by hand and chemical intuition. Furthermore, the size of these metallopeptides is small enough to address their energetic and structural properties by means of quantum chemical calculations. For larger metallopeptides, however, the computational study of the folding of the peptide around the metal cation becomes more challenging due to many possible metal coordinations and peptide configurations. For these systems, initial plausible 3D models can be built by using a computational approach that combines HM with quantum-mechanics-based methods, as illustrated for $Cu^{2+}$-A$\beta_{1-16}$. From these initial structures, the conformational space can be explored through classical MD simulations. Obtained 3D structures can be refined by running quantum chemical calculations on representative models of the most populated clusters. For larger metallopeptides, such as $Cu^{2+}$-A$\beta_{1-42}$, QM simulations are computationally too costly, and so, classical MD simulations using enhanced sampling techniques, such as the GaMD approach, are the techniques to be applied to identify the low-energy states of the complex conformational landscape. This kind of simulation (full atom with explicit solvent) provides essential insights into the structural and dynamic features of metallopeptides. However, they do not address critical issues, such as the metal binding process, which would require accounting for metal desolvation or the effects of subtle electronic changes on the metal during the folding process.

## References

**1** Gomes, C.M. and Wittung-Stafshede, P. (eds.) *Protein Folding and Metal Ions.* Taylor & Francis; 2011.

**2** Lykourinou, V. and Ming, L.-J. Metalloproteins and metallopeptides – natural metallofoldamers. In *Metallofoldamers: Supramolecular Architectures from*

*Helicates to Biomimetics* (ed. Maayan, G.G. and Albrecht, M. (Markus)). Wiley-VCH; 2013, pp. 1–50.

**3** Schweitzer-Stenner, R. (ed.) *Protein and Peptide Folding, Misfolding, and Non-Folding*. John Wiley & Sons; 2012.

**4** Gamba, I., Rama, G., Ortega-Carrasco, E. et al. The folding of a metallopeptide. *Dalt. Trans.* 2016;*45*:881–885.

**5** Belmonte, L., Rossetto, D., Forlin, M. et al. Cysteine containing dipeptides show a metal specificity that matches the composition of seawater. *Phys. Chem. Chem. Phys.* 2016;*18*:20104–20108.

**6** Belmonte, L. and Mansy, S.S. Metal catalysts and the origin of life. *Elements* 2016;*12*:413–418.

**7** Ball, Z.T. Designing enzyme-like catalysts: a rhodium(II) metallopeptide case study. *Acc. Chem. Res.* 2013;*46*:560–570.

**8** Lewis, J.C. Artificial metalloenzymes and metallopeptide catalysts for organic synthesis. *ACS Catal.* 2013;*3*:2954–2975.

**9** Hou, Y. and Gochin, M. Artificial ion channel biosensor in human immunodeficiency virus Gp41 drug sensing. *Anal. Chem.* 2008;*80*:5924–5929.

**10** Borghesani, V., Alies, B., and Hureau, C. Cu II binding to various forms of amyloid-$\beta$ peptides: are they friends or foes? *Eur. J. Inorg. Chem.* 2018;*2018*:7–15.

**11** Rana, M. and Sharma, A.K. Cu and Zn interactions with A$\beta$ peptides: consequence of coordination on aggregation and formation of neurotoxic soluble A$\beta$ oligomers. *Metallomics* 2019;*11*:64–84.

**12** Viles, J.H. Metal ions and amyloid fiber formation in neurodegenerative diseases. Copper, zinc and iron in Alzheimer's, Parkinson's and Prion diseases. *Coord. Chem. Rev.* 2012;*256*:2271–2284.

**13** Wright, J.A. and Brown, D.R. Alpha-synuclein and its role in metal binding: relevance to Parkinson's disease. *J. Neurosci. Res.* 2008;*86*:496–503.

**14** Alí-Torres, J., Maréchal, J.-D., Rodríguez-Santiago, L. et al. Three dimensional models of $Cu^{2+}$-A$\beta$(1–16) complexes from computational approaches. *J. Am. Chem. Soc.* 2011;*133*:15008–15014.

**15** Mujika, J.I., Rodríguez-Guerra Pedregal, J., Lopez, X. et al. Elucidating the 3D structures of Al(III)-A$\beta$ complexes: a template free strategy based on the pre-organization hypothesis. *Chem. Sci.* 2017;*8*:5041–5049.

**16** Dunbar, R.C., Berden, G., and Oomens, J. How does a small peptide choose how to bind a metal ion? IRMPD and computational survey of CS versus iminol binding preferences. *Int. J. Mass Spectrom.* **2013**;*354–355*:356–364.

**17** Dunbar, R.C., Steill, J.D., and Oomens, J. Conformations and vibrational spectroscopy of metal-ion/polylalanine complexes. *Int. J. Mass Spectrom.* 2010;*297*:107–115.

**18** Martens, J.K., Compagnon, I., Nicol, E. et al. Globule to helix transition in sodiated polyalanines. *J. Phys. Chem. Lett.* 2012;*3*:3320–3324.

**19** Semrouni, D., Balaj, O.P., Calvo, F. et al. Structure of sodiated octa-glycine: IRMPD spectroscopy and molecular modeling. *J. Am. Soc. Mass Spectrom.* 2010;*21*:728–738.

**20** Wang, P., Wesdemiotis, C., Kapota, C. et al. The sodium ion affinities of simple di-, tri-, and tetrapeptides. *J. Am. Soc. Mass Spectrom.* 2007;*18*:541–552.

21 Wong, C.H.S., Ma, N.L., and Tsang, C.W. A theoretical study of potassium cation binding to glycylglycine (GG) and alanylalanine (AA) dipeptides. *Chem. - A Eur. J.* 2002;*8*:4909–4918.

22 Wyttenbach, T., Bushnell, J.E., and Bowers, M.T. Salt bridge structures in the absence of solvent? The case for the oligoglycines. *J. Am. Chem. Soc.* 1998;*120*:5098–5103.

23 Ye, S.J. and Armentrout, P.B. Absolute thermodynamic measurements of alkali metal cation interactions with a simple dipeptide and tripeptide. *J. Phys. Chem. A* 2008;*112*:3587–3596.

24 Dunbar, R.C., Steill, J.D., Polfer, N.C. et al. Metal cation binding to gas-phase pentaalanine: divalent ions restructure the complex. *J. Phys. Chem. A* 2013;*117*:1094–1101.

25 Balaj, O.-P., Kapota, C., Lemaire, J. et al. Vibrational signatures of sodiated oligopeptides (GG–$Na^+$, GGG–$Na^+$, AA–$Na^+$ and AAA–$Na^+$) in the gas phase. *Int. J. Mass Spectrom.* 2008;*269*:196–209.

26 Balaj, O.P., Semrouni, D., Steinmetz, V. et al. Structure of sodiated polyglycines. *Chem. A Eur. J.* 2012;*18*:4583–4592.

27 Benzakour, M., Mcharfi, M., Cartier, A. et al. Interactions of peptides with metallic cations. I. Complexes glycylglycine–$M^+$ (M = Li, Na). *J. Mol. Struct. THEOCHEM* 2004;*710*:169–174.

28 Cerda, B.A., Hoyau, S., Ohanessian, G. et al. $Na^+$ binding to cyclic and linear dipeptides. Bond energies, entropies of $Na^+$ complexation, and attachment sites from the dissociation of $Na^+$-bound heterodimers and ab initio calculations. *J. Am. Chem. Soc.* 1998;*120*:2437–2448.

29 Dunbar, R.C., Steill, J.D., Polfer, N.C. et al. Peptide length, steric effects, and ion solvation Govern Zwitterion stabilization in barium-chelated di- and tripeptides. *J. Phys. Chem. B* 2009;*113*:10552–10554.

30 Kish, M.M., Wesdemiotis, C., and Ohanessian, G. The sodium ion affinity of glycylglycine. *J. Phys. Chem. B* 2004;*108*:3086–3091.

31 Kohtani, M., Kinnear, B.S., and Jarrold, M.F. Metal-ion enhanced helicity in the gas phase. *J. Am. Chem. Soc.* 2000;*122*:12377–12378.

32 Dunbar, R.C., Martens, J., Berden, G. et al. Complexes of Ni(II) and Cu(II) with small peptides: deciding whether to deprotonate. *Phys. Chem. Chem. Phys.* 2016;*18*:26923–26932.

33 Constantino, E., Rimola, A., Rodríguez-Santiago, L. et al. Coordination properties of glycylglycine to $Cu^+$, $Ni^+$ and $Co^+$. Influence of metal cation electronic configuration. *New J. Chem.* 2005;*29*: 1585.

34 Constantino, E., Rimola, A., Sodupe, M. et al. Coordination of (Glycyl) n glycine (n = 1–3) to $Co^+$ and $Co^{2+}$. *J. Phys. Chem. A* 2009;*113*:8883–8892.

35 Rimola, A., Constantino, E., Rodríguez-Santiago, L. et al. Binding properties of $Cu^{+/2+}$–(Glycyl) n glycine complexes (n = 1–3). *J. Phys. Chem. A* 2008; *112*:3444–3453.

36 Shoeib, T., Rodriquez, C.F., Michael Siu, K.W. et al. A comparison of copper(I) and silver(I) complexes of glycine, diglycine and triglycine. *Phys. Chem. Chem. Phys.* 2001;*3*:853–861.

**37** Saunders, M., Houk, K.N., Wu, Y.D. et al. Conformations of cycloheptadecane. A comparison of methods for conformational searching. *J. Am. Chem. Soc.* 1990;*112*: 1419–1427.

**38** Ren, P. and Ponder, J.W. Consistent treatment of inter- and intramolecular polarization in molecular mechanics calculations. *J. Comput. Chem.* 2002;*23*:1497–1506.

**39** Chattopadhyaya, R., Meador, W.E., Means, A.R. et al. Calmodulin structure refined at 1.7 Å resolution. *J. Mol. Biol.* 1992;*228*:1177–1192.

**40** Kim, B.-R. and Kim, H.-T. A DFT study of the $[Cu^{2+}$—(GlyGlyHis − $3H^{+})]^{-1}$ ion complex structure. *J. Mol. Struct. Theochem* 2008;*868*:109–112.

**41** Dunbar, R.C., Martens, J., Berden, G. et al. Transition metal(II) complexes of histidine-containing tripeptides: structures, and infrared spectroscopy by IRMPD. *Int. J. Mass Spectrom.* 2018;*429*:198–205.

**42** Dunbar, R.C., Berden, G., Martens, J.K. et al. Divalent metal-ion complexes with dipeptide ligands having Phe and His side-chain anchors: effects of sequence, metal ion, and anchor. *J. Phys. Chem. A* 2015;*119*:9901–9909.

**43** Shoeib, T. and Sharp, B.L. A structural and free energy analysis of $Ag^{+}$ complexes to five small peptides. *Inorg. Chim. Acta* 2009;*362*:1925–1934.

**44** Dunbar, R.C., Steill, J.D., and Oomens, J. Encapsulation of metal cations by the PhePhe ligand: a cation–π ion cage. *J. Am. Chem. Soc.* 2011;*133*:9376–9386.

**45** Shin, J.-W. Structure characterization of electrosprayed [A·PheGlyGly-H]+ (A = Ca, Sr, Ba, Cu, Fe, Zn, Mn, and Eu) clusters through collision-induced dissociation and extended DFT studies. *Int. J. Mass Spectrom.* 2014;*372*:39–45.

**46** Polfer, N.C., Oomens, J., and Dunbar, R.C. Alkali metal complexes of the dipeptides PheAla and AlaPhe: IRMPD spectroscopy. *ChemPhysChem.* 2008;*9*: 579–589.

**47** Prell, J.S., Demireva, M., Oomens, J. et al. Role of sequence in salt-bridge formation for alkali metal cationized GlyArg and ArgGly investigated with IRMPD spectroscopy and theory. *J. Am. Chem. Soc.* 2009;*131*:1232–1242.

**48** Abirami, S., Wong, C.H.-S., Tsang, C.-W. et al. A theoretical study of potassium cation binding to prolylglycine (PG) and glycylproline (GP) dipeptide. *J. Mol. Struct. THEOCHEM* 2005;*729*:193–202.

**49** Baldauf, C., Pagel, K., Warnke, S. et al. How cations change peptide structure. *Chem. - A Eur. J.* 2013;*19*:11224–11234.

**50** Mena, S., Mirats, A., Caballero, A.B. et al. Drastic effect of the peptide sequence on the copper-binding properties of tripeptides and the electrochemical behaviour of their copper(II) complexes. *Chem. - A Eur. J.* 2018;*24*:5153–5162.

**51** Faller, P., Hureau, C., and Berthoumieu, O. Role of metal ions in the self-assembly of the Alzheimer's amyloid-β peptide. *Inorg. Chem.* 2013;*52*:12193–12206.

**52** Drew, S.C. and Barnham, K.J. The heterogeneous nature of $Cu^{2+}$ interactions with Alzheimer's amyloid-β peptide. *Acc. Chem. Res.* 2011;*44*:1146–1155.

**53** Gómez-Balderas, R., Raffa, D.F., Rickard, G. et al. Computational studies of Cu(II)/Met and Cu(I)/Met binding motifs relevant for the chemistry of Alzheimer's disease. *J. Phys. Chem. A* 2005;*109*:5498–5508.

**54** Raffa, D.F., Gómez-Balderas, R., Brunelle, P. et al. Ab initio model studies of copper binding to peptides containing a His–His sequence: relevance to the

$\beta$-Amyloid peptide of Alzheimer's disease. *JBIC J. Biol. Inorg. Chem.* 2005;*10*:887–902.

55 Alí-Torres, J., Rodríguez-Santiago, L., Sodupe, M. et al. Structures and Stabilities of $Fe^{2+}/^{3+}$ complexes relevant to Alzheimer's disease: an ab initio study. *J. Phys. Chem. A* 2011;*115*:12523–12530.

56 Furlan, S., Hureau, C., Faller, P. et al. Modeling copper binding to the amyloid-$\beta$ peptide at different pH: toward a molecular mechanism for Cu reduction. *J. Phys. Chem. B* 2012;*116*:11899–11910.

57 Raffa, D.F., Rickard, G.A., and Rauk, A. Ab initio modelling of the structure and redox behaviour of copper(I) bound to a His–His model peptide: relevance to the $\beta$-amyloid peptide of Alzheimer's disease. *JBIC J. Biol. Inorg. Chem.* 2007; *12*:147–164.

58 Furlan, S., Hureau, C., Faller, P. et al. Modeling the $Cu^{+}$ binding in the 1–16 region of the amyloid-$\beta$ peptide involved in Alzheimer's disease. *J. Phys. Chem. B* 2010;*114*:15119–15133.

59 Naganathan, A.N. and Muñoz, V. Scaling of folding times with protein size. *J. Am. Chem. Soc.* 2005;*127*:480–481.

60 Webb, B. and Sali, A. Comparative protein structure modeling using MODELLER. *Curr. Protoc. Bioinform.* 2016;*54*:5.6.1–5.6.37.

61 Martí-Renom, M.A., Stuart, A.C., Fiser, A. et al. Comparative protein structure modeling of genes and genomes. *Annu. Rev. Biophys. Biomol. Struct.* 2000; *29*:291–325.

62 Rost, B. Twilight zone of protein sequence alignments. *Protein Eng. Des. Sel.* 1999;*12*:85–94.

63 Srinivasan, N. and Blundell, T.L. An evaluation of the performance of an automated procedure for comparative modelling of protein tertiary structure. *Protein Eng. Des. Sel.* 1993;*6*:501–512.

64 Mackerell, A.D. Empirical force fields for biological macromolecules: overview and issues. *J. Comput. Chem.* 2004;*25*:1584–1604.

65 Braun, W. and Gō, N. Calculation of protein conformations by proton-proton distance constraints. *J. Mol. Biol.* 1985;*186*:611–626.

66 Clore, G.M., Brünger, A.T., Karplus, M. et al. Application of molecular dynamics with interproton distance restraints to three-dimensional protein structure determination. *J. Mol. Biol.* 1986;*191*:523–551.

67 Jones, T.A. and Thirup, S. Using known substructures in protein model building and crystallography. *EMBO J.* 1986;*5*:819–822.

68 Moult, J. and James, M.N.G. An algorithm for determining the conformation of polypeptide segments in proteins by systematic search. *Proteins Struct. Funct. Genet.* 1986;*1*:146–163.

69 van Vlijmen, H.W.T. and Karplus, M. PDB-based protein loop prediction: parameters for selection and methods for optimization. *J. Mol. Biol.* 1997; *267*:975–1001.

70 Shen, M. and Sali, A. Statistical potential for assessment and prediction of protein structures. *Protein Sci.* 2006;*15*:2507–2524.

**71** Zirah, S., Kozin, S.A., Mazur, A.K. et al. Structural changes of region 1–16 of the Alzheimer disease amyloid beta-peptide upon zinc binding and in vitro aging. *J.Biol.Chem.* 2006;*281*:2151–2161.

**72** Alí-Torres, J., Mirats, A., Maréchal, J.-D. et al. 3D structures and redox potentials of Cu(2$^+$)-A$\beta$(1–16) complexes at different PH: a computational study. *J. Phys. Chem. B* 2014;*118*:4840–4850.

**73** Alí-Torres, J., Mirats, A., Maréchal, J.-D.D. et al. Modeling $Cu^{2+}$-A$\beta$ complexes from computational approaches. *AIP Adv.* 2015;*5*: 092402.

**74** Miao, Y., Sinko, W., Pierce, L. et al. Improved reweighting of accelerated molecular dynamics simulations for free energy calculation. *J. Chem. Theory Comput.* 2014;*10*:2677–2689.

**75** Roldan, L. Impact of Cu(II) and Al(III) on the conformational landscape of amyloidβ1-42. *Phys. Chem. Chem. Phys.* 2021;*23*:13023–13032.

**76** Maier, J.A., Martinez, C., Kasavajhala, K. et al. Ff14SB: improving the accuracy of protein side chain and backbone parameters from Ff99SB. *J. Chem. Theory Comput.* 2015;*11*:3696–3713.

**77** Boopathi, S., Dinh Quoc Huy, P., Gonzalez, W. et al. Zinc binding promotes greater hydrophobicity in Alzheimer's A$\beta$42 peptide than copper binding: molecular dynamics and solvation thermodynamics studies. *Proteins Struct. Funct. Bioinform.* 2020;1–18.

**78** Huy, P.D.Q., Vuong, Q.V., La Penna, G. et al. Impact of Cu(II) binding on structures and dynamics of A$\beta$42 monomer and dimer: molecular dynamics study. *ACS Chem. Neurosci.* 2016;*7*:1348–1363.

**79** Liao, Q., Owen, M.C., Olubiyi, O.O. et al. Conformational transitions of the amyloid-$\beta$ peptide upon copper(II) binding and PH changes. *Isr. J. Chem.* 2017;*57*:771–784.

**80** Barz, B., Buell, A.K., and Nath, S. Compact fibril-like structure of amyloid $\beta$-peptide (1–42) monomers. *Chem. Commun.* 2021;*57*:947–950.

**81** Xiao, Y., Ma, B., McElheny, D. et al. A$\beta$(1–42) Fibril structure illuminates self-recognition and replication of amyloid in Alzheimer's disease. *Nat. Struct. Mol. Biol.* 2015;*22*:499–505.

**82** Wälti, M.A., Ravotti, F., Arai, H. et al. Atomic-resolution structure of a disease-relevant A$\beta$(1–42) amyloid fibril. *Proc. Natl. Acad. Sci. USA.* 2016;113:E4976–E4984.

**83** Gremer, L., Schölzel, D., Schenk, C. et al. Fibril structure of amyloid-$\beta$(1–42) by cryo–electron microscopy. *Science.* 2017;*358*:116–119.

**84** Exley, C., Price, N.C., Kelly, S.M. et al. An interaction of $\beta$-amyloid with aluminium in vitro. *FEBS Lett.* 1993;*324*:293–295.

# 9

# Computational Studies of Metallodrugs

*James A. Platts and Matthew Turner*

*School of Chemistry, Cardiff University, Park Place, Cardiff, United Kingdom*

## 9.1 Introduction

Metallodrugs have a long history, with traditional formulations based on mercury, gold, and antimony dating back to antiquity [1]. This article will set out some of the ways that computational modeling can be used to explore the structure, properties, and function of metal-based drugs. There are numerous ways such an article could be organized, including by the identity of the metal involved, by the disease or biomolecular target of interest, or by mode of action or binding mode. However, for pedagogical reasons, we choose to arrange this work by the computational model used, since this helps to clarify some of the challenges in this field and the reasons for choices commonly made.

This article therefore contains sections on (i) purely quantum mechanical (QM) studies of metallodrugs, with and without solvation effects; (ii) QM studies of metallodrug interactions with biomolecular species; (iii) hybrid QM/molecular modeling (QM/MM) of drug-biomolecule interactions; (iv) purely classical molecular mechanics (MM) description of drugs and their interactions; and (v) docking and quantitative structure-activity relation (QSAR) studies. Throughout, we will place particular focus on platinum-based anticancer drugs and their binding to DNA since these are probably those most commonly studied computationally, but where appropriate will illustrate key concepts with other metals and/or biomolecular targets.

## 9.2 QM Studies of Metallodrugs and Their Solvation

In principle, *ab initio* QM methods are able to find approximate solutions to the time-independent Schrödinger equation for a given molecule with great accuracy. A hierarchy of methods exists, from Hartree–Fock, through Møller–Plesset perturbation theory (MP2) and configuration interaction, to coupled cluster methods; coupled with

*Computational Bioinorganics: From Description to Prediction*, First Edition. Edited by Jean-Didier Maréchal.

systematically improved orbital basis sets, these converge towards exact solutions. In practice, however, the size and complexity of many metallodrugs, coupled with the presence of heavy transition or main group elements, means that significant approximations must be made in order for practical computations to be carried out.

An early study from Wysokiński and Michalska [2] makes this clear. Here, the performance of numerous QM methods for the calculation of the structures and vibrational spectra of cisplatin and carboplatin was assessed, using experimental data as well as second-order MP2 as a benchmark. This showed the inadequacy of Hartree–Fock for this purpose, with density functional theory (DFT) yielding the best combination of accuracy and computational efficiency. In particular, hybrid methods that mix in HF exchange into DFT predictions are shown to perform better than pure (also known as generalized gradient approximation, GGA) ones, and the mPW1PW functional is highlighted as reproducing benchmark data particularly well. Application of this method then allowed unambiguous assignment of the spectra of cisplatin that had previously been disputed in the literature, for instance in the low-frequency N—Pt—N mode in IR, and the IR/Raman signatures of $NH_3$ vibrations.

This chapter also highlights a further issue with QM calculations on Pt, and indeed with all elements in the lower half of the periodic table. The large nuclear charge leads to strong relativistic effects for core electrons, and hence to relativistic contraction and ultimately to the difference in appearance and chemistry of silver and gold, for instance. A full description of such relativistic effects *via* the Dirac equation is possible but not often practical for "real" molecules. Instead, many researchers choose to use effective core potentials (ECPs, also known as pseudopotentials), in which core electrons are not explicitly included in the QM calculation, but instead are represented by a model potential that mimics the full relativistic solution. This approach has numerous advantages: calculations are simplified, as fewer electrons/basis functions are required, while relativistic effects are implicitly included, at least approximately, within the normal framework of Hartree–Fock or Kohn–Sham DFT. Michalska's work tested several common choices of ECP, including those from Hay and Wadt, the "Stuttgart-Dresden" set, and that from Stevens, Basch, and Krauss, all of which explicitly include $n = 5$, $n = 6$ electrons in the calculation (so-called small-core ECPs), with implicit effects of inner shells. The effect of adding polarization functions to heavy atoms was also tested, but found not to be large in this case.

Around the same time, Zhang et al. [3] took a similar approach to model the interaction and reaction of cisplatin and analogs with water. This is vital, since it is recognized that cisplatin as normally formulated is actually a prodrug, and that replacement of chloride ligands by water or hydroxide is a key step in the mechanism of action. Moreover, since essentially all biology takes place in an aqueous environment, the specifics of how drugs interact with solvents are an important part of achieving full understanding of their chemistry. This work tested numerous DFT methods against the experimental geometry of cisplatin, as well as $H_2O$ and $[PtCl_4]2^-$, concluding that mPW1PW91/SDD was the best of those tested. The authors also recognized the importance of bulk solvation, including this through the self-consistent reaction field (SCRF) approach. This model was then used to characterize all stationary points on the potential energy surface for the replacement of both chlorides by water. The nature of these stationary points as minima (i.e. reactant, intermediates, and product) or first-order saddle points (i.e. transition states [TS]) was confirmed through the use of harmonic frequency calculation, which also allowed the estimation of thermal contributions to overall energies.

Using this approach, they were able to characterize all minima as four-coordinate square planar species linked by five-coordinate trigonal bipyramidal TS along an associative reaction path, in agreement with available experimental data. The level of detail available from DFT calculation allowed specific comparison of key geometrical details, such as metal to leaving/entering ligand bond lengths and angles at TS, and also to compare these quantities between cisplatin and $[Pt(dien)Cl]^+$ (dien = diethylenetriamine) to estimate steric effects. Thermodynamic data from these calculations indicate $\Delta H^{\ddagger}$ of 23 and 20 kcal $mol^{-1}$ (96 and 84 kJ $mol^{-1}$) for replacement of the first and second chloride, again in accord with the experiment. Natural bond orbital (NBO) analysis was also used to track the changes in atomic charge across the reaction path, with Pt and Cl exhibiting the largest variations in charge. It is also shown that proper inclusion of bulk solvation effects is key to the success of this approach: gas phase estimates of barrier heights are as much as 10 kcal $mol^{-1}$ higher than aqueous ones.

## 9.3 QM Studies of Metallodrug–Biomolecule Interactions

Computational studies of metallodrugs alone can yield useful insight, but ultimately their interaction with potential biological targets is key to understanding their pharmacology. There have been many such studies using purely QM methods, almost exclusively DFT: we focus here on a small number of these to exemplify key aspects of the necessary calculations.

An early example of this approach was reported by Burda et al. [4] in which the hybrid DFT method B3LYP was combined with MWB ECP/basis set on Pt and Cl and 6-31G(d) on all other atoms, and used to predict Pt binding to guanine and adenine (G and A) DNA bases and the effect of this on tautomeric forms of these bases. No solvent was included, but comparison of neutral, +1 and +2 charged systems allowed the estimation of gross electrostatic effects. Geometry optimization of Pt adducts yielded square planar coordination and short Pt—$NH_2$—H…$O_6$/$N_6$ hydrogen bonds when canonical tautomers were examined, while the *syn* enol tautomer contained $O_6$—H…Cl hydrogen bonding interaction. Direct Pt—$O_6$ binding was also examined, but found to be much (53 kJ $mol^{-1}$) less stable than Pt—$N_7$ coordination. The authors conclude that the platination of $N_7$ in purine bases should not promote the formation of mispairs.

Following this, Baik et al. [5] used DFT to compare the binding of cisplatin to G and A to explain the clear preference for the former nucleobase in DNA adducts. They used the hybrid B3LYP functional, coupled with Los Alamos ECP/basis set on Pt and conventional 6-31G(d,p) or cc-pVTZ basis sets on remaining atoms. The smaller basis was used to locate geometries of all relevant stationary points in the gas phase, at which single point energies using the larger basis were calculated, including solvent effects through the SCRF method. They also used an energy decomposition method to obtain deeper insight into the origin of energy changes and barriers located.

This approach predicted gas phase electronic binding energies, $\Delta H$(SCF), for mono-functional binding of −80.7 and −65.5 kcal mol$^{-1}$ to G and A, respectively. These values fall to $\Delta G$(Sol) = −16.1 and −8.8 kcal mol$^{-1}$ on inclusion of solvent effects and thermodynamic (zero-point energy, enthalpy, and entropy) corrections. Hydrogen bonding from ammine N—H to $O_6$ of G or $N_6$ of A contributes to the preference for G: on the basis of N—H and H...O/H...N distances and comparison to analogous compounds, their strength is estimated at between 5 and 8 kcal mol$^{-1}$. While significant, these interactions are not sufficient to explain all the binding energy difference, reflecting the inherently greater basicity of N7 in G over that in A. It is also stressed that the compensatory effects of solvation and entropy are vital in obtaining realistic binding energies and hence description of the thermodynamics of binding.

To assess the kinetics of cisplatin-DNA binding, several possible TS for the ligand exchange reaction $[Pt(NH_3)_2(OH_2)Cl]^+ + X \rightarrow [Pt(NH_3)_2(X)Cl]^+ + OH_2$ (where X = guanine or adenine) were located, taking care that the eliminated water was properly described. All such TS are trigonal bipyramidal in geometry, corresponding to associative reaction pathways, with Pt—N7 and Pt—$OH_2$ bonds that are formed and broken approximately equal in length. In the case of G, hydrogen bonding patterns do not change significantly at the TS, but in A the situation is more complex with two almost isoenergetic TS geometries that differ in H-bond pattern. These searches yield free-energy barriers, $\Delta G^{\dagger}$(Sol), of 24.6 and 30.2 kcal mol$^{-1}$ for G and A, the former in good agreement with a value derived from $^{195}$Pt NMR of 23.4 kcal mol$^{-1}$. Thus, kinetics as well as thermodynamics contribute to cisplatin's preference for binding at guanine over adenine. The origin of the difference is traced to the difference in electronic energy, since although solvent and thermodynamic corrections are important, much of their effect cancels out when comparing G with A. This is further traced to approximately equal contributions of orbital interaction and of hydrogen bonding.

DNA is not the only biomolecule that can interact with metallodrugs: amino acids and peptides also contain numerous nucleophilic sites that can bind to metals. Michera et al. [6] examined the binding of cisplatin to the simplest amino acid, glycine. They used B3LYP with a 6-31+G(d) basis set on light atoms, MWB-60 ECP on Pt and MWB-10 on Cl: the latter were augmented with diffuse and polarization functions, which are deemed important for the description of anionic species to locate relevant stationary points. The larger basis 6-311++G(2df, 2pd), with similar extensions to those used for Pt and Cl, was then used to obtain single-point energies. The polarizable continuum model version of implicit solvation was used, and care was taken to properly treat acid-base equilibria. Electronic wavefunctions obtained with the larger basis set were analyzed with NBO and atoms in molecules (AIM) methods, which are discussed in more detail later in this chapter. This approach showed the pH dependence of key steps in metal–biomolecule interactions: at pH values close to 7, glycine exclusively forms a chelating complex through both N and O, a reaction that is exoergic by around 23 kcal mol$^{-1}$ (if displacing chloride and water) or 33 kcal mol$^{-1}$ (if displacing two water molecules).

The same group has also used a similar approach to explore the interaction of cisplatin and related molecules with sulfur-containing amino acids cysteine and methionine [7]. These relatively soft ligands are expected to be well suited for coordination to the soft Pt(II) ion. They show that cysteine sulfur exhibits the strongest thermodynamic binding to cisplatin, in accord with the expected soft nature of both ligand and metal, with a bond energy of over 80 kcal $mol^{-1}$. Methionine S is predicted to form much weaker Pt—S bonds, but for both amino acids a chelating $\kappa^2$(S, N)-binding mode is found to give the greatest stabilization. This is of more than academic interest, given that the S-rich tripeptide glutathione is believed to play an important role in cisplatin metabolism, toxicity, and resistance.

Aside from cisplatin and related species, DFT has been used to examine how Pt(II) complexes might bind to models of amyloid-$\beta$ (A$\beta$), the peptide that is closely associated with the onset of Alzheimer's disease [8]. This peptide is known to bind endogenous metals, such as Cu, Zn, and Fe through its N-terminal sequence, and in particular His6, His13, and His14 residues. Pt(II), especially when bound to large, planar ligands such as 1,10-phen-anthroline, has been shown to compete with native metals and inhibit aggregation of A$\beta$ and the generation of reactive oxygen species. DFT, alongside numerous experimental techniques, was used to probe the detailed structure of possible adducts: the MPW1PW91 functional with SDD basis set was chosen on the basis of comparison against X-ray geometries of related species. Rather than simply optimizing structures at this level, these were used as the starting point for refinement against X-ray absorption fine structure (EXAFS) data for the entire peptide to arrive at a model of Pt(BPS) (BPS = 4,7-diphenyl-[1,10]phenanthroline disulfonate) binding to A$\beta$ 1–42 through the N-terminal histidines. This is in contrast to cisplatin binding, which binds predominantly through Met35, an observation that might help to explain the requirement for large, planar ligands for antiaggregation activity.

Similar models were considered by Novato et al. [9] who used DFT to examine the binding of two Pt(phenanthroline)-based complexes to a series of amino acids. They used the B3LYP functional with a basis set consisting of 6-31G(d) on light atoms and Lanl2DZ on Pt, within the IEFPCM description of aqueous solvation to locate and characterize minima and TS, then obtained improved energetic profiles through single-point calculations using LC-$\omega$PBE and CAM-B3LYP functionals along with 6-31+G(2df) basis on light atoms and an augmented SDD basis on Pt. This approach was used to estimate barriers for ligand exchange reactions, including $H_2O$ and various amino acids replacing chloride, both of which occur *via* an associative mechanism with a distorted trigonal bipyramidal TS. The fastest and most exergonic reaction is predicted to occur with Met, followed by Lys (if modeled with a neutral side chain) and His, with only small differences between N$\varepsilon$ and N$\delta$ of His. Oxygen-binding residues, such as Glu and Asp were found to exhibit markedly higher barriers to reaction.

Of course, platinum is by no means the only metal used in metallodrugs: our group has carried out calculations on a set of ruthenium-arene structures [10] first proposed by Sadler et al. and their binding to DNA. A key feature of these systems is that the arene ligands are designed to intercalate between DNA base pairs: this interaction is driven largely by dispersion forces, which typically are not well modeled by DFT. Through the use of a specific function that had previously been shown to model $\pi$–$\pi$ stacking well, namely BHandH, we were able to properly model the interaction of a series of Ru-arene complexes with guanine, and to probe the relative importance of $\pi$–$\pi$ stacking and hydrogen bonding to their relative stability. Since publishing that work, DFT treatment of dispersion interactions has become a hot topic, and these effects are now routinely used as part of such studies. This is typically achieved using a so-called empirical dispersion correction, i.e. adding an extra term to standard DFT to account for the $R^{-6}$ dependence of dispersion interactions. Several such schemes have been proposed, varying in the functional form, parameters used, and in damping functions used to prevent the dispersion term exploding at short distances. At the time of writing, a popular version of this approach is "DFT-D3(BJ)," which applies Grimme's third-generation dispersion correction along with Becke and Johnson's damping function [11], although this is an area of rapid theoretical development. One example of this approach is a recent study examining the noncovalent interaction of cisplatin with models of graphene [12].

The studies reported so far have concentrated largely on the potential energy surface of metallodrugs and their interaction with biomolecules. Location of minima and TS, and their relative energies, are of course key to the understanding of such systems. In addition, though, the results of QM calculations such as Kohn–Sham DFT can also be interpreted through examination of the orbitals and electron density that result. One approach to doing so is the Quantum Theory of Atoms in Molecules (QTAIM, or often just AIM), in which the electron density distribution across a molecule is partitioned into distinct atomic basins *via* a set of zero-flux surfaces [13]. The intersection of these surfaces with the bond paths that link bonded nuclei give rise to bond critical points in the electron density, which have been shown to be an excellent means to probe the bonding interactions within and between molecules. Our group used this approach to monitor the hydrogen bonding that is present when cisplatin binds to adenine and guanine [14], showing that the strong Pt—N—H...$O_6$=C interaction formed with guanine forms a significant part of the energy preference for binding to guanine over adenine. Another popular analysis method is the NBO scheme, recently used by Tai and Nhat [15] to examine the hydrogen bonds present when cisplatin and two derivatives bind to guanine.

## 9.4 QM/MM Modeling of Drug–Biomolecule Interactions

Hybrid QM/MM methods are increasingly attractive in studies of metallodrugs. In these, a relatively small fraction of the whole system is described with QM methods, often but not exclusively DFT, while the remainder is treated with MM. This allows calculation of the properties of much larger systems that cannot be tackled by QM methods alone, while retaining the attractive features of DFT for a description of metal centers. The suitability of QM methods for bond breaking and formation has meant that much effort in this field has been focused on enzymatic reaction mechanisms, but applications to metallodrugs have also been published.

An early study by Spiegel et al. [16] used Car-Parrinello molecular dynamics (MD) coupled to AMBER-style force fields to probe the binding of cisplatin to three different DNA sequences, two taken from experimental X-ray structures and the third from an MD-equilibrated simulation. Key aspects of QM/MM calculations include the choice of atoms to be included in the QM region, and the means by which QM and MM regions are joined. In this chapter, $[Pt{-}(NH_3)_2]^{2+}$, along with coordinated guanines and, in two cases, a phosphate group that is in close proximity to the drug were treated at the QM level. Extension of this with neighboring thymine and cytosine was reported not to significantly change the electronic structure of Pt. Guanine and phosphate groups were truncated with capping hydrogen atoms. DFT calculations used the GGA BLYP functional with a plane wave basis set and included electrostatic effects of the MM region *via* a hierarchical method that explicitly calculates such effects for close atoms but approximates them by a multipole model for more distant ones. Calculations of HOMO and LUMO, along with $^{195}$Pt chemical shifts, were also performed on selected snapshots of the QM/MM trajectory. Significant changes in the geometry about Pt were observed over the course of 5–7 ps simulation: planarity around Pt and roll angle between coordinated guanines move to values closer to those found from solution-phase NMR than from X-ray experiments. The Pt—N—H… O=P hydrogen bond that is observed in the X-ray structure is lost over the course of the trajectory, whereas the Pt—N—H…O6 is fully conserved. Watson–Crick pairing of coordinated guanines with cytosine is also found to be highly flexible. The same protocol was used to perform QM/MM docking of $[Pt{-}(NH_3)_2]^{2+}$ to canonical B-DNA, showing the decrease in N7—Pt—N7 angle and concomitant increase in roll angle between guanines, associated with relaxation energy of around 150 kcal mol$^{-1}$.

Our group took a similar approach to compare the binding of five different Pt-based anticancer drugs to DNA [17]. Here, the focus was on comparing the effects on DNA binding and distortion between cisplatin, oxaliplatin, heptaplatin, and two isomers of lobaplatin. DFT, in the form of BH and H with the 6-31+G(d,p)/SDD basis set, was used to model Pt, the ligand and two coordinated guanines, with capping hydrogens replacing glycosidic N—C bonds.

The AMBER (parm96) MM approach was used for the remainder of a DNA octamer taken from an NMR structure of cisplatin-DNA (Protein Data Bank [PDB] entry 1AU5). All the drugs considered were found to disrupt the native structure of DNA in similar ways, with subtle differences observed for the larger ligands used in oxaliplatin, heptaplatin, and lobaplatin resulting in larger rise and roll values compared to cisplatin. Binding energies are also found to be greater for the larger drugs than for cisplatin, despite very similar properties of Pt—N7 bonds (probed using AIM methods) – an observation that was ascribed to more significant noncovalent interactions in the larger drugs, facilitated by the axis bend induced by Pt coordination. In a follow-up study, we also examined how the proposed anticancer complex kiteplatin, cis-$[PtCl_2$(cis-1,4-DACH)] (DACH = diaminocyclohexane), compares to the more established drugs [18]. These calculations indicate that kiteplatin behaves more like cisplatin than oxaliplatin, despite being an isomer of the latter.

## 9.5 Classical MM Description of Drugs and Their Interactions

While QM-based methods, such as DFT, offer the means to describe metals and their interaction with biomolecules from first principles, the computational overhead involved with many of these, even when applied only to a subset of the whole system, means that they can limit the length- or timescales that are realistically available for simulation. In many other areas of biomolecular simulation, fully classical MM is the method of choice, since this allows simulation of entire biomolecules over biologically relevant timescales of nano- or microseconds, often within a simulated environment that is made up of water, ions, and/or lipid bilayers. MD is widely used in this field, as the biological molecules and processes of interest are inherently dynamic in nature and require sampling over conformations that are accessible at body temperature. Enhanced sampling techniques, such as metadynamics, umbrella sampling, or accelerated MD, allow simulations to efficiently explore the free-energy landscape of flexible systems.

Despite their popularity in the wider biomolecular simulation field, classical MD studies are still relatively rare in the field of metal-based drugs. A study from 1999 by Scheeff et al. [19] recognized the need for, and potential of, such an approach, and aimed to examine the differences between cisplatin and oxaliplatin binding to DNA through classical MD. A prerequisite of this approach is a set of parameters that properly describes all atoms in the system: new atom types for Pt, coordinated ammine/amine groups, and for N7 in guanine coordinated to Pt were added to the standard AMBER set, with force constants for Pt—N bond stretching and N—Pt—N angle bending drawn by comparison with available X-ray crystallographic structures. Importantly, both proper and improper torsions for atoms bound to Pt were required to retain the expected square-planar geometry and prevent puckering of the bound structure. Partial charges for all groups bound to Pt were obtained by a combination of comparison to previous work [20] and analogy with established charges within the AMBER approach. Application of these parameters resulted in a root mean square deviation (RMSD) of 0.84–0.92 Å relative to the known crystal structure of cisplatin bound to a dodecamer DNA sequence, depending on the specifics of the charge scheme used. Having validated their parameterization, MD simulations were carried out *in vacuo* with the positions of phosphorus atoms at the 5′ and 3′ termini fixed to prevent chain unraveling. Importantly, given the poly-electrolyte nature of DNA, no cutoffs were applied in the calculation of non-bonded interactions. MD was run for 500 ps after 20 ps of equilibration, with a time step of 1 fs, and average structures extracted from snapshots taken at regular intervals throughout these trajectories.

After MD, the averaged cisplatin adduct structure had an RMSD of 1.37 Å from the crystal structure, and retained the expected square-planar coordination geometry around Pt. Differences from the crystal structure were noted in the hydrogen bond pattern: the coordinated ammine N—H are strong donors, and a persistent interaction with O6 of guanine 7 was noted throughout the simulation. A more expected Watson–Crick base pairing pattern for thymine 8 was also observed after the MD simulation, compared to the unusual form seen in the crystal structure. These observations lent further weight to the suitability of the MM parameters and MD protocol used, and also showed how simulations can complement experimental studies. Comparison with the equivalent simulation of oxaliplatin revealed broadly similar structures, with only subtle differences due to the extra bulk of the chair cyclohexane in the latter molecule. Hydrogen bonding patterns were one point of difference, with coordinated N—H now found to interact preferentially with O4 of thymine 8 rather than guanine 7. Oxaliplatin was also found to cause a small but significant narrowing of the major groove of DNA, whereas the minor grooves showed very similar properties. The authors discuss their findings in the light of proposed mechanisms of cisplatin resistance, noting that their findings are consistent with the alteration of DNA mismatch repair binding to cisplatin adducts. They also propose that explicit inclusion of aqueous solvent and counterions, and particle mesh Ewald (PME) treatment of electrostatics would be necessary for deeper insight: these techniques were new at the time of writing, but are now an essential part of MD simulations.

A more recent study took just this approach: Sharma et al. [21] reported MD studies designed to understand the differences in the DNA intrastrand cross-links formed by cisplatin and oxaliplatin. The authors point out that the various experimental reports of these structures are far from ideal for direct comparison due to differences in the techniques used (X-ray crystallography vs. solution state NMR), DNA sequence, and conditions used. In contrast, molecular simulation allows computational "experiments" under strictly comparable conditions. Starting from NMR structures of cisplatin and oxaliplatin complexes with the DNA sequence 5′-d(CCTCA<u>GG</u>CCTCC)-3′, as well as undamaged DNA with the same sequence in ideal B-DNA form, MD studies were carried out. Most parameters were taken from Scheeff's work (*vide supra*), but partial charges for Pt and coordinated N atoms were recalculated for all systems using the Mulliken scheme based on B3LYP calculations. Five simulations of 10 ns each were run from the different starting points, differing only in the initial velocities assigned after heating. Unlike previous work, no restraints on any atoms were applied, but 5′ and 3′ terminal base pairs were excluded from analysis, as is standard practice for MD studies of nucleic acids, since these bases undergo much larger fluctuations than the remainder of the sequence.

The resulting analysis showed that the final structural ensembles converged, whether from the same starting point with different initial velocities or different starting points, and that they did so within 4 ns of unrestrained MD. The converged structures agreed slightly better with NMR data than with crystallography, and largely reproduced inter-proton distances obtained from NMR. Principal component analysis indicates that key motions of the systems correspond to bending, twisting, and winding of the DNA double helix, and that these do not differ significantly between cisplatin and oxaliplatin. Both Pt-drugs are shown to disrupt the standard Watson–Crick hydrogen bonding pattern at platinated bases, and also to the 5′ side of this, while the 3′ base pairs remain essentially intact. Hydrogen bonding from Pt-coordinated N—H donors of cisplatin was observed to O6 of G7 and N7 of A5, which were present for 47 and 74% of the trajectory, respectively. This pattern is more complex for oxaliplatin due to the asymmetry of donors, altering the occupancy of the contacts noted above to 58 and 79%. Statistically significant differences in the frequency of DNA helical parameters were observed between free and platinated DNA, and also in the propeller, shift, and slide values of central GC base pairs between cisplatin and oxaliplatin. The authors use their findings to postulate that cisplatin binding leads to DNA spending more time in conformations that are likely to be recognized by mismatch repair proteins, whereas oxaliplatin coordination is more likely to be bypassed, hence accounting for the ability of oxaliplatin to circumvent resistance to chemotherapy with cisplatin.

One issue with the approach used in the studies discussed above is that the standard expression used to evaluate MM energies and forces is not well suited for the description of metal complexes. Energies of bond stretch and angle bend terms require reference parameters, typically consisting of harmonic force constants and equilibrium lengths or angles. Such references are not always simply defined: a homoleptic $ML_4$ square-planar complex would contain reference L—M—L angles of both 90° and 180°, while in heteroleptic complexes M—L lengths may differ depending on whether ligands are located *cis* or *trans* to one another. Such effects were circumvented in the above studies by assigning different atom types to chemically equivalent centers: for instance, ammine N on the 5′ and 3′ sides of coordinated cisplatin were assigned differently. While allowing regular MD studies to be performed, this approach can be impractical and limit the transferability of forcefields between chemically related systems. An alternative is to build in the electronic effects of *d*-orbitals on transition metal coordination into the standard MM expression. Ligand field molecular mechanics (LFMM) is one approach to doing so: here, an extra term for ligand field stabilization energy is added to the regular MM energy equation, thus capturing key *d*-orbital effects in a few transferable parameters [22]. This approach has been applied to many types of systems, including metallodrugs. Tai et al. [23] used it to examine the binding of cisplatin and a novel, photoactive complex based on *trans*-$[\mathrm{Pt(II)(pyridine)_2}]^{2+}$ to a DNA dodecamer. The suitability of LFMM for the description of square-planar Pt(II) was confirmed by comparing it to geometrical and vibrational data obtained from DFT calculations, as well as against geometries obtained from X-ray crystallography. Pt-DNA complexes were solvated within cuboidal boxes containing 3000–4000 water molecules and subjected to MD using a version of the DL-POLY package adapted for use with LFMM. These MD trajectories were found to converge within 2.5 ns, allowing the analysis of several further ns of equilibrated trajectory. The effects of intrastrand cisplatin coordination are shown to be similar to those discussed above and to experiment, leading to a bent double-helical structure with essentially intact Watson–Crick GC base pairing. 1,2-Interstrand coordination of cisplatin was also examined and found to yield a structure similar to the experiment, in which a cytosine residue is rotated out of the helical structure. Similar coordination of *trans*-$[\mathrm{Pt(II)(pyridine)_2}]^{2+}$ leads to a quite different structure, in which the double helix is preserved, but the GC pairing of coordinated guanines is completely destroyed, and significant propeller twist of nearby residues is induced, leading to a bend in the double helix of 67° and helical unwinding of 20°.

Apart from Pt-DNA complexes, our group has used the LFMM approach to model the interaction of Pt(II) with A$\beta$. The size and flexibility of this peptide mean that the speed of MM methods is required: QM or QM/M methods may be able to treat single, static structures but cannot easily model the conformational freedom of such peptides. To check the suitability of LFMM for the description of such systems, we first chose to study a subset of the whole peptide, including key metal binding residues: His6-Asp-Ser-Gly-Tyr-Glu-Val-His-His14, taken from the N-terminus of the whole peptide [24]. This was chosen to retain sufficient flexibility to make a realistic testbed for methods while remaining sufficiently small to carry out QM comparisons. This peptide was bound to Pt(II)L, where L = biphenyl or phenanthroline. Since His13 or His14 can bind metals through either N$\delta$ or N$\varepsilon$, this leads to eight possible coordination modes for each ligand. Each was built, and an exhaustive conformational search carried out using LFMM. Fortunately, parameters developed for Pt-guanine binding were transferable to this case, although atomic partial charges were recalculated for Pt, L, and coordinated His using the Merz–Kollman scheme, with all other atoms treated using AMBER94. Conformational searching resulted in several thousand possible structures for each ligand. Selected cases were fully reoptimized using DFT methods, and with the semi-empirical PM7 method. Comparison of resulting structures clearly showed that LFMM results in comparable geometries to the QM methods examined. However, only a rather weak correlation between LFMM and QM energies was observed, leading us to rely on PM7 data to identify specific low-energy complexes that, in the case of L = phenanthroline, agreed with experimental reports of the preferred coordination mode.

Having determined that LFMM properly describes Pt—A$\beta$ binding correctly, we applied this approach to carry out MD simulations of these flexible systems, concentrating on Pt(phen) bound to the 1–16 N-terminal fragment, which is known to bind metals such as Cu, Zn, and Fe, as well as the full 1–42 sequence. All simulations were performed in a reaction field model of implicit aqueous solvation, since we wanted to avoid the friction effects of explicit solvent [25]. Five simulations of 200 ns each starting from different geometries, taken from conformational searching, were carried out for the 1–16 fragment [26], which were found to reach equilibration after 25–50 ns. This is an order of magnitude greater than the times needed for the DNA studies discussed above, reflecting the less rigid nature of the short peptide compared to the double helix of DNA. Averaging data across equilibrated portions of the whole simulations, and comparing against equivalent trajectories for the metal-free peptide, allowed us to draw conclusions on the effect of Pt on peptide structure. In general, residues in the Pt—A$\beta$ complex are found to be more mobile than in the free peptide, with the exception of the histidines bound directly to the metal. Pt-binding does not drastically affect the secondary structure of the peptide, although a greater propensity for helical structures towards the C-terminus of the 1–16 fragment is observed. However, it does induce major differences in the pattern of salt bridges within the peptide, disrupting the main contacts in the free peptide and inducing new ones, most obviously between Glu11 and Lys16. Similar patterns are observed in the full 1–42 Pt—A$\beta$ complex [27], although here multiple simulations of length 1 μs each were required to reach pseudo-equilibration. Over the course of such a long timescale, Pt(phen) was found to induce rather flexible and extended structures, quite different from those induced by Cu(II) binding.

## 9.6 Docking and QSAR Studies

In the field of organic drug discovery and development, QSAR and molecular docking techniques are very widely used, often in early screens to find promising lead molecules that might then be developed into more drug-like compounds. These techniques have not yet had the same level of impact on metal-based drugs, probably because the software tools required do not automatically account for metal complexes. There are, however, some reports of these techniques in the literature. QSAR aims to relate the known biological activity of a set of compounds to measured or calculated molecular properties, and hence to predict those properties that affect the activity and to propose novel molecules with favorable properties. Sarmah and Deka [28] reported efforts to use molecular properties calculated by DFT to construct models of the cytotoxic activity of 24 Pt(II) and Pt(IV) complexes, including the widely used A2780 human ovarian adenocarcinoma cell line and its cisplatin-resistant subline. Linear relationships between cytotoxicity and properties such as LUMO and LUMO+1 energies and philicity indices were established, and the importance of including solvent effects in DFT calculations emphasized. Further models of the octanol/water partition coefficient were also explored, since this is a molecular property widely used in drug design as a surrogate for the ability to cross biological membranes. More recently, Varbanov et al. [29] showed that DFT-calculated properties of 53 platinum(IV) complexes can be used to model activity against two different cell lines (CH1 and SW480). Properties such as molecular size and weight, hydrogen bond donor and acceptor counts, and oxidation/reduction potentials were found to be important for predicting activity. The importance of different functional groups was also examined using Free–Wilson analysis, in which the independent variables within QSAR analysis are either 1 or 0, depending on whether a given group is present in the structure.

Molecular docking aims to locate the optimal way a potential drug molecule might interact with its biological target and often assumes that this interaction is purely noncovalent in nature. Coordination bonds formed between the metal and nucleophilic sites in a biomolecule are not typically well described by this approach, and would be better treated using QM, QM/MM, or MD methods described above. One recent study used standard docking methods to estimate how an azido-bridged dinuclear copper(II) thiosemicarbazide complex might interact with DNA as well as human and bovine serum albumin (HSA and BSA) [30]. They used the CDOCKER algorithm, part of the popular Discovery Studio package, to find 10 possible docking poses for each target, then energy-minimized the resulting complexes. All interactions within and between molecules were treated using the CHARMM force field. Given the complexity of the molecule and the metal–ligand and metal–metal interactions in the target molecule, it is not clear how well CHARMM models these effects, as it was originally designed to treat proteins and nucleic acids only. On this basis, the dinuclear copper complex is predicted to intercalate into DNA and form weaker noncovalent complexes with HSA and BSA. Similarly, Zhao et al. [31] used the same CDOCKER/CHARMM approach to model the interaction of a platinum pyrithione complex with two deubiquitinating enzymes (DUBs). As is standard practice, the geometry of the receptors was obtained from the PDB and kept rigid at the experimental geometry, while the Pt complex was allowed torsional flexibility. This suggested that the complexes of interest were able to interact with the catalytic cores of the DUBs, with particularly strong interactions between Pt and the S of a key cysteine residue.

One of the major complications of the application of docking to metallodrugs is the fact that most such drugs are expected to form coordination bonds with nucleophilic sites within their biomolecular targets. This is different from the typical case for organic drugs, which usually bind to their receptors through noncovalent interactions. Recent work has shown how conventional docking algorithms can be adapted for coordinative binding. Initial application to $V^{IV}O$ complexes binding to lysozyme used the GOLD software, and described metal–ligand bonds using a modification of the potential used for hydrogen bonding [32], using a dummy hydrogen atom to "activate" the metal–protein interaction. The approach was validated against known structures of V bound to chymotrypsin A and several phosphatases: redocking of metal complexes to these receptors resulted in RMSD values between 0.2 and 1.0 Å, and favorable values of the GoldScore scoring function used to rank possible binding modes. The docking algorithm was then used to predict the binding of different stereoisomers of *cis*-$[VO(pic)_2(H_2O)]$ to hen egg white lysozyme, indicating that binding through Asp52 or Glu35 is favored, but not through His15 that had been proposed previously, in agreement with EPR spectral data. The preference for the OC-6–23-Δ isomer binding was further supported by QM/MM calculations and NCIPlot analysis of the resulting electron density. A subsequent report from the same group confirmed that this approach is applicable to a much wider range of metals, coordination geometries, and ligand types. Thirty-nine known complexes containing 15 different metals were selected from the PDB, resulting in a mean RMSD of less than 1 Å, with over 90% of simulated poses having RMSD less than the experimental resolution [33].

## 9.7 Conclusion

We have set out some of the ways that computational chemistry and molecular simulation have been used in the field of metallodrugs. The majority of these studies use QM methods, especially DFT, to explicitly calculate the electronic structure of relevant metal–ligand complexes and their interaction with biomolecules. Popular methods, such as B3LYP, potentially with correction for dispersion effects, are widely used and successfully capture key aspects of metallodrug properties. Care must be taken to account for relativistic effects, especially for heavier elements such as Pt: this is typically performed with ECPs that replace inner electrons.

The computational overhead associated with DFT methods means that interactions with biomolecules must be modeled using relatively small parts of the whole nucleic acid or protein of interest. Hybrid QM/MM methods are therefore of interest, since they retain the accuracy of QM methods for the metal and coordinated groups but reduce the resource requirement for more distant groups. Purely MM description of metal-based drugs bound to biomolecules are also popular, but here specific parameterization of metal and ligands is required. This has been achieved for Pt(II) and the atoms coordinated to it in drugs, such as cisplatin, allowing a detailed study of the dynamics of their interactions and their effects on biomolecular structure.

Application of more empirical methods, such as QSAR and docking are less common than in the modeling of more conventional organic drugs. This may partly be due to lack of capability in software packages to deal with the challenges of metal complexes, or may stem from the specifics of metallodrug chemistry, particularly the tendency to form coordination bonds with biomolecules rather than interacting solely through noncovalent interactions as many organic drugs do. Recent applications of both methods to different classes of metallodrugs have been published, demonstrating how these challenges may be met through proper understanding of the theoretical and chemical nature of the difference between organic and metallodrugs. In our view, the future of such virtual screening of metal-based therapeutics is highly promising, and we expect an increase in published studies in this field in the coming years.

## References


1 Rehder, D. *Bioinorganic Chemistry*. Oxford: Oxford University Press; 2014.

2 Wysokiński, R. and Michalska, D. The performance of different density functional methods in the calculation of molecular structures and vibrational spectra of platinum (II) antitumor drugs: cisplatin and carboplatin. *J. Comput. Chem.* 2001;*22*:901–912.

3 Zhang, Y., Guo, Z., and You, X.-Z. Hydrolysis theory for cisplatin and its analogues based on density functional studies. *J. Am. Chem. Soc.* 2001;*123*:9378–9387.

4 Burda, J.V., Šponer, J., and Leszczynski, J. The interactions of square platinum (II) complexes with guanine and adenine: a quantum-chemical ab initio study of metalated tautomeric forms. *JBIC J. Biol. Inorg. Chem.* 2000;*5*:178–188.

5 Baik, M.-H., Friesner, R.A., and Lippard, S.J. Theoretical study of cisplatin binding to purine bases: why does cisplatin prefer guanine over adenine? *J. Am. Chem. Soc.* 2003;*125*:14082–14092.

6 Michera, L., Nekadová, M., and Burda, J.V. Reactions of cisplatin and glycine in solution with constant pH: a computational study. *Phys. Chem. Chem. Phys.* 2012;*14*: 12571.

7 Zimmermann, T., Chval, Z., and Burda, J.V. Cisplatin interaction with cysteine and methionine in aqueous solution: computational DFT/PCM study. *J. Phys. Chem. B* 2009;*113*:3139–3150.

8 Streltsov, V.A., Chandana Epa, V., James, S.A. et al. Structural insights into the interaction of platinum-based inhibitors with the Alzheimer's disease amyloid-β peptide. *Chem. Commun.* 2013;*49*: 11364.

9 Novato, W.T.G., Stroppa, P.H.F., Da Silva, A.D. et al. Reaction between the Pt(II)-complexes and the amino acids of the β-amyloid peptide. *Chem. Phys. Lett.* 2017;*667*:4–8.

**10** Gkionis, K., Platts, J.A., and Hill, J.G. Insights into DNA binding of ruthenium arene complexes: role of hydrogen bonding and π stacking. *Inorg. Chem.* 2008;*47*:3893–3902.

**11** Grimme, S., Ehrlich, S., and Goerigk, L. Effect of the damping function in dispersion corrected density functional theory. *J. Comput. Chem.* 2011; *32*:1456–1465.

**12** Cuevas-Flores, M. del R., Garcia-Revilla, M.A. et al. Noncovalent interactions between cisplatin and graphene prototypes. *J. Comput. Chem.* 2018;*39*:71–80.

**13** Bader, R.F.W. *Atoms in Molecules: a Quantum Theory*. Oxford [England], New York: Clarendon Press, Oxford University Press; 1994.

**14** Robertazzi, A. and Platts, J.A. Hydrogen bonding and covalent effects in binding of cisplatin to purine bases: ab initio and atoms in molecules studies. *Inorg. Chem.* 2005;*44*:267–274.

**15** Tai, T.B. and Nhat, P.V. A DFT investigation on interactions between asymmetric derivatives of cisplatin and nucleobase guanine. *Chem. Phys. Lett.* 2017;*680*:44–50.

**16** Spiegel, K., Rothlisberger, U., and Carloni, P. Cisplatin binding to DNA oligomers from hybrid car-parrinello/molecular dynamics simulations. *J. Phys. Chem. B* 2004;*108*:2699–2707.

**17** Gkionis, K., Mutter, S.T., and Platts, J.A. QM/MM description of platinum–DNA interactions: comparison of binding and DNA distortion of five drugs. *RSC Adv.* 2013;*3*: 4066.

**18** Mutter, S.T., Margiotta, N., Papadia, P. et al. Computational evidence for structural consequences of kiteplatin damage on DNA. *JBIC J. Biol. Inorg. Chem.* 2015; *20*:35–48.

**19** Scheeff, E.D., Briggs, J.M., and Howell, S.B. Molecular modeling of the intrastrand guanine-guanine DNA adducts produced by cisplatin and oxaliplatin. *Mol. Pharmacol.* 1999;*56*:633–643.

**20** Yao, S., Plastaras, J.P., and Marzilli, L.G. A molecular mechanics AMBER-type force field for modeling platinum complexes of guanine derivatives. *Inorg. Chem.* 1994;*33*:6061–6077.

**21** Sharma, S., Gong, P., Temple, B. et al. Molecular dynamic simulations of cisplatin- and oxaliplatin-d (GG) intrastand cross-links reveal differences in their conformational dynamics. *J. Mol. Biol.* 2007;*373*:1123–1140.

**22** Deeth, R.J., Anastasi, A., Diedrich, C. et al. Molecular modelling for transition metal complexes: dealing with d-electron effects. *Coord. Chem. Rev.* 2009; *253*:795–816.

**23** Tai, H.-C., Brodbeck, R., Kasparkova, J. et al. Combined theoretical and computational study of interstrand DNA guanine–guanine cross-linking by *trans* -[Pt(pyridine)$_2$ ] derived from the photoactivated prodrug *trans,trans,trans* -[Pt(N$_3$ )$_2$ (OH)$_2$ (pyridine)$_2$ ]. *Inorg. Chem.* 2012;*51*:6830–6841.

**24** Turner, M., Platts, J.A., and Deeth, R.J. Modeling of platinum-aryl interaction with amyloid-β peptide. *J. Chem. Theory Comput.* 2016;*12*:1385–1392.

**25** Shen, M. and Freed, K.F., Long time dynamics of met-enkephalin: comparison of explicit and implicit solvent models. *Biophys. J.* 2002;*82*:1791–1808.

**26** Turner, M., Mutter, S.T., Deeth, R.J. et al. Ligand field molecular dynamics simulation of Pt(II)-phenanthroline binding to N-terminal fragment of amyloid-β peptide. Lakshmana MK, editor. *PLoS One* 2018;*13*:e0193668.

**27** Mutter, S.T., Turner, M., Deeth, R.J. et al. Metal binding to amyloid-$\beta_{1-42}$: a ligand field molecular dynamics study. *ACS Chem. Neurosci.* 2018;*9*:2795–2806.

**28** Sarmah, P. and Deka, R.C. DFT-based QSAR and QSPR models of several cis-platinum complexes: solvent effect. *J. Comput. Aided. Mol. Des.* 2009; *23*:343–354.

**29** Varbanov, H.P., Jakupec, M.A., Roller, A. et al. Theoretical investigations and density functional theory based quantitative structure–activity relationships model for novel cytotoxic platinum(IV) complexes. *J. Med. Chem.* 2013; *56*:330–344.

**30** Biswas, N., Khanra, S., Sarkar, A. et al. One new azido bridged dinuclear copper ( II ) thiosemicarbazide complex: synthesis, DNA/protein binding, molecular docking study and cytotoxicity activity. *New J. Chem.* 2017;*41*:12996–13011.

**31** Zhao, C., Chen, X., Zang, D. et al. Platinum-containing compound platinum pyrithione is stronger and safer than cisplatin in cancer therapy. *Biochem. Pharmacol.* 2016;*116*:22–38.

**32** Sciortino, G., Sanna, D., Ugone, V. et al. Elucidation of binding site and chiral specificity of oxidovanadium drugs with lysozyme through theoretical calculations. *Inorg. Chem.* 2017;*56*:12938–12951.

**33** Sciortino, G., Rodríguez-Guerra Pedregal, J., Lledós, A. et al. Prediction of the interaction of metallic moieties with proteins: an update for protein-ligand docking techniques. *J. Comput. Chem.* 2018;*39*:42–51.

# 10

# Computational Design of Artificial Metalloenzymes

*Laura Tiessler-Sala, Maria Fatima Lucas, and Emanuele Monza*

*Zymvol Biomodeling SL, Barcelona, Spain*

## 10.1 Introduction

Enzymes present significant advantages compared to organometallic catalysts due to their well-defined structures around the reactants, which can boost specificity, selectivity, and turnover frequencies (TOFs). While industrial chemical catalysts have TOFs in the $10^{-2}$–$10^{2}$ $s^{-1}$ range [1], certain enzymes work in a diffusion-controlled regime, achieving ranges of $10^{5}$–$10^{6}$ $s^{-1}$ [2]. Unlike organometallic catalysts, which mostly work in organic solvents and are modified chemically, enzymes are soluble in water, highly specific and selective, optimized genetically, biodegradable, and do not produce toxic waste.

These characteristics make enzymes ideal templates for green chemistry. However, natural enzymes need to be redesigned for a number of reasons: (i) *conditions in a reactor can significantly differ from nature*. Enzymes are sensitive to temperature, pH, and pressure variations, and industrial processes often require conditions that are more extreme than what these enzymes can naturally tolerate; (ii) *a target reaction might lack a natural enzyme*. Often, the reactions sought for industrial processes do not exist in nature, and enzymes need to be modified to accommodate non-natural substrates; (iii) *nature of the solvent*. Industrial reactions frequently occur in organic solvents or other nonaqueous media where natural enzymes may have limited solubility or activity; (iv) *reaction kinetics might be unsatisfying*. The last point is quite an issue considering that the rate constant ($k_{cat}$) of the average enzyme is around 10 $s^{-1}$ [2], stating clearly that these biocatalysts evolved to just accomplish their natural task fairly, leaving room for improvement. This is achieved experimentally with rational design and directed evolution.

In rational design, site-directed mutations are generally introduced based on structural and mechanistic knowledge of the enzyme. The main limitation of this approach is the difficulty of predicting the effect of a point mutation, even if the system is very well known. Opposite to rational design, directed evolution randomly introduces mutations in an enzyme, not requiring a thorough understanding of the

*Computational Bioinorganics: From Description to Prediction*, First Edition. Edited by Jean-Didier Maréchal.

system [3]. Although directed evolution is a revolutionary technique (potential discovery of unintuitive beneficial mutations) and has led to significant breakthroughs, its random nature and the limited sequence space sampling (no more than $10^6$ variants) keep the rate of beneficial mutations low. A compromise is a semi-rational design that focuses mutagenesis to regions where, according to structural and/or computational analysis, the chances of discovering beneficial mutations are higher [4]. Computational enzyme design is an appealing alternative because most sequence space screening can be performed using *in silico* sequence- and structure-based simulations. This approach dramatically reduces the number of candidates for experimental validation [5, 6]. Previously, computational design relied heavily on the quality and availability of the structural data. However, recent machine-learning advances have made it possible to obtain accurate and reliable protein models [7].

Notwithstanding all the advantages of biocatalysis, organometallic catalysis offers a wider, potentially infinite, reaction toolkit. Artificial Metalloenzymes (ArMs) combine the best of both worlds, expanding the reactivity space of biocatalysts through organometallic moieties while unlocking great potential for high specificity, selectivity, and TOFs. In ArMs, the catalytic reaction is controlled by the first-coordination sphere of the metal, which refers to the ligands or residues directly bonded to the metal. Meanwhile, all other variables regarding activity, specificity, and selectivity are controlled by the second sphere of coordination provided by the protein, including all residues interacting with the first-coordination sphere.

Although both directed evolution and rational designs have led to several achievements in recent years [8–16], computer simulations promise to boost the development of ArM by sensibly reducing experimental screening, therefore saving time and resources. Indeed, the impact of computer-driven design is increasing, and protocols are becoming more robust [17–20]. In this chapter, the principles of computational ArM design are explored, with a focus on hurdles and frontiers rather than available software and methodologies, which has been covered in many reviews [17, 19] and are extensively presented in other chapters of this book.

## 10.2 Typical Computational Pipeline

A typical computational chemogenetic optimization pipeline is depicted in Figure 10.1. Chemogenetic means that both the metallic catalyst and the protein host are optimized through chemical and genetic methods, respectively. In this process, the catalytic reaction of the ArM can be tuned by altering the metallic cofactor nature, while residues from the second sphere of coordination can be mutated genetically to improve the selectivity or enantioselectivity [21].

The first step toward a novel ArM is the choice of an appropriate homogeneous catalyst for the target reaction. The catalyst-reactant potential energy surface is explored with quantum chemical methods [22]. The reaction path of the initial catalyst-reactant system or theozyme (catalyst-reactant and key amino acids for reaction) is calculated with quantum mechanics (QM) in water solvent. In particular, it is

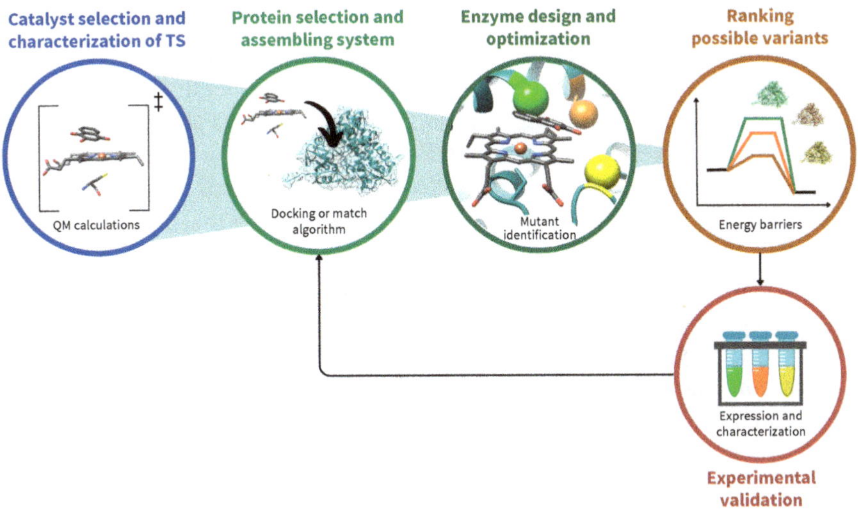


**Figure 10.1** ArM chemogenetic computer-lab iterative optimization pipeline.

important to identify the transition state (TS) of the rate-determining step of the reaction. Afterward, the ligands of the metal atom(s) can be altered to tune the catalytic efficiency and selectivity of the desired reaction. Nowadays, this can be performed with automatic pipelines that modify the ligand at the stationary points, which are refined accordingly [23].

Next, the three-dimensional model of the most relevant TS or other intermediates is introduced to potential protein scaffolds, searching for one that is more capable of hosting and stabilizing it [17]. At this stage, metal prediction software, such as BioMetAll, metalNet, or Metal3, can help identify possible metal binding sites within these scaffolds [24–26]. The TS or intermediate can be placed into the protein with plain protein–ligand docking [19, 27] or using matching algorithms, such as Rosetta [28]. The second method has the advantage that a theozyme or another minimal active site model [29] can be docked in the scaffold, explicitly taking into account the effect of potential coordinating amino acids instead of the plain catalyst. The protein host could be either a natural protein or one created *de novo* for the task at hand. In the case of *de novo* design, recent advances in deep learning (DL) are transforming traditional approaches in the field, as will be described in detail later.

The generated ArM scaffold is then optimized with computational enzyme design techniques, usually improving the interaction energy between the host protein and the reactive moiety (catalyst plus reactant, often in a TS configuration) through virtual mutagenesis [30]. Importantly, not only the first-coordination sphere but also the second shell around the reactive moiety must be optimized, maximizing structural and electrostatic preorganization [31]. The virtual variants are then reevaluated with more complex techniques, like molecular dynamics (MD) and QM-molecular mechanics (QM-MM) schemes [32]. The idea behind this screening is to validate the design by using orthogonal concepts among them and using the computational design strategy itself.

For instance, MD simulations are used to assess the dynamics of catalytic contacts in the active site of directed evolution variants [30]. Remarkable success in tuning selectivity has been achieved through MD simulations by identifying crucial positions [8, 33, 34]. Generally speaking, running multiple replicas provides better sampling and accuracy of results. In enzyme reactions, studies have revealed that high-throughput MD (HTMD) is more effective than traditional MD for near attack conformations (NAC) sampling, which are conformations that resemble the TS [35, 36]. HTMD consists of running many short replicas instead of a few longer runs, thereby accumulating statistics on NAC frequency. Additionally, MD simulations can also shed light on the importance of protein dynamics and how these can affect the activity of the ArM [37]. Nevertheless, modeling metallic centers with classical force fields is not always straightforward; depending on how exotic the metal center may be, this approach becomes less feasible.

On the other hand, QM-MM is used to model possible changes in the electronic and coordination states of the catalyst [19], and to approximate reaction barriers within the enzyme. For example, once the reaction coordinate is determined for the wild type, usually by constraining the reactive coordinate in a geometry optimization run, mutations can then be introduced in the stationary points to approximate the reaction barrier [38, 39]. The most promising mutants could then be rescored with a higher-level QM-MM calculation along the reaction coordinate to estimate the energy barrier [40]. The main drawback of this methodology is its high computational cost, especially for high-throughput screening of virtual variants. Once the final list of selected enzymes has been identified, the final candidates are expressed in the lab and characterized. The experimental validation can then be used to refine the predictions and design improved virtual enzymes in an iterative computer-lab feedback loop [41].

## 10.3 Critical Aspects and Frontiers of ArM Design

### 10.3.1 Challenges in Modeling Metallic Centers

MD simulations greatly depend on the accuracy of the force field in describing the parameters of the biomolecules. While organic molecules are well represented in force fields available in biomolecular simulation software, most standard force fields lack robust metal parameters. Although dedicated force fields have been developed [42, 43], they are usually incompatible with protein-focused force fields. As a result, metals and nonstandard organometallic cofactors usually lack reliable parameters. The parameterization of metals is challenging due to their unique characteristics, like their capacity to change their coordination number during binding processes [44]. The lack of experimental data further complicates the development of reliable parameters. Some progress has been reported regarding machine-learning methods, but their application is still limited [45, 46]. These factors pose some limitations and challenges in running classical simulations with metals.

There are different models to represent metals in MM methods, and their applicability depends on the system's characteristics. However, each approach has disadvantages that need to be considered. Although detailed in other chapters of this book, here we briefly summarize the most significant aspects. Nonbonded models account for ligand exchange or geometry changes, but they may lead to lower accuracy, especially for transition metals, making them less ideal for ArM design. Bonded models are effective when the ligand exchange process does not occur at the computational timescale. However, they require careful parameterization derived from a combination of experimental data and QM calculations [44]. This type of parameterization can now be sped up with scripts like *MCPB.py* [47], but it still requires expertise, has low transferability, and is time-consuming due to the requirement of QM methods. In practice, instead of using bonded parameters, sometimes fixed geometries obtained from QM or QM-MM geometry optimization are used to represent metal–ligand interactions. However, this approach can misrepresent the conformational ensemble of the catalyst within the protein. This highlights the importance of implementing QM conformational analysis and using appropriate bonded parameterization protocols that consider the flexibility of the catalyst. Furthermore, selecting the best starting points for calculating reaction pathways is also crucial to ensure that the TSs and intermediates are properly docked in the protein host and modeled.

MD simulations provide valuable insights for ArM design, but more precise methods, such as QM/MM, are crucial to accurately capture the electronic properties of metals. However, the main limitation of QM/MM methods is their slow speed in screening hundreds of variants in a reasonable time. To overcome this, shortcuts that limit computational time without jeopardizing accuracy and predictive reliability are needed. While machine learning has accelerated QM/MM calculations [48–50], no application to metalloenzymes has been reported yet. An approach suited for metalloenzymes is QM-discrete MD (QM-DMD), which uses a simplified version of MD [39]. DMD [51] approximates the dynamics as purely collision-based by using discrete (square-well) potential energy functions. As a result, fewer calculations are performed, allowing DMD to access much longer timescales although some accuracy is sacrificed. In QM-DMD, the metal center is periodically modeled (with geometry optimization) using QM calculations and is kept frozen during the DMD phase. In such a way, the mutual catalyst-protein influence is taken into account without the need for extensive parameterization and at a fraction of the cost of standard QM/MM calculations [52]. This methodology has been successfully applied for the study of the catalytic mechanism of $Co^{2+}$ acireductone dioxygenase [53] and the roles of each metal in Cu,Zn-superoxide dismutase [54].

Another promising methodology is the metal molecular fractionation with conjugate caps (metal-MFCC) approach, which divides the metalloprotein into fragments. This allows for the calculation of potential energies and atomic forces as the linear combination of fragment properties. This approach could potentially be used to perform *ab initio* MDs; its high computational cost has limited its application to structural optimization of small zinc-finger metalloproteins [55]. Other fragmentation schemes, such as fragment combination ranges, have also been recently applied to the study of metalloenzymes to investigate substrate-protein interactions [56]. Recently, multiscale quantum refinement methods combining experimental data (X-ray diffraction) and several computational schemes have been applied to metalloproteins and metallopeptides. Specifically, it was found that using ONIOM [57] with QM/SE/MM approaches presented good results with low computational costs, as the metal binding site is described with high accuracy by QM methods and the remaining structures/interactions with low-level methods (SE or MM) [58].

To recapitulate, quantum calculations present a significant challenge when dealing with ArMs. For MD simulations, precise parameterization is required, relying on time-consuming QM-deriving methods. Streamlining the parameterization or integrating machine-learning approaches could reduce time and improve accuracy. Furthermore, when multiple QM/MM calculations are required, other low-accuracy approaches need to be considered, but there is still a need to benchmark these approaches and establish specific guidelines for ArMs.

### 10.3.2 Unnatural Amino Acids Can Unlock New Reactivity

The last decade has seen the successful adoption of unnatural amino acids to incorporate metals in protein cavities, yielding new reactivity [59, 60]. These unnatural amino acids contain catalytic moieties in their side chains that allow expanding the range of reactions. The potential of incorporating unnatural amino acids in ArM is very high and, in combination with computational design, could yield a broad range of reactions; however, due to the complexity involved, only a few computational efforts have been reported using (2, 2′-dipyridin-5yl)alanine (Bpy-Ala) (Figure 10.2).

Mills et al. were able to capture divalent cations with the artificial amino acid Bpy-Ala through computational design, specifically following a theozyme approach [61]. The initial theozyme attempt included the substrate, the unnatural amino acid, and two ligation spots that could be filled by either histidine (His) or tyrosine (Tyr) (Figure 10.3a). The theozyme was optimized computationally by QM methods and introduced to a protein scaffold with RosettaMatch [62], in which Bpy-Ala was treated as a regular amino acid. The obtained matches were redesigned with RosettaDesign [63], introducing specific interactions to stabilize the theozyme. The virtual enzymes were then filtered based on geometric constraint satisfaction.

Nonetheless, X-ray diffraction proved that the loop containing Bpy-Ala was greatly displaced from the design position, making the designed variant ineffective. As the

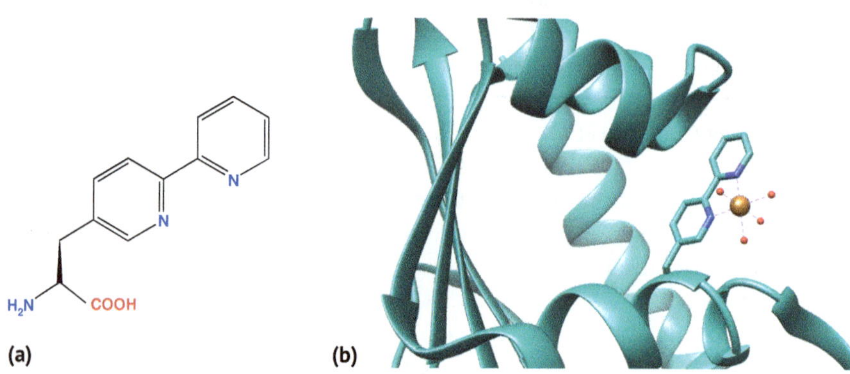


**Figure 10.2** (a) Structure unnatural amino acid (2, 2′-dipyridin-5yl)alanine (Bpy-Ala). (b) Structure of human sterol carrier protein incorporating unnatural amino acid Bpy-Ala coordinating to Cu (PDB = 8AF2).

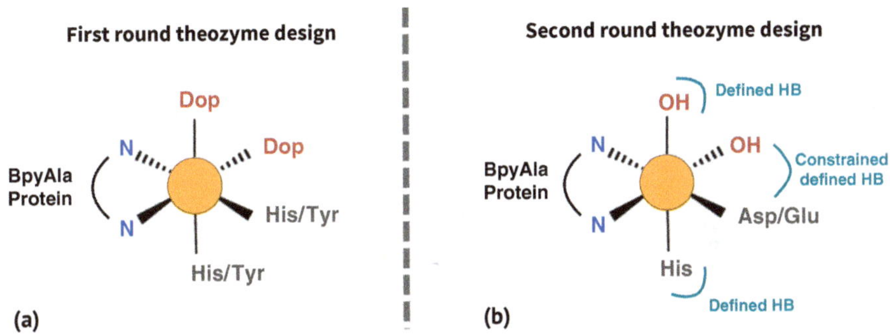


**Figure 10.3** Theozyme model and criteria utilized for (a) first and (b) second round of computational design of ArM based on Bpy-Ala. *Source*: Adapted from [61].

author underlines in the paper, it was not checked (computationally) whether the metal ligands were properly arranged prior to the incorporation of the metal. Additionally, the orientation of the free substrate within the active site was not simulated.

A second round of design was carried out, with many variations in the theozyme and in the RosettaMatch search criteria (Figure 10.3b): (i) the system was simplified replacing the substrate with two water molecules; (ii) Bpy-Ala was forced to be part of stable secondary structure elements; (iii) aspartic (Asp) and glutamic (Glu) acid were included as ligating residues in addition to His, aiming to increase the geometric agreement between the best matches and the theozymes; (iv) Tyr was eliminated, as this residue is often observed in trivalent cation-binding sites; (v) Asp and Glu were placed so as to make hydrogen bonds to the bound water ligands; (vi) the metal-binding His and bound water molecules were required to form at least one hydrogen bond with the remainder of the protein. In total, 15 independent theozyme models were input in RosettaMatch, mimicking alternative coordination motifs. These modifications in the protocol resulted in variants with the desired geometry, as shown by X-ray diffraction, with structures bound to divalent metals with only small deviations from the designed models. Due to the high affinity obtained for $Zn^{2+}$, this was a starting point for future hydrolases and other related enzymes.

Bpy-Ala is also the unnatural amino acid used in a completely different computer-based approach to the design of an enantioselective artificial metallohydratase enzyme [64]. To achieve the target reaction, enantioselective hydration of an enone, an integrative computational workflow was applied (Figure 10.4a). The authors tackled the problem with QM cluster calculations (Figure 10.4b) to realize that the interaction of the enone carbonyl group with a metal and of the nucleophilic water with a general base yielded more favorable energy barriers. Then, they characterized the substrate–metal–protein interactions with protein–ligand dockings. Finally, they used MD to quantify the structural integrity of reactive conformations in the active site.

This protocol led to the following results: (i) the native amino acid base that was supposed to activate water was not optimally positioned; (ii) the active site was allowing many unproductive conformations; (iii) water accessibility was not symmetric on the two-substrate prochiral faces, leading to selectivity. With this information, the authors

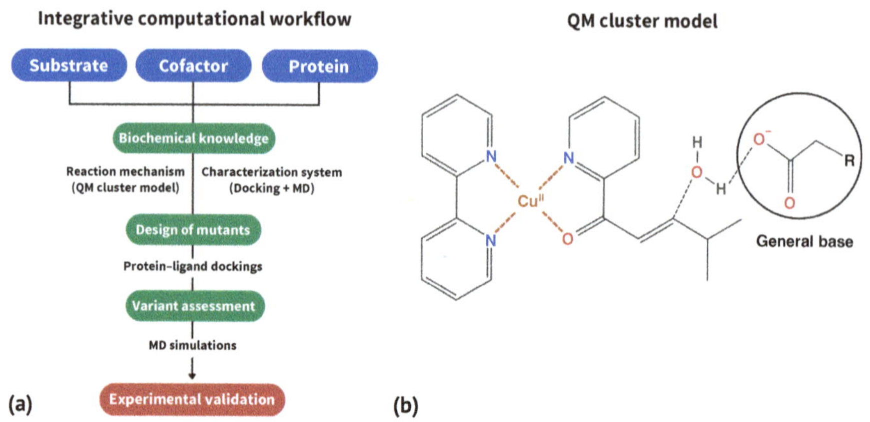


**Figure 10.4** (a) Integrative computational workflow and (b) QM cluster model followed for enantioselective artificial metallo hydratase. *Source*: Adapted from [64].

redesigned the active site, searching for a better positioning of the catalytic base. Docking and MD were used to test a few mutants, in a rational fashion, leading to a satisfactory activity and enantioselectivity. In particular, MD scored both active site integrity and enantiomeric excess. It is worth noting that although one single 100 ns MD was utilized to score mutants, a HTMD scheme can be envisioned for brute force computational mutagenesis, switching from rational design to computational evolution.

The take-home messages of these works are: (i) the need to ensure a high degree of preorganization of the first metal coordination sphere by optimizing second sphere packing and/or hydrogen bond pattern; (ii) the importance of deeply studying the wild type system to highlight system flaws to be smoothened during design; (iii) that it is nowadays possible to introduce unnatural amino acid capable of coordinating metals through computer simulations.

### 10.3.3 Beyond TS Stabilization

Traditionally, the primary factor considered in computational ArM design has been on TS stabilization [30]. However, over the last decade, it has been acknowledged that other factors should also be considered, such as noncatalytic residues, distal mutations, substrate entrance, and protein dynamics.

Early work by Siegel et al. highlighted this shift. They expressed and characterized 100 mutants of a glycosyl hydrolase, and then they used computational design algorithms to calculate several features, including the theozyme-protein interaction energy [65]. Finally, they used simple machine-learning algorithms to single out which features were informative of the kinetic constants. Three key findings arose from such analysis: (i) no feature strongly correlates with the experimental data; (ii) the machine-learning algorithm selected some features and discarded others to decently reproduce the results; (iii) the theozyme–protein interaction energy was not selected by the algorithm. These findings, although limited to a single enzyme, suggest that enzymatic function is shaped by many different molecular properties and that TS stabilization is only one of them. Furthermore, they highlight the importance of the entire active site in catalysis.

Most current efforts have relied on computational rational design (dockings and MD simulations) or directed evolution to identify mutations in the active site [66]. However, methods have also been developed to automatically generate more efficient enzymes through mutation in active sites [67]. Machine learning is now being used to complement experimental or computational approaches, accelerating these processes [68, 69]. Despite these advances, all of these new methodologies have yet to be applied in the ArM field. So far, only Vornholt et al. [70, 71] have reported the use of machine learning to enhance sequence-activity mapping in ArM.

Dynamical effects are often underrepresented in computational design pipelines. A few studies have used MD simulations to successfully characterize long-range mutation effects on active site catalysis [72–76]. Silvia Osuna's group developed the tool Shortest Path Map (SPM), which elucidates the underlying allosteric network of interactions through extended MD simulations and identifies distal positions for the first time [77].

Despite these successes, they all require long and computationally expensive MD simulations, and no reliable tool has been available for fast detection up until recently. Studies using Zymspot, a software based only on bioinformatics and structure-based approaches, have recently demonstrated its ability to identify distal mutations that improve the activity or stability of both artificial and natural enzymes [78, 79]. Moreover, the impact of distal mutations can now be predicted with machine-learning algorithms, such as EnzyACT [80].

Enzyme efficiency is also influenced by how the substrate (and the metal cofactor) accesses the active site or how the global protein dynamics accompany the catalysis. These two factors can sometimes be the key to designing selective ArMs, with computational approaches providing valuable insights. Although a few computer-driven enzyme designs focused on substrate entrance have been successfully carried out in recent times [14, 81–83], this is far from being the rule. For example, MD simulations have shown that a tightly bound heme in the protein LmrR can catalyze cyclopropanation due to the flexibility of its helices, which open up to facilitate substrate entry and reach the precatalytic stage [14]. In another approach involving an ArM based on POP-$Rh_2$, the inclusion of substrate pathway calculations as a last step of the multiscale workflow revealed how the entrance and diffusion of the substrates can also be the origin of the ArM's enantioselectivity [83].

The previous studies highlight the importance of taking into account possible conformational changes. Enhanced sampling methods could be used to overcome the transitions between relevant states [72]. For example, replica-exchange methods and accelerated MD do not rely on collective variables, allowing them to extensively sample the conformational space with less time and even predict kinetic parameters [84]. These technologies have yet to be explored in ArM design, but they could be incorporated when the system's dynamics may play a crucial role.

### 10.3.4 The New Era: *De Novo* Protein Design with DL

In the *de novo* computational design of ArM, one of the biggest challenges lies in the fact that the designed protein should fold into a specific 3D structure capable of accommodating a metallic cofactor. While computational *de novo* ArMs have been reported over the years, so far most computational ArMs have been designed by repurposing natural protein scaffolds.

*De novo* designs are often guided by known structural or sequence motifs, relying on simple scaffolds, like 4-helical bundles or TIM barrels. Rosetta [28] is one of the main software tools used for *de novo* design, relying on force fields to identify structures with low-energy conformations. Examples of successful computationally designed *de novo* ArMs include 4-helix bundles that mimic natural Due Ferri proteins, tuned to perform oxidase, oxygenase, or ferroxidase activities [85]; heme binding protein (dnHEM1) with peroxidase and carbene transferase activity [86]; and TIM barrels engineered for cerium photoredox catalysis [87].

The main problem with these scaffolds is their limited diversity, they rely on simple and symmetric structural motifs, which do not provide enough structural variability to support all types of reactions. Therefore, there is a need to expand to more complex scaffolds. This challenge can now be overcome with the recent introduction of DL techniques, like AlphaFold [7] or RoseTTAFold [88], which are able to predict protein structures from sequences with high accuracy. These advancements have opened the field for the development of *de novo* software, which, in the future, may enable creating novel structures with completely different folds or modification of natural structures to incorporate metal catalysts.

The first steps toward the design of metalloenzymes using DL techniques involved inpainting (filling in missing parts of protein) and hallucination (generating structure from scratch) methods, both derived from RoseTTAFold. These approaches have been applied to successfully design a luciferase [89], but early attempts included work with metal moieties. The first successful metal-binding proteins were generated with *RFjoint* [90], which generates sequences and structures to create a desired structural motif, using as input a partial structure or sequence. Successful designs of metal-binding proteins were reported, such as di-iron proteins or calcium-binding EF-hand motifs, and the active site of carbonic anhydrase II. These results demonstrated the feasibility to create metalloenzymes, but these methods lacked precision, and the number of possible outputs was limited. This laid the groundwork for more advanced tools like RFDiffusion [91], which is constructed upon RoseTTAFold architecture, but uses diffusion models.

In RFDiffusion, protein backbones are generated from scratch through iterative denoising processes, either unconditionally or by imposing conditions. This program can be used in combination with ProteinMPNN [92], which generates sequences that will fold into a specific given backbone, whether from native proteins or designed proteins from RFDiffusion. The combination of both tools can be used to generate a vast pool of structures, which can be screened against a metallic catalyst of interest, but can also be used to improve the characteristics of an ArM. A clear example of its potential is the combination of *RFjoint* and ProteinMPNN that resulted in designing a myoglobin structure with increased solubility, thermostability, and functional stability (Figure 10.5) [93].

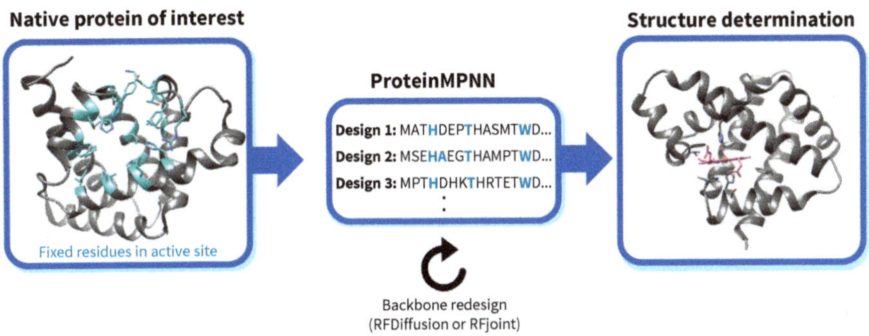


**Figure 10.5** Workflow used for the *de novo* design of myoglobin with ProteinMPNN *Source*: Adapted from [93].

Due to the relevance of heme enzymes, work with hemoproteins has also been reported with RFDiffusion All Atom, which is a version that takes into consideration cofactors. In this version, the design of a heme-binding protein obtained by diffusion around the heme-binding site was reported [94]. Furthermore, specific diffusion models, such as P450Diffusion, have been developed to design the catalytic pockets of P450 [95]. All these recent works with metalloproteins suggest the future development of ArMs may increasingly include *de novo* folds created using DL tools.

## 10.4 Concluding Remarks

Computational metalloenzyme design is still in its early stages compared to the more developed field of metal-free enzymes. This is the consequence of the higher difficulty in modeling metal centers, which requires a good knowledge of QM and careful parameterization. Furthermore, the majority of the open-source MM software landscape does not include metal modeling. In any case, even if QM-based solutions are well known, the available techniques are often too slow for high-throughput virtual screening. We believe that machine-learning QM methods and chemical intuition in applying shortcuts will significantly speed up these tools and render them useful for brute-force metalloenzyme design.

Regarding unnatural amino acids, these need to be systematically incorporated into computational design toolboxes to fully unlock the potential of ArMs. Furthermore, the TS stabilization-centric vision should be complemented with more comprehensive approaches to enzyme design (not only metalloenzymes). This passes through a synergistic effort of the lab and computer communities in producing and analyzing data, the adoption of available ligand migration tools, and the continuous research in what may be the major challenge in computational enzyme engineering: predicting the effect of mutations far from the active site in affordable timeframes and computational costs.

Finally, DL technologies mark a major step forward in *de novo* design by expanding the range of available structures. This progress opens up exciting possibilities for incorporating metals into unexplored protein folds. While initial successes in the incorporation of metals into these *de novo* structures have been achieved, the precise design of ArMs with novel functions still remains a challenge. As DL methods continue to improve, we can anticipate new methodologies, especially those focused on ArMs.

## References


1 Hagen, J. *Industrial Catalysis: A Practical Approach*, 1e. Wiley; 2015.

2 Bar-Even, A., Noor, E., Savir, Y. et al. The moderately efficient enzyme: evolutionary and physicochemical trends shaping enzyme parameters. *Biochemistry* 2011;*50*(21):4402–4410.

3 Romero, P.A. and Arnold, F.H. Exploring protein fitness landscapes by directed evolution. *Nat. Rev. Mol. Cell Biol.* 2009;*10*(12):866–876.

4 Chica, R.A., Doucet, N., and Pelletier, J.N. Semi-rational approaches to engineering enzyme activity: combining the benefits of directed evolution and rational design. *Curr. Opin. Biotechnol.* 2005;*16*(4):378–384.

5 Monza, E., Acebes, S., Lucas, F. et al. *Molecular Modeling in Enzyme Design, Toward In Silico Guided Directed Evolution*. Springer; 2017, pp. 257–284.

6 St-Jacques, A.D., Gagnon, O., and Chica, R.A. *Computational Enzyme Design: Successes, Challenges, and Future Directions*; 2018.

7 Jumper, J., Evans, R., Pritzel, A. et al. Highly accurate protein structure prediction with AlphaFold. *Nature* 2021;*596*(7873):583–589.

8 Christoffel, F., Igareta, N.V., Pellizzoni, M.M. et al. Design and evolution of chimeric streptavidin for protein-enabled dual gold catalysis. *Nat. Catal.* 2021;*4*(8):643–653.

9 Gu, Y., Bloomer, B. J., Liu, Z. et al. Directed evolution of artificial metalloenzymes in whole cells. *Angew. Chem. Int. Ed.* 2022;*61*(5):e202110519.

10 Hestericová, M., Heinisch, T., Lenz, M. et al. Ferritin encapsulation of artificial metalloenzymes: engineering a tertiary coordination sphere for an artificial transfer hydrogenase. *Dalton Trans.* 2018;*47*(32):10837–10841.

11 Mirts, E.N., Petrik, I.D., Hosseinzadeh, P. et al. A designed Heme-[4Fe-4S] metalloenzyme catalyzes sulfite reduction like the native enzyme. *Science (New York, N.Y.)* 2018;*361*(6407):1098.

12 Serrano-Plana, J., Rumo, C., Rebelein, J.G. et al. Enantioselective hydroxylation of benzylic C($sp^3$)–H bonds by an artificial iron hydroxylase based on the biotin–streptavidin technology. *J. Am. Chem. Soc.* 2020;*142*(24):10617–10623.

13 Upp, D.M., Huang, R., Li, Y. et al. Engineering dirhodium artificial metalloenzymes for diazo coupling cascade reactions. *Angew. Chem. Int. Ed.* 2021;*60*(44):23672–23677.

14 Villarino, L., Splan, K.E., Reddem, E. et al. An artificial heme enzyme for cyclopropanation reactions. *Angew. Chem. Int. Ed.* 2018;*57*(26):7785–7789.

15 Yang, H., Swartz, A.M., Park, H.J. et al. Evolving artificial metalloenzymes via random mutagenesis. *Nat. Chem.* 2018;*10*(3):318–324.

16 Yu, K., Zou, Z., Igareta, N.V. et al. Artificial metalloenzyme-catalyzed enantioselective amidation via nitrene insertion in unactivated C($sp^3$)–H bonds. *J. Am. Chem. Soc.* 2023;*145*(30):16621–16629.

**17** Akcapinar, G.B. and Sezerman, O.U. Computational approaches for de novo design and redesign of metal-binding sites on proteins. *Biosci. Rep.* 2017; *37*(2):BSR20160179.

**18** Alonso-Cotchico, L., Rodríguez-Guerra, J., Lledós, A. et al. Molecular modeling for artificial metalloenzyme design and optimization. *Acc. Chem. Res.* 2020; *53*(4):896–905.

**19** Muñoz Robles, V., Ortega-Carrasco, E., Alonso-Cotchico, L. et al. Toward the computational design of artificial metalloenzymes: from protein–ligand docking to multiscale approaches. *ACS Catal.* 2015;*5*(4):2469–2480.

**20** Rodríguez-Guerra, J., Alonso-Cotchico, L., Sciortino, G. et al. Computational studies of artificial metalloenzymes: from methods and models to design and optimization. In *Artificial Metalloenzymes and MetalloDNAzymes in Catalysis.* John Wiley & Sons, Ltd.; 2018, pp. 99–136.

**21** Rosati, F. and Roelfes, G. Artificial metalloenzymes. *ChemCatChem* 2010;*2*(8):916–927.

**22** Clot, E. and Norrby, P.-O. Modeling in homogeneous catalysis: a tutorial. In *Innovative Catalysis in Organic Synthesis.* John Wiley & Sons, Ltd.; 2012, pp. 167–191.

**23** Guan, Y., Ingman, V.M., Rooks, B.J. et al. AARON: an automated reaction optimizer for new catalysts. *J. Chem. Theory Comput.* 2018;*14*(10):5249–5261.

**24** Cheng, Y., Wang, H., Xu, H. et al. Co-evolution-based prediction of metal-binding sites in proteomes by machine learning. *Nat. Chem. Biol.* 2023;*19*(5):548–555.

**25** Dürr, S.L., Levy, A., and Rothlisberger, U. Metal3D: a general deep learning framework for accurate metal ion location prediction in proteins. *Nat. Commun.* 2023;*14*(1):2713.

**26** Sánchez-Aparicio, J.-E., Tiessler-Sala, L., Velasco-Carneros, L. et al. BioMetAll: identifying metal-binding sites in proteins from backbone preorganization. *J. Chem. Inf. Model.* 2021;*61*(1):311–323.

**27** Martins, F.L., Pordea, A., and Jäger, C.M. Computationally driven design of an artificial metalloenzyme using supramolecular anchoring strategies of iridium complexes to alcohol dehydrogenase. *Faraday Discuss.* 2022;*234*:315–335.

**28** Richter, F., Leaver-Fay, A., Khare, S.D. et al. De novo enzyme design using Rosetta3. *PloS One* 2011;*6*(5):e19230.

**29** Tantillo, D.J., Jiangang, C., and Houk, K.N. Theozymes and compuzymes: theoretical models for biological catalysis. *Curr. Opi. Chem. Biol.* 1998; *2*(6):743–750.

**30** Kiss, G., Çelebi-Ölçüm, N., Moretti, R. et al. Computational enzyme design. *Ang. Chem. Int. Ed.* 2013;*52*(22):5700–5725.

**31** Smith, A.J.T., Müller, R., Toscano, M.D. et al. Structural reorganization and preorganization in enzyme active sites: comparisons of experimental and theoretically ideal active site geometries in the multistep serine esterase reaction cycle. *J. Am. Chem. Soc.* 2008;*130*(46):15361–15373.

**32** Senn, H.M. and Thiel, W. QM/MM methods for biomolecular systems. *Angew. Chem. Int. Ed.* 2009;*48*(7):1198–1229.

**33** Oohora, K., Kagawa, Y., Nishiura, T. et al. Rational design of an artificial ethylbenzene hydroxylase using a molecular dynamics simulation to enhance enantioselectivity. *Chem. Lett.* 2024;*53*(2):upad042.

**34** Wijma, H.J., Floor, R.J., Bjelic, S. et al. Enantioselective enzymes by computational design and in silico screening. *Angew. Chem.* 2015;*127*(12):3797–3801.

**35** Wang, P., Zhang, J., Zhang, S. et al. Using high-throughput molecular dynamics simulation to enhance the computational design of kemp elimination enzymes. *J. Chem. Inf. Model.* 2023;*63*(4):1323–1337.

**36** Wijma, H.J., Marrink, S.J., and Janssen, D.B. Computationally efficient and accurate enantioselectivity modeling by clusters of molecular dynamics simulations. *J. Chem. Inf. Model.* 2014;*54*(7):2079–2092.

**37** Alonso-Cotchico, L., Rodríguez-Guerra Pedregal, J., Lledós, A. et al. The effect of cofactor binding on the conformational plasticity of the biological receptors in artificial metalloenzymes: the case study of LmrR. *Front. Chem.* 2019;*7*:211.

**38** Lonsdale, R., Harvey, J.N., and Mulholland, A.J. A practical guide to modelling enzyme-catalysed reactions. *Chem. Soc. Rev.* 2012;*41*(8):3025–3038.

**39** Sparta, M., Shirvanyants, D., Ding, F. et al. Hybrid dynamics simulation engine for metalloproteins. *Biophys. J.* 2012;*103*(4):767.

**40** Zheng, F., Yang, W., Ko, M.-C. et al. Most efficient cocaine hydrolase designed by virtual screening of transition states. *J. Am. Chem. Soc.* 2008;*130*(36):12148–12155.

**41** Privett, H.K., Kiss, G., Lee, T.M. et al. Iterative approach to computational enzyme design. *Proc. Natl. Acad. Sci.* 2012;*109*(10):3790–3795.

**42** Brodbeck, R. and Deeth, R.J. Extending ligand field molecular mechanics to modelling organometallic $\pi$-bonded systems: applications to ruthenium–arenes. *Dalton Trans.* 2011;*40*(42):11147–11155.

**43** Tubert-Brohman, I., Schmid, M., and Meuwly, M. Molecular mechanics force field for octahedral organometallic compounds with inclusion of the trans influence. *J. Chem. Theory Comput.* 2009;*5*(3):530–539.

**44** Li, P. and Merz, K.M. Jr. Metal ion modeling using classical mechanics. *Chem. Rev.* 2017;*117*(3):1564–1686.

**45** Fracchia, F., Del Frate, G., Mancini, G. et al. Force field parametrization of metal ions from statistical learning techniques. *J. Chem. Theory Comput.* 2018; *14*(1):255–273.

**46** Juraskova, V., Tusha, G., Zhang, H. et al. Modelling ligand exchange in metal complexes with machine learning potentials. *Faraday Discuss.* 2024;*256*:156.

**47** Li, P. and Merz, K.M. Jr. MCPB.py: a Python based metal center parameter builder. *J. Chem. Inf. Model.* 2016;*56*(4):599–604.

**48** Shen, L., Wu, J., and Yang, W. Multiscale quantum mechanics/molecular mechanics simulations with neural networks. *J. Chem. Theory Comput.* 2016;*12*(10):4934–4946.

**49** Zeng, J., Giese, T.J., Ekesan, Ş. et al. Development of range-corrected deep learning potentials for fast, accurate quantum mechanical/molecular mechanical simulations of chemical reactions in solution. *J. Chem. Theory Comput.* 2021;*17*(11):6993–7009.

**50** Zhou, B., Zhou, Y., and Xie, D. Accelerated quantum mechanics/molecular mechanics simulations via neural networks incorporated with mechanical embedding scheme. *J. Chem. Theory Comput.* 2023;*19*(4):1157–1169.

**51** Proctor, E.A., Ding, F., and Dokholyan, N.V. Discrete molecular dynamics. *WIREs Comput. Mol. Sci.* 2011;*1*(1):80–92.

**52** Nechay, M.R., Valdez, C.E., and Alexandrova, A.N. Computational treatment of metalloproteins. *J. Phys. Chem. B* 2015;*119*(19):5945–5956.

**53** Valdez, C.E., Gallup, N.M., and Alexandrova, A.N. $Co^{2+}$ acireductone dioxygenase: $Fe^{2+}$ mechanism, $Ni^{2+}$ mechanism, or something else? *Chem. Phys. Lett.* 2014;*604*:77–82.

**54** Nedd, S., Redler, R.L., Proctor, E.A. et al. Cu,Zn-superoxide dismutase without Zn is folded but catalytically inactive. *J. Mol. Biol.* 2014;*426*(24):4112–4124.

**55** Xu, M., He, X., Zhu, T. et al. A fragment quantum mechanical method for metalloproteins. *J. Chem. Theory Comput.* 2019;*15*(2):1430–1439.

**56** Hellmers, J., Hedegård, E.D., and König, C. Fragmentation-based decomposition of a metalloenzyme–substrate interaction: a case study for a lytic polysaccharide monooxygenase. *J. Phys. Chem. B* 2022;*126*(29):5400–5412.

**57** Chung, L.W., Sameera, W.M.C., Ramozzi, R. et al. The ONIOM method and its applications. *Chem. Rev.* 2015;*115*(12):5678–5796.

**58** Yan, Z., Li, X., and Chung, L.W. Multiscale quantum refinement approaches for metalloproteins. *J. Chem. Theory Comput.* 2021;*17*(6):3783–3796.

**59** Drienovská, I. and Roelfes, G. Expanding the enzyme universe with genetically encoded unnatural amino acids. *Nat. Catal.* 2020;*3*(3):193–202.

**60** Yu, Y., Hu, C., Xia, L. et al. Artificial metalloenzyme design with unnatural amino acids and non-native cofactors. *ACS Catal.* 2018;*8*(3):1851–1863.

**61** Mills, J.H., Khare, S.D., Bolduc, J.M. et al. Computational design of an unnatural amino acid dependent metalloprotein with atomic level accuracy. *J. Am. Chem. Soc.* 2013;*135*(36):13393–13399.

**62** Zanghellini, A., Jiang, L., Wollacott, A.M. et al. New algorithms and an in silico benchmark for computational enzyme design. *Protein Sci. Publ. Protein Soc.* 2006;*15*(12):2785.

**63** Kuhlman, B. and Baker, D. Native protein sequences are close to optimal for their structures. *Proc. Natl. Acad. Sci.* 2000;*97*(19):10383–10388.

**64** Drienovská, I., Alonso-Cotchico, L., Vidossich, P. et al. Design of an enantioselective artificial metallo-hydratase enzyme containing an unnatural metal-binding amino acid. *Chem. Sci.* 2017;*8*(10):7228–7235.

**65** Carlin, D.A., Caster, R.W., Wang, X. et al. Kinetic characterization of 100 glycoside hydrolase mutants enables the discovery of structural features correlated with kinetic constants. *PLoS One* 2016;*11*(1):e0147596.

**66** Van Stappen, C., Deng, Y., Liu, Y. et al. Designing artificial metalloenzymes by tuning of the environment beyond the primary coordination sphere. *Chem. Rev.* 2022;*122*(14):11974–12045.

**67** Khersonsky, O., Lipsh, R., Avizemer, Z. et al. Automated design of efficient and functionally diverse enzyme repertoires. *Mol. Cell* 2018;*72*(1): 178.

**68** Wu, Z., Kan, S.B.J., Lewis, R.D. et al. Machine learning-assisted directed protein evolution with combinatorial libraries. *Proc. Natl. Acad. Sci.* 2019:*116*(18): 8852–8858.

**69** Yang, J., Li, F.-Z., and Arnold, F.H. Opportunities and challenges for machine learning-assisted enzyme engineering. *ACS Central Sci.* 2024;*10*(2):226–241.

**70** Vornholt, T., Christoffel, F., Pellizzoni, M.M. et al. (2021). Systematic engineering of artificial metalloenzymes for new-to-nature reactions. *Sci. Adv.*, *7*(4):eabe4208.

71 Vornholt, T., Mutný, M., Schmidt, G.W. et al. Enhanced sequence-activity mapping and evolution of artificial metalloenzymes by active learning. *ACS Central Sci.* 2024;*10*(7):1357–1370.

72 Maria-Solano, M., Serrano-Hervás, E., Romero-Rivera, A. et al. Role of conformational dynamics in the evolution of novel enzyme function. *Chem. Commun.* 2018;*54*(50):6622–6634.

73 Hong, N.-S., Petrović, D., Lee, R. et al. The evolution of multiple active site configurations in a designed enzyme. *Nat. Commun.* 2018;*9*(1):3900.

74 Petrović, D., Risso, V.A., Kamerlin, S.C.L. et al. Conformational dynamics and enzyme evolution. *J. R. Soc. Interface* 2018;*15*(144):20180330.

75 Romero-Rivera, A., Garcia-Borràs, M., and Osuna, S. Role of conformational dynamics in the evolution of retro-aldolase activity. *ACS Catal.* 2017;*7*(12):8524–8532.

76 Romero-Rivera, A., Iglesias-Fernández, J., and Osuna, S. Exploring the conversion of a d-sialic acid aldolase into a l-KDO aldolase. *Eur. J. Org. Chem.* 2018;*2018*(20–21): 2603–2608.

77 Osuna, S. The challenge of predicting distal active site mutations in computational enzyme design. *WIREs Comput. Mol. Sci.* 2021;*11*(3):e1502.

78 Casilli, F., Canyelles-Niño, M., Roelfes, G. et al. Computation-guided engineering of distal mutations in an artificial enzyme. *Faraday Discuss.* 2024;*252*:262–278.

79 De Simone, M., Alonso-Cotchico, L., Lucas, M.F. et al. Distal mutations enhance efficiency of free and immobilized NOV1 dioxygenase for vanillin synthesis. *J. Biotechnol.* 2024:*391*:92–98.

80 Li, G., Zhang, N., Dai, X. et al. EnzyACT: a novel deep learning method to predict the impacts of single and multiple mutations on enzyme activity. *J. Chem. Inf. Model.* 2024;*64*(15):5912–5921.

81 Acebes, S., Fernandez-Fueyo, E., Monza, E. et al. Rational enzyme engineering through biophysical and biochemical modeling. *ACS Catal.* 2016;*6*(3):1624–1629.

82 Giacobelli, V.G., Monza, E., Lucas, M.F. et al. Repurposing designed mutants: a valuable strategy for computer-aided laccase engineering – the case of POXA1b. *Catal. Sci. Technol.* 2017;*7*(2):515–523.

83 Sánchez-Aparicio, J.-E., Sciortino, G., Mates-Torres, E. et al. Successes and challenges in multiscale modelling of artificial metalloenzymes: the case study of POP-Rh2 cyclopropanase. *Faraday Discus.* 2022;*234*:349–366.

84 Chen, J., Wang, W., Sun, H. et al. Roles of accelerated molecular dynamics simulations in predictions of binding kinetic parameters. *Mini Rev. Med. Chem.* 2024;*24*(14):1323–1333.

85 Lombardi, A., Pirro, F., Maglio, O. et al. De novo design of four-helix bundle metalloproteins: one scaffold, diverse reactivities. *Acc. Chem. Res.* 2019;*52*(5):1148–1159.

86 Kalvet, I., Ortmayer, M., Zhao, J. et al. Design of heme enzymes with a tunable substrate binding pocket adjacent to an open metal coordination site. *J. Am. Chem. Soc.* 2023;*145*(26):14307–14315.

87 Klein, A.S., Leiss-Maier, F., Mühlhofer, R. et al. A de novo metalloenzyme for cerium photoredox catalysis. *J. Am. Chem. Soc.* 2024;*146*(38):25976–25985.

**88** Baek, M., DiMaio, F., Anishchenko, I. et al. Accurate prediction of protein structures and interactions using a three-track neural network. *Science* 2021;*373*(6557):871–876.

**89** Yeh, A.H.-W., Norn, C., Kipnis, Y. et al. De novo design of luciferases using deep learning. *Nature* 2023;*614*(7949):774–780.

**90** Wang, J., Lisanza, S., Juergens, D. et al. Scaffolding protein functional sites using deep learning. *Science* 2022;*377*(6604):387.

**91** Watson, J.L., Juergens, D., Bennett, N.R. et al. De novo design of protein structure and function with RFdiffusion. *Nature* 2023;*620*(7976):1089–1100.

**92** Dauparas, J., Anishchenko, I., Bennett, N. et al. Robust deep learning–based protein sequence design using ProteinMPNN. *Science* 2022;*378*(6615):49–56.

**93** Sumida, K.H., Núñez-Franco, R., Kalvet, I. et al. Improving protein expression, stability, and function with ProteinMPNN. *J. Am. Chem. Soc.* 2024;*146*(3):2054–2061.

**94** Krishna, R., Wang, J., Ahern, W. et al. Generalized biomolecular modeling and design with RoseTTAFold All-Atom. *Science* 2024;*384*(6693):eadl2528.

**95** Wang, Q., Liu, X., Zhang, H. et al. Cytochrome P450 enzyme design by constraining the catalytic pocket in a diffusion model. *Research* 2024;*7*: 0413.

# 11

# How Computational Chemistry Can Contribute to the Understanding of the Effects of Metal Ions in Biological Systems: Aluminum as a Case Study

*J.I. Mujika*[1,2] *and X. Lopez*[1,2]

[1] *Faculty of Chemistry, Universidad del País Vasco/Euskal Herriko Unibertsitatea (UPV/EHU), Manuel Lardizabal 3, Donostia-San Sebastian 20018, Spain*
[2] *DIPC, Universidad del País Vasco/Euskal Herriko Unibertsitatea (UPV/EHU), Manuel Lardizabal 4, Donostia-San Sebastian 20018, Spain*

## 11.1 Introduction

Living organisms require many chemical elements for their normal function. These elements are considered essential for the survival of the organisms. Quite surprisingly, however, the number of essential elements is really small: from the 118 chemical elements found in the periodic table (see Figure 11.1), only six elements (C, H, N, O, P, and S) account for approximately 97% in the mass of a human body. Other elements that are essential in significant quantities are Na, K, Mg, Ca, and Cl. The lack or insufficient amount of any of these elements leads inevitably to a malfunction of the organism, or even to its death.

Some other elements are also essential for the organisms. Nevertheless, they are defined as essential trace elements because they are required in low concentrations so that either an excess or deficiency can be harmful to biological organisms. Examples of these trace elements are Fe, Zn, or Cu. By contrast, other elements are highly toxic to humans and to most living organisms. The ingestion of elements, such as lead (Pb) or arsenic (As) leads very rapidly to death. As can be seen in Figure 11.1, up to eight elements are included in this fateful list. Other elements are also harmful due to their radioactivity.

Less clear is the role of the remaining elements. Most of them do not exhibit any biological function, and they are not required by the human body. However, ingesting these elements does not trigger (in principle) any noxious effect, at least for a short period, and therefore they cannot be considered toxic. But are they harmless elements? Do we know the consequences of their presence in the human body over a long period?

*Computational Bioinorganics: From Description to Prediction*, First Edition. Edited by Jean-Didier Maréchal.
 Published 2026 by John Wiley & Sons Ltd.

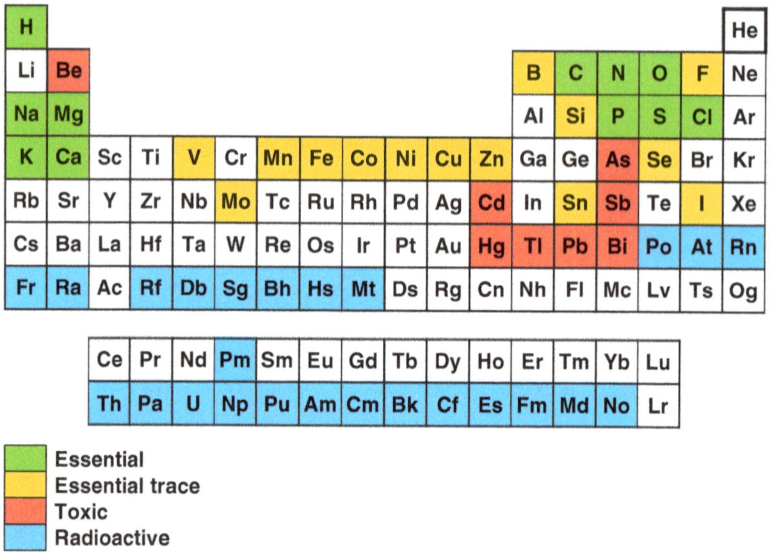


**Figure 11.1** Periodic table of elements and their biological roles in the human body.

Due to the complexity of biological systems, in which many molecules of different physical and chemical characteristics are close, the answers to these questions are complicated to address. We must realize that any of these elements might interact in different ways with biological materials like proteins, lipids, DNA, etc. Therefore, the effects of these elements are of enormous complexity, and can only be studied by well-defined strategies. Combinations of techniques from different fields are likely to allow the best understanding of how nonbiological elements impact biological systems.

Among a large number of scientific techniques available out there, computational chemistry is one of the most promising ones. In computational chemistry, well-established physically sound methods are applied to realistic models of biological molecules. With these models, the interactions between different molecules can be described at an atomistic level, providing very valuable information that is challenging to obtain by experimental techniques. Moreover, the energies of these systems can be evaluated, and the interaction energies between the elements and main biological building blocks determined, thus, establishing from a thermodynamic point of view, which could be the preferential binding target of the element under study.

Of course, computational chemistry has its inherent limitations. Computational chemistry employs models that are quite simplified regarding real biological systems. Although, conclusions achieved by approaches need to be confirmed by experiments; these approaches can be highly valuable in predicting and guiding experimental research. A critical advantage of computational chemistry is that almost any system can be studied (although simplified) without additional expense. Therefore, the results obtained may rationalize the preferential targets of a chemical element and rule out others.

As shown in Figure 11.1, there are many elements with unknown biological consequences. This chapter focuses on one of these elements: aluminum and some of the studies carried out in our group by means of computational chemistry to understand the biophysics and chemistry of this nonessential metal. After a brief introduction to aluminum and its potential risks to human health, we introduce the complex biophysics problem ascribed to this element, known as aluminum speciation. Then, we describe in more detail three case studies in which we combined different computational methods to tackle complex problems and address plausible answers to some specific questions. We believe these studies are good examples to illustrate the robustness and capabilities of computational chemistry to study bioinorganic issues.

## 11.2 Aluminum

Aluminum is the third most abundant element in Earth's crust (8.1%), following oxygen (46.6%) and silicon (27.7%). However, as shown in Figure 11.1, this element is not essential in the human organism, and therefore, it is considered an exogenous element. But does this mean that aluminum is a toxic element? The answer, at least for a short time, is that it is not: if a high dose of this element is introduced in a human organism, that person would not die. That would not be the fate with highly toxic elements, such as Pb or As. Nevertheless, the long-term effects of aluminum are not so straightforward, and a large body of evidence suggests harmful effects when a significant concentration of aluminum is accumulated in the organism.

Regarding aluminum exposure, the industrial revolution triggered a big change for humans. Before that revolution, aluminum remained in Earth's crust, and complex biogeochemical cycles prevented direct contact between this metal and humans. However, in the last decades, this scenario has changed because aluminum is widely used in many different products and processes. The physical and chemical properties of aluminum, along with its low cost, make this element very attractive for many uses, such as flocculent in drink water treatment, in vaccines, in food staff, in deodorants, etc. Consequently, this element is not out of the biotic cycle anymore, and significant traces of this metal are introduced into the human bodies, with unknown consequences. Nevertheless, it must be pointed out that the human organism contains efficient mechanisms to throw aluminum away and guarantee a low absorption of this element, as for example: Aluminum is "sequestrated" by the transferrin protein to reduce the concentration of free aluminum in blood serum. The blood–brain barrier prevents a high concentration of aluminum in the brain. However, despite all these mechanisms to keep aluminum out of the most vital parts of the body, it is well established that the body's burden of aluminum increases with aging and that this element accumulates in several different compartments, including skin, brain, or bones. So, the question is straightforward: does the accumulation of aluminum in the body harms the organism? This question has no definitive answer, but many studies have linked aluminum with several diseases.

Concern about aluminum toxicity started in the 1970s. At that time, aluminum-containing dialysis baths were administered to patients with chronic renal failure. It was discovered that these patients developed dialysis encephalopathy disease or dialysis dementia, and it was confirmed that aluminum was the cause of the disease. The role of aluminum in other diseases is unclear, but more diseases have been related to aluminum. Of particular controversy raised about aluminum and Alzheimer's disease (AD) [1], the main neurodegenerative disease. Nowadays, there is a consensus considering aluminum a risk factor in AD development.

### 11.2.1 Aluminum Speciation in Water

The unbound aluminum is present in an aqueous environment as a trivalent cation ($Al^{3+}$). Due to its physicochemical features, aluminum is a very reactive element that may interact with many different molecules to form stable complexes. In addition, to make the problem more complex, many of these molecules may present different protonation states. Consequently, there exists a myriad of different possible combinations to form the complexes. To understand the biological effects of aluminum, however, it is of paramount importance to identify the main targets of aluminum, which are their most likely protonation states, and in the end, how aluminum is distributed in the body. This is known as the speciation chemistry of aluminum.

As an illustrative example of the complexity of aluminum speciation, the main aqueous mononuclear complexes (but not all of them) formed by the cation in an aqueous solution are shown in Figure 11.2. At each pH regime, the predominant hydrolytic species differs, with different protonation states. For instance, at pH >6.5 conditions, $[Al(OH)_4]^-$ is predominant, whereas at pH <5.5 $[Al(H_2O)_6]^{3+}$ prevails. However, at middle pH conditions $[Al(OH)_2(H_2O)_4]^+$ and $[Al(OH)(H_2O)_6]^{2+}$ are more likely; the $[Al.(OH)_3]$ species is also formed, but this compound precipitates.

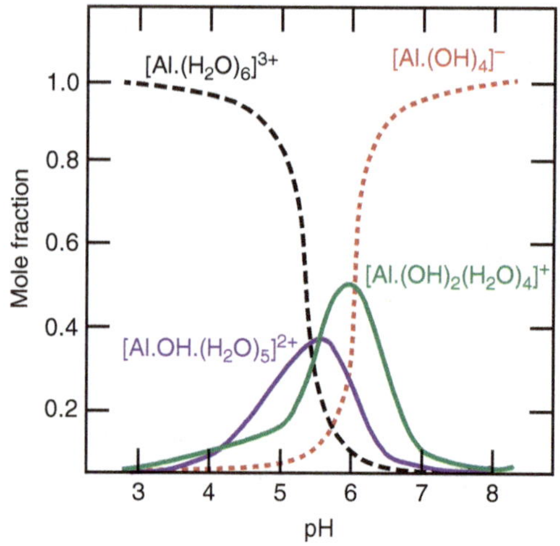


**Figure 11.2** Distribution upon pH of main soluble mononuclear aluminum ion species in aqueous solution.

### 11.2.2 Mapping Aluminum Speciation with Biological Building Molecules

The scenario is even more complex when other molecules emerge. According to the HSAB principle, aluminum is a hard Lewis acid, and therefore, it shows a preference for hard Lewis bases, such as oxygen-containing biomolecules (phosphates, lipids, DNA, proteins, etc.). Smaller molecules, such as citrate, oxalate, or phosphorylated peptides, are also identified as preferential targets for aluminum. Besides, catecholamine-based neurotransmitters (L-DOPA, dopamine, adrenaline, catechol, etc.) are well-established potential aluminum ligands. In conclusion, the speciation of this metal is a vast combinational problem with an unaffordable number of possibilities to form aluminum-containing complexes. In this sense, computational chemistry emerges as a relevant field that can shed light on the speciation problem by determining the formation energies of the most likely complexes and mapping their thermodynamic stability. This information does not entirely clarify how aluminum is distributed in the human body, but at least it may point to the most likely molecule targets of the cation.

As pointed out above, in this chapter, we are not going to describe all these studies. The reader can find elsewhere the studies about aluminum and its interactions with different biomolecules conducted by our group [2–17]. Instead, we prefer to explain in detail three case studies, hoping that they can help the reader understand the strength of computational chemistry to address some of the fundamental questions that are difficult to answer by other techniques.

## 11.3 The Pro-oxidant Activity of Aluminum

In biological environments, regardless of with which molecule is interacting, aluminum always presents a +3 oxidation state. Due to this unique oxidation state, aluminum should not be involved in reduction/oxidation (redox) reactions, where metal's valence changes. However, quite surprisingly, some *in vitro* and *in vivo* experiments demonstrated that aluminum promotes the biological oxidation of certain molecules, a fact difficult to explain due to the nonredox character of aluminum. As a plausible explanation for this contra-intuitive evidence, Prof. Exley suggested [18] that aluminum could favor the oxidation reaction by increasing the lifetime of the superoxide radical ($\bullet O_2^-$). But that was a mere hypothesis, difficult to demonstrate by experimental techniques. By contrast, the study of whether aluminum can stabilize the radical is affordable by means of computational chemistry, and in fact, this is an excellent example to illustrate how computational chemistry can shed light on questions challenging to address by other techniques [2–4].

Radical species are very reactive, and they react with neighboring molecules, such as lipids, proteins, DNA, etc. They are naturally produced, and they perform essential tasks inside the organisms, such as apoptosis, signaling, or homeostasis. Nevertheless, an overproduction of these species, known as oxidative stress, cause irreversible damage to cells. Oxidative stress has been related to several diseases, including neurodegenerative diseases.

The Haber–Weiss reaction is the main source of radical oxygen species (ROS) in cells. This reaction includes the Fenton reaction, a prolonged reaction that takes place only when iron catalyzes it. The overall Haber–Weiss reaction can be split into two reactions

$$Fe^{3+} + \bullet O_2^- \leftrightarrow Fe^{2+} + O_2 \tag{11.1}$$

$$Fe^{2+} + H_2O_2 \leftrightarrow Fe^{3+} + \bullet OH + OH^- \tag{11.2}$$

where the first reaction is the Fenton reaction. Thus, the overall Haber–Weiss reaction is

$$H_2O_2 + \bullet O_2^- \leftrightarrow O_2 + \bullet OH + OH^- \tag{11.3}$$

The superoxide ($\bullet O_2^-$) species is used to produce $\bullet OH$ radicals, the most reactive ROS. But how can aluminum promote this reaction?

To understand aluminum's role, we must first mention a nice experimental study by Fukuzumi et al. [19]. In this study, the authors evaluated the interaction strengths between several metals and the superoxide radical by measuring the electron spin resonance (ESR) g-tensor in several metal superoxide complexes. They demonstrated a linear relationship between the oxidant activity of the metals and the interaction strength with the superoxide anion: the stronger the metal–superoxide interaction, the higher the oxidative activity of the metal. Moreover, the interaction strength increases with metal charge (monovalent cations < divalent cations < trivalent cations). In addition, for metals with the same charge, the strength increases by decreasing the ion radius. Aluminum was not included in this study, but based on these trends and due to its high charge (+3) and small radius, it was reasonable to think that aluminum can interact strongly with the superoxide and therefore can exhibit a high pro-oxidant activity.

For computational chemists, the concept that the metal–superoxide interaction strength is a reliable indicator of the pro-oxidant activity of a given metal is appealing because that interaction can be calculated precisely by first-principle quantum methods. Hence, the first step in the computational study was to compute the g-tensor values and compare them with the experimental values determined by Fukuzumi et al. A good agreement between the computed and experimental g-tensor values would indicate that the methodology was adequate to study the system and that it could also be used to compute the g-tensor of the Al. $\bullet$ $O_2^-$ complex, thus quantifying the pro-oxidant capability of this metal.

The g-tensor is related to the energy splitting of the antibonding $\pi_g^*$ levels of $\bullet O_2^-$. As shown in Figure 11.3, the two antibonding $\pi_g^*$ orbitals of the superoxide $\bullet O_2^-$ radical do not degenerate, so there is an energy gap ($\Delta\varepsilon_{\pi g}$) between them. This energy difference is directly related to the g-tensor ($g_{zz}$):

$$g_{zz} = g_e + 2\lambda/\Delta\varepsilon_{\pi g} \text{ for } \Delta\varepsilon_{\pi g} >> \lambda \tag{11.4}$$

where $g_e$ is the ESR g-tensor of the free electron spin value ($g_e$ = 2.0023) and $\lambda$ the spin-orbit coupling constant of oxygen ($\lambda$ = 0.014 eV). Therefore, in order to calculate the g-tensor of the metal. $\bullet O_2^-$ complexes we needed to calculate accurately $\Delta\varepsilon_{\pi g}$, that is, the energy splitting between the two $\pi_g^*$ orbitals of the $\bullet O_2^-$ radical.

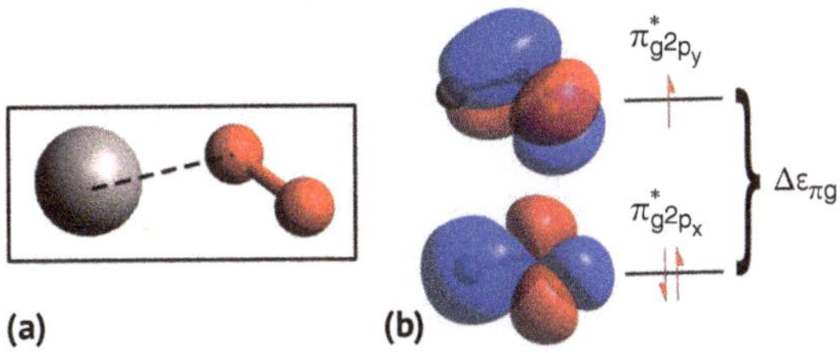


**Figure 11.3** (a) Metal–• $O_2^-$ superoxide interaction. (b) Energy splitting of the antibonding $\pi_g$ orbitals ($\Delta\varepsilon_{\pi g}$), which is proportional to the metal–superoxide interaction strength.

For that problem, we chose the complete active space self-consistent field (CASSCF) method. In short, CASSCF is a multiconfigurational method; that is, multiple electronic configurations are taken into account to correctly represent a molecule's reference state. The particularity of this method is that not all orbitals of the molecule are included in the calculations, but only the chemically most relevant ones, referred to as *active space*. Applying this method, ESR g-tensor values, and the associated interaction energies were computed for selected gas-phase metal–superoxide complexes (M = $Na^+$, $K^+$, $Mg^{2+}$, $Ca^{2+}$), and the computed values compared with the experimental values determined by Fukuzumi et al. The results were amazingly encouraging, with an excellent correlation between experimental and computational values, so that the $\Delta\varepsilon_{\pi g}$ values increased by increasing the metal charge: 0.31 eV ($K^+$), 0.35 eV ($Na^+$), 0.56 eV ($Ca^{2+}$), and 0.65 eV ($Mg^{2+}$). More importantly, these results confirm that the methodology used was adequate to calculate the metal–superoxide interaction strength, an essential step in any computational study to provide meaningful conclusions.

The next step was to calculate the $\Delta\varepsilon_{\pi g}$ value and g-tensor for the Al. • $O_2^-$ complex, not measured experimentally. Interestingly, a very high $\Delta\varepsilon_{\pi g}$ value was computed, namely 1.11 eV. According to the experimental trend, this result confirms the strong interaction between aluminum and superoxide. They provided the first solid evidence to explain the pro-oxidant activity of aluminum through superoxide radical stabilization.

Nevertheless, both the experiments conducted by Fukuzumi et al. and the CASSCF calculations were performed in the gas phase, that is, the metal and the radical were the only molecules in the system. But these conditions are not given in a biological environment, where at least an aqueous solvent molecular surrounds the complex. Moreover, as shown in Figure 11.2, aluminum can form different hydrolytic species in which the metal coordination shell contains water and/or hydroxide molecules in an aqueous environment. Consequently, we investigated whether the formation of the Al. • $O_2^-$ complex is favorable not only in the gas phase but also in the solution. In the new models, explicit water molecules were included to complete the metal first coordination shell (for guaranteeing more accurate results, the second solvation sphere was also explicitly added, see Figure 11.4). The effect of the bulk solvent was accounted for by including a homogeneous polarizable medium with a dielectric constant, known as continuum solvation models or implicit solvent. These structures were optimized by a first-principle quantum mechanical method, namely a density functional theory (DFT) functional.

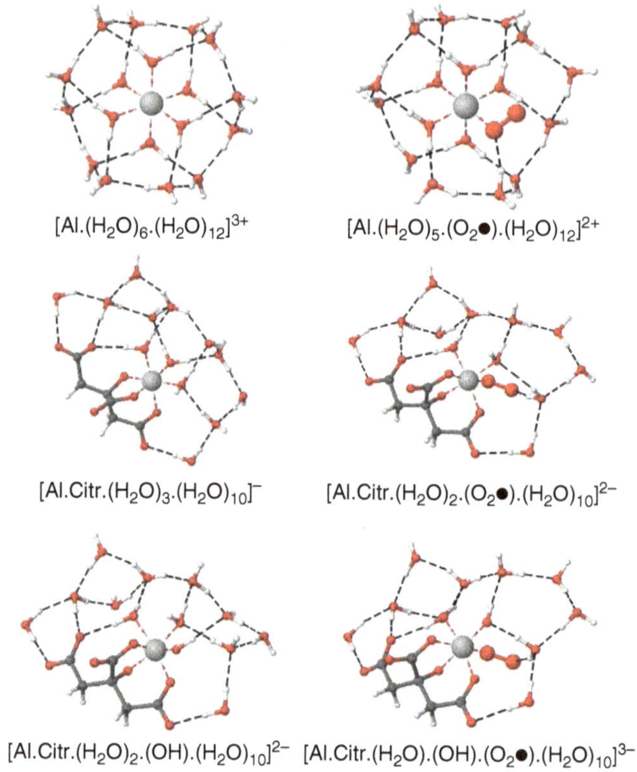


**Figure 11.4** As illustrative example, some microsolvated complex structures characterized in solution by first-principle QM methods and implicit solvent.

The most likely hydrolytic species (see Figure 11.4) of aluminum were used to study whether the formation of the $Al.\bullet O_2^-$ complex is favorable in solution. Different numbers of water or/and hydroxide molecules are placed in the metal's first coordination shell in these structures. Therefore, the formation of a microsolvated $Al.\bullet O_2^-$ complex requires the substitution of either a water molecule or a hydroxide by the superoxide at the metal first shell:

$$Al.(H_2O)_n.(OH)_m + \bullet O_2^- \leftrightarrow Al.(H_2O)_{n-1}.(OH)_m.(\bullet O_2^-) + H_2O \qquad (11.5)$$

$$Al.(H_2O)_n.(OH)_m + \bullet O_2^- \leftrightarrow Al.(H_2O)_n.(OH)_{m-1}.(\bullet O_2^-) + OH^- \qquad (11.6)$$

where the superoxide radical substitutes either a water molecule (Reaction (11.5)) or a hydroxide (Reaction (11.6)) at the metal's first coordination shell. The energies of these reactions were computed for different hydrolytic species with different numbers of water molecules ($n$) or hydroxides ($m$). The results confirmed that the reaction is thermodynamically favorable when the superoxide radical replaces a water molecule but not when it does a hydroxide. This information can help understand the superoxide's role in terms of pH, which determines the concentrations of water and hydroxide. Again, we remark that this information is difficult to get by techniques other than computational chemistry.

Once the formation and stability of the aluminum–superoxide complex were confirmed, the next step was to study whether aluminum could promote the formation of ROS radicals and subsequent oxidation of biomolecules, as it was proposed. It was hypothesized that the $Al \bullet O_2^-$ complex could be involved in the following reaction

$$Fe^{3+} + [Al.(\bullet\, O_2)]^{2+} \leftrightarrow Fe^{2+} + Al^{3+} + O_2 \qquad (11.7)$$

where iron is reduced, and the $\bullet O_2^-$ radical oxidized. Hence, aluminum favors the reaction by forming a stable complex with the radical and increasing its lifetime. In this manner, $Fe^{2+}$ is produced, which in turn produces OH radicals through the Haber–Weiss reaction (Reaction (11.3)). But that was a mere hypothesis without solid evidence.

Computationally, the question of whether the reaction is thermodynamically favorable can be addressed by calculating the reaction (free) energy of this reaction. For that purpose, all the species included in Reaction (11.7) were characterized by the methodology used to characterize the microsolvated species described above; that is, they were optimized by a first-principle quantum method in the presence of a continuum solvation model. The models included the metal second solvation layer (see Figure 11.4) to guarantee a more accurate description of the strong short-range interactions present in the complex, and again, alternative species, with different numbers of water and hydroxide molecules were considered for the first solvation sphere of the metal-superoxide complex. The free energies computed confirm that Reaction (11.7) is favorable when the $Al. \bullet O_2^-$ complexes contain up to two hydroxide ions. All in all, the results confirm that (i) aluminum increases the lifetime of the superoxide radical, forming a stable $Al. \bullet O_2^-$ complex, and (ii) that this complex may be oxidized and favors the reduction of $Fe^{+3}$ to form $Fe^{+2}$, which ultimately promotes the production of ROS species.

Nevertheless, biological systems are highly complex, as many molecules coexist within the same biological compartment. Consequently, a vast number of different complexes can be potentially formed, including binary complexes by combining two molecules, ternary complexes by combining three molecules, or even more complex compounds. In the case of aluminum, citrate is a molecule that has to be taken into account for the formation of these complexes. Citrate is a small organic molecule made of three carboxylic groups and an alcohol group, so these chemical structures make citrate a prominent target for aluminum. The concentration of citrate is very significant in many biological environments; in fact, this organic molecule is aluminum's primary low molecular mass ligand. Similarly, citrate also presents a high affinity toward iron, so a realistic description of the aluminum pro-oxidant activity requires the inclusion of citrate in the Reaction (11.7). Since the distribution of citrate is not homogeneous in all biological compartments, several scenarios were contemplated: (i) citrate is not included in the reaction (this is the situation described above), (ii) citrate interacts only with iron, (iii) citrate interacts only with aluminum, and (iv) citrate interacts with both aluminum and iron.

The computed free energy values show that the ternary aluminum–citrate–superoxide complexes are stable, so even with the presence of citrate; aluminum is able to stabilize the superoxide, which in turn may promote the Fenton reaction.

In fact, when citrate interacts only with aluminum, the reduction of Fe (II) to Fe (III) (Reaction (11.7)) becomes even more favorable than in the absence of citrate. So, these results point to a more pro-oxidant activity of aluminum when the metal interacts with citrate, a likely situation in biological conditions.

However, citrate can also interact with iron. The affinity of citrate to this metal is higher than to aluminum. Interestingly, the computed free energy values for reducing iron when this metal is chelated to citrate demonstrated that the reaction is thermodynamically unfavorable irrespective of whether aluminum interacts with citrate or not (see Figure 11.5). In other words, the chelation of iron by citrate prevents the pro-oxidant activity of aluminum. The protective role of citrate can be even larger if we take into account that in excess of citrate, the [Al.Citr$_2$] complex can be formed, so far, the most stable complex among all the complexes of aluminum modeled by our group. All its coordination positions are occupied if aluminum interacts with two citrate molecules. There would not be changes for binding the superoxide radical to the cation and subsequent promotion of the Fenton reaction.

In summary, the computational study described herein provides evidence for the pro-oxidant role of aluminum. It shows how this element may increase the lifetime of the $\bullet O_2^-$ superoxide radical, which may contribute to the Fenton reaction. The

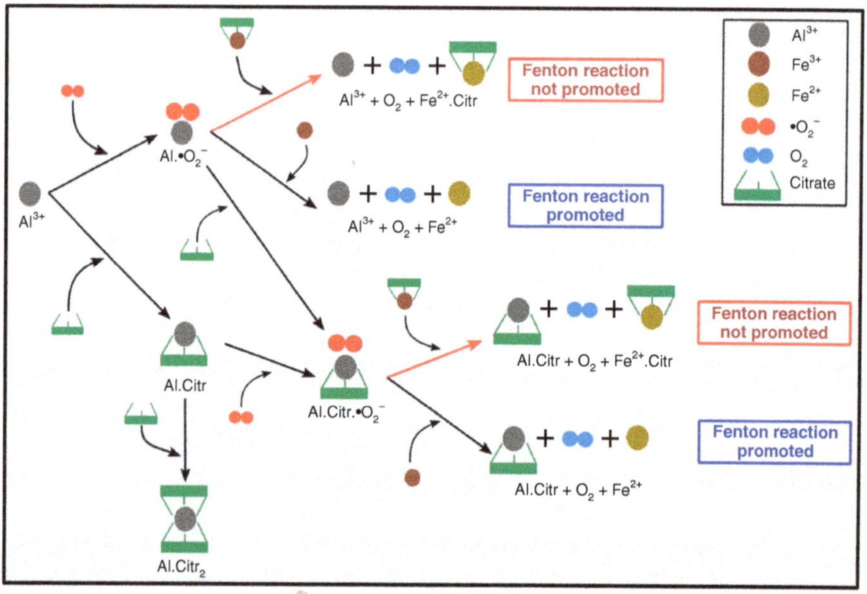


**Figure 11.5** Scheme of all the reactions studied to determine the pro-oxidant activity of aluminum. The molecules included are: aluminum (III) in solution (Al$^{3+}$), iron (III) in solution (Fe$^{3+}$), iron (II) in solution (Fe$^{2+}$), superoxide radical ($\bullet O_2^-$), oxygen molecule (O$_2$), and citrate. Note that black arrows refer to thermodynamically favorable reactions, and red arrows to thermodynamically unfavorable ones. The label at the right part of the four reaction pathways studied indicates whether that particular pathway does promote the Fenton reaction or does not.

studies, however, put in context the process and conclude that the presence of citrate may be crucial in the process since it may either alleviate or enhance the pro-oxidant activity of aluminum.

## 11.4 Unveiling the Al(III)–Aβ Complex

AD has become the most prevalent neurodegenerative disease in the world. It has been estimated that by 2040 more than 80 million people will suffer from this disease. From a biophysical point of view, two main hallmarks have been associated with AD: extracellular senile plaques and intraneuronal neurofibrillary tangles.

Senile plaques mainly contain β amyloids (Aβ), small peptide fragments typically made of 40 or 42 amino acids. These fragments are produced by the γ-secretase enzyme, which cleaves the amyloid precursor protein membrane protein. The formation of these peptides is a normal process in healthy people. Still, for reasons not understood yet the concentration of Aβ increases substantially in patients with AD, inducing their aggregation to form soluble amyloids first and ultimately insoluble aggregates. Even though the mechanism by which these aggregates grow has not been elucidated yet; several factors have been associated with this process, including oxidative stress or abnormal concentration of metals such as Fe, Zn, Cu, Al, etc.

Regarding the metals, two prominent roles have been attributed to them: (i) to speed up Aβ aggregation process and (ii) oxidative stress promotion. In the case of aluminum, the relationship between this element and AD is still under debate since a large body of evidence either supports or refutes the participation of aluminum in AD's development [1]. Despite this controversy, nowadays, aluminum is accepted as a risk factor in disease.

To better understand how metals may contribute to the disease, it is essential to unveil their binding patterns to the Aβ peptide, an issue that computational chemistry can afford. However, predicting metal-binding processes to biomolecules is one of the most challenging exercises that computation could answer, even more challenging when the target biomolecule is a small peptide such as Aβ. The Aβ peptide is defined as an intrinsically disordered protein, which implies that the peptide does not show a preference for any secondary structure (alpha helix, β-sheets, etc.), so the variability of the first coordination sphere of the metal upon binding of Aβ is enormous, with numerous possible intermediates. Due to this myriad of possibilities, an initial template is advisable to conduct a robust computational study and obtain physically sound results.

For metals such as Zn(II) and Cu(II), there is a large body of experimental data about their binding pattern to the Aβ peptide. Techniques such as EPR have allowed the establishment of the preferential binding modes of Zn(II) and Cu(II), and the nuclear magnetic resonance (NMR) structures of the Zn(II) bound Aβ peptides have been solved. With all this information, the 1 : 1 stoichiometric complexes formed by Aβ peptide and Zn(II) and Cu(II) are well established, and residues such as His6, His13, and His14 are identified as the main ligands for these two cations. All these experimental data were beneficial in conducting various computational studies on the complexes formed by these two metals.

Unfortunately, to the best of our knowledge, the number of experimental structures concerning the Al.Aβ complex is very scarce, and there is not any reliable experimental structure that could be used as an initial template for running computational studies. As an alternative, it could be tempting the use of the complexes formed by Zn(II) or Cu(II) as the initial template. However, one must bear in mind that the chemical nature of the metals determines their coordination modes to Aβ. According to the HSAB theory, Zn(II) and Cu(II) are acids of intermediate strength, whereas Al(III) is defined as a hard acid. Consequently, the chemistry of the latter differs from the chemistry of the other two metals; Al(III) shows a preference for negatively charged oxygen-containing groups, such as phosphates and carboxylates, whereas Zn(II) and Cu(II) tend to interact with N-containing groups. This different nature of the metals will be reflected in their coordination sites at the Aβ peptide and rules out the use as templates of structures solved with these metals.

Hence, we were compelled to design a novel computational methodology to characterize the Al.Aβ complex precisely, in which multiple approaches of different scales were combined to tackle the problem [5]. The critical step in this novel protocol relies on identifying pre-organized structures of the metal-free peptide that could host the ion in its best possible coordination environments. At room temperature, a flexible molecule such as an Aβ peptide may present many different conformations (see Figure 11.6). The basic idea of the protocol is to identify among all these different conformations those that are suitable to accommodate a metal, that is, that the peptide is pre-organized to interact efficiently with the cation. This concept is recurrent in many molecular fields, such as protein–ligand interactions, *de novo* enzyme design, or metal binding to proteins and cyclic peptides. In the next subsection, we describe the different approaches combined in this novel computational protocol in order to propose for the first time realistic three-dimensional (3D) structures for the Al.Aβ complex.

Asp1-Ala2-Glu3-Phe4-Arg5-His6-Asp7-Ser8-Gly9-Tyr10-Glu11-Val12-His13-His14-Gln15-Lys16-Leu17-Val18-Phe19-Phe20-Ala21-Glu22-Asp23-Val24-Gly25-Ser26-Asn27-Lys28-Gly29-Ala30-Ile31-Ile32-Gly33-Leu34-Met35-Val36-Gly37-Gly38-Val39-Val40-Ile41-Ala42

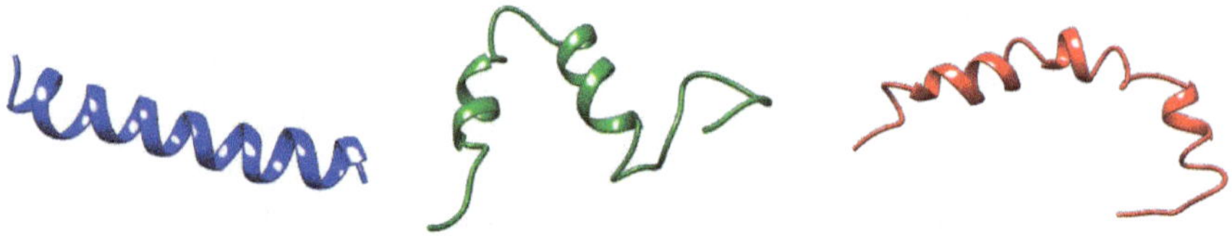

**Figure 11.6** Top: The sequence of the $A\beta_{42}$ peptide. Bottom: three conformations of the Aβ peptide solved by NMR. A total of 157 NMR structures of the Aβ peptides in solution were used to find in them those conformations pre-organized to interact efficiently with aluminum.

### 11.4.1 Step 1: Most Favorable First Coordination Spheres of Aluminum

As already pointed out above, the lack of any reliable template for constructing a reasonable model for the Al.Aβ complex forced us to search for pre-organized structures of the peptide. These conformations can be considered to "readily" interact with the cation. But, the initial question is, what does it mean to be pre-organized? How can we consider a conformation suitable to interact with the cation? The short answer to these questions is that the conformation of the peptide should be the one that provides an optimum binding site to the cation. Among the 40 or 42 amino acids present in the Aβ peptide (sequence shown in Figure 11.6), a specific combination of amino acids should provide an optimum coordination shell that strengthens the interaction between the peptide and the cation, thus increasing the stability of the complex. Therefore, the first step in this computational protocol was to establish which should be the preferential, or more precisely, the optimum coordination shell that the Aβ peptide can provide to the cation to form, in terms of thermodynamics, the most favorable complex. Once such a binding site is defined, this information will be determinant to build 3D models of the Al.Aβ complex.

With the aim of defining such optimum metal-binding site, we constructed a number of small models, also known as cluster models, to mimic possible metal-binding sites. Each of these clusters contains an Al(III) ion surrounded by several amino acid side chains, mimicking a metal-binding site. As pointed out above, due to aluminum's hard Lewis acid character, electron-donating ligands, such as O-containing amino acids, are the most likely candidates to interact with the metal. In addition, based on the coordination modes established for Cu(II) and Zn(II), histidine could be one of the metal ligands, and in many metalloproteins, the metal coordination shell is fulfilled by peptide bonds (either a carbonyl group or a terminal amine group), or by water molecules. Moreover, even though Al(III) prefers an octahedral coordination mode, coordination numbers five or four should also be considered. Therefore, due to all these considerations, the next chemical groups were selected to build the cluster models (presented in Figure 11.7): (i) acetate (C) to mimic aspartic or glutamic acid (mono- or bidentate interaction modes), (ii) phenol for tyrosine (T), (iii) acetamide for a peptide bond carbonyl amide (A), (iv) water (W), (v) 4(5)-methylimidazole for histidine (H), and (vi) methylamine for the N-terminal (N). In order to reduce the number of possibilities, at least two nonwater ligands were included in the models, from which at least two side chains are carboxylic groups.

Each of the cluster model structures were optimized combining a DFT functional with an implicit solvent model. In order to establish the stability of each cluster model, the next reaction was defined:

$$[\mathrm{Al.}\ (\mathrm{H_2O})_6] + m.\ [\mathrm{Lig}^{ni}]^p + n.\ [\mathrm{Lig}^i]^q \leftrightarrow [\mathrm{Al.}\ \mathrm{Lig}_{m+n}\cdot(\mathrm{H_2O})_{6-m-n}]^{3+p+q} + m.\mathrm{H_2O} + n.\ \mathrm{H_3O^+} \tag{11.8}$$

where $m$ refers to the number of ligands that are not ionized ($\mathrm{Lig}^{ni}$) upon their complexation to Al(III), and $n$ refers to the number of ligands that are deprotonated ($\mathrm{Lig}^i$) when they interact with Al(III); $p$ and $q$ refer to the total charge of $\mathrm{Lig}^{ni}$ and $\mathrm{Lig}^i$, respectively. The distinction between ionized and non-ionized ligands is due to the fact

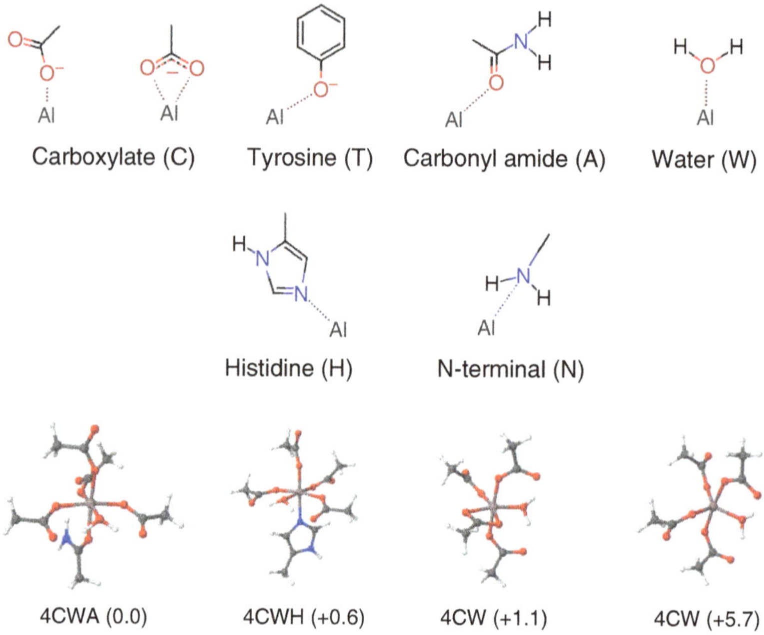


**Figure 11.7** Top: The ligands combined to build the cluster models. Bottom: the four most stable models encountered.

that the acidity of a molecule changes in solution and interacting with a metal. For instance, it was calculated that the $pK_a$ of a tyrosine drops from 10.1 in solution (at these conditions the hydroxyl group remains protonated) to 4.1 when interacting with Al(III) (the group is deprotonated). By contrast, the $pK_a$ of a carboxylic group is around 4.0, so the group is deprotonated both in solution and interacting with the metal. In addition to tyrosine, the amine group is the other group that was ionized upon the complexation ($pK_a = 12.6$ in solution). Note that an octahedral Al(III) with six water molecules is taken as referenced in Reaction (11.8), and that the reaction can be considered as a substitution reaction in which the water molecules are replaced by the ligands. In other words, the thermodynamics of the reaction refer to the difference in energy between aluminum in solution and buried in a metal-binding site model.

It is worth noting that the aim of this part of the study was to determine the preferential coordination shell of aluminum, so the results should be analyzed qualitatively and, from them, draw an overall pattern. Interestingly, in all of the most stable cluster models (the four most stable models are shown in Figure 11.7; more details can be found in the original article) [5], at least three carboxylic groups appear. Moreover, among the 10 most stable ones (except one), four carboxylic groups are present. This is not unexpected since the carboxylic group is a negatively charged O-ligand that neutralizes the cation's high positive charge. Note that tyrosine is the other negatively charged O-ligand. Still, its coordination requires being deprotonated first with a subsequent deprotonation energy penalty, making its inclusion in the metal first shell-less favorable. In fact, tyrosine does not appear in any of the most stable binding site models.

In summary, the characterization and sorting of the cluster models indicate that the most stable binding site of Al(III) should provide either three or four carboxylates. The nature of the remaining ligands has a more negligible effect, with a water molecule, a carbonyl group, or a histidine being the best candidates. However, it must be pointed out that even though the characterization of the metal-binding sites using cluster models provides valuable information about the preferred binding site of Al(III) to an Aβ peptide, these types of calculations do not take into account some acute effects, such as the influence of the residues located beyond the first coordination shell, the peptide's conformation, or entropic effects. The whole Al(III)–Aβ structure is required to account for these effects.

### 11.4.2 Step 2: The Building of Preliminary Al.Aβ Complexes

The next step was to build physically sound Al.Aβ complex models. As pointed out above, building these structures requires filtering among the available Aβ peptide's NMR structures that present a conformation suitable to accommodate an Al(III) atom. Thanks to the cluster model calculations, we can now establish what a pre-organized conformation implies: a peptide conformation that provides at least three carboxylic groups to the metal-binding site.

The geometric of the Aβ peptide solved by NMR were downloaded from the Protein Data Bank. In particular, 13 NMR Protein Data Bank files were extracted, containing 157 different conformations of the peptide (three of them shown in Figure 11.6). In all these structures, the Aβ peptide presents a large variety of conformations. To identify among them, those geometries pre-organized to host aluminum, a two-step search was performed using in-home Python scripts.

In the first part of the search, a 3D grid was positioned on top of each Aβ NMR structure (see Figure 11.8a). The search program measured the distances between a given grid point and the $C_\alpha$ atoms of the six aspartate or glutamate amino acids present in the Aβ peptide (sequence shown in Figure 11.6). This search aimed to

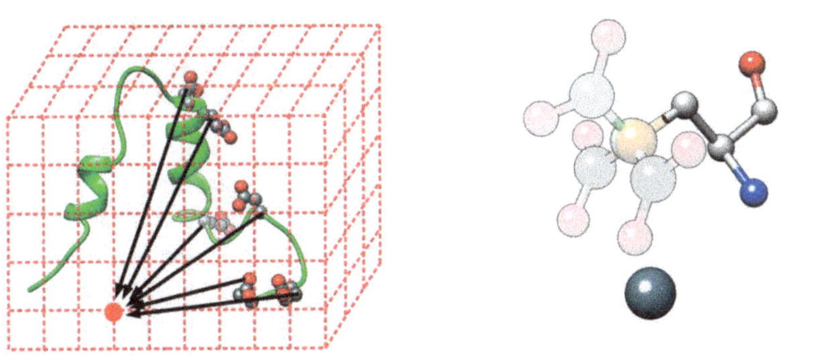

(a) (b)

**Figure 11.8** (a) The 3D grid constructed to find pre-organized NMR structures of the Aβ peptide. At each grid point, the distances to the $C_\alpha$ atom of the six carboxylic groups hosted by the Aβ peptide were measured. (b) Three of the possible arrangements (rotamers) that a carboxylic group can adopt. The MOGA algorithm, among other factors, evaluated which one is the most favorable one to interact with Al(III).

find those grid points where at least three carboxylic groups (aspartate or glutamate) are in the vicinity. In other words, to find spatial positions suitable to place aluminum in concordance with the criteria established by the cluster model calculations. Note that in the search $C_\alpha$ atoms and not the carboxylic side chains were taken as reference. The reason for this geometric criterion is the high flexibility of the carboxylic side chains. In this manner, the side chain has more changes to adopt an optimum conformation, and it is not confined to the configuration solved by NMR. In the next step, a more specific calculation was performed to determine the precise arrangements of the carboxylic groups.

The 3D grid screening located a total of 194 positions in which three carboxylic groups were in the vicinity. Interestingly, only O-containing amino acids (or water molecules) are found surrounding these grid points, and no nitrogen atoms from the backbone or side chains were identified in them. However, the number of tentative locations to place Al(III) was still too large. This number was drastically reduced in the second step of the screening, in which the Multi-Objective Genetic Algorithm (MOGA) method was applied. In this calculation, the 194 3D models of the Al.Aβ complex were scored according to a scoring function than basically evaluated two factors: (i) quality in terms of geometric factors of the first coordination sphere of the metal and (ii) the energetic stabilization of the rotameric conformation of the side chain with respect to the metal environment.

Regarding the first point, the quality of the metal coordination, three geometrical variables were analyzed in each model: (i) metal–ligand distance, (ii) ligand-metal-ligand angles, and (iii) dihedral between the metal and the three atoms of the coordinating group. For the second point, the MOGA algorithm uses rotameric libraries in order to search for the most favorable rotameric state of each amino acid side chain (see Figure 11.8b). Besides, steric clashes were asked to be minimized during the exploration. A scoring function that contains all these factors was defined to filter the 194 Al-Aβ model candidates, from which 13 structures were selected as the most representative 3D models of the Al-Aβ complex. The next step involves the refinement of the structures of these models and the evaluation of their energetic stability in order to propose the most likely coordination mode of Al(III) by an Aβ peptide.

### 11.4.3 Step 3: Refinement of the Al.Aβ Complexes

The 13 models selected in the previous step are not yet realistic structures of the Al.Aβ complex. They were constructed based on a 3D grid and optimum rotamer search, so the conformation of the peptide is not "adapted" for interacting with Al(III). In computational chemistry terminology, these structures have not been optimized yet. Therefore, they do not lay on a minimum on the potential energy surface associated with each structure. To obtain such representative geometries, first, the structures were equilibrated by molecular dynamics (MD) simulations, and then the geometries were optimized by a hybrid QM/MM method. Finally, the 13 structures were conveniently sorted according to their thermodynamic stability.

MD simulation is a prevalent technique to explore the potential energy surface of a system and sample as much as possible the conformations that the system can adopt, and if the simulation is long enough, reach an equilibrated structure.

The MD simulations require evaluation of the energy of the system for a given set of atom coordinates at every simulation step. Different methods can be used to compute the energy: a first-principle QM method, a molecular mechanics (MM) force field, or even a hybrid QM/MM method. For simulating the Al.Aβ complexes, an MM force field, namely the CHARMM36 force field (as implemented in the CHARMM suite of programs) was chosen to calculate the system's energy. As standard in these simulations, the system was solvated with explicit water molecules, and periodic boundary conditions were applied in all directions. The simulation models the Al.Aβ system in an aqueous environment. A production run of 10 ns was carried out for each of the 13 systems under canonical thermodynamic ensemble (NVT).

Once the MD simulations equilibrated the structures, three different geometries were picked up for optimizations from each trajectory generated. By selecting three structures, the final result does not rely on a single conformation. The geometries were optimized using a hybrid QM/MM method so that the most relevant part of the system, in this case, aluminum and its first coordination shell, is described by a QM method, which guarantees an accurate description of this part of the system. By contrast, the remaining part of the system, i.e. the rest of the peptide and the solvent, are described by a MM force field. In this manner, the influence of this part is also taken into account, but using a less rigorous methodology, thus speeding up the calculations. The pDynamo, an open-source program written by Professor M. Field, was chosen to perform these optimization calculations, in which the Orca quantum package was called to describe the QM part with a first-principle DFT functional, whereas the MM part was described by the CHARMM36 force field.

The geometries optimized by QM/MM hybrid methods provide valuable information about the complex's structure and the metal coordination shell. These calculations accurately describe the electronic structure of the metal and its first coordination shell since they are treated by a QM method. Nevertheless, the energies of the different Al.Aβ complexes cannot be compared directly. On the one hand, the total number of atoms is different in each system, mainly because different numbers of solvent molecules were added. On the other hand, the QM regions differ on the 13 Al.Aβ complexes because the metal coordination shells vary in each complex, with different ligands in them. Thus, since the QM/MM optimized structures are minimal on the potential energy surface, their geometries are meaningful, but nevertheless, the relative stability between them cannot be compared directly.

Hence, we decided to evaluate the interaction energies between the Aβ peptide and Al(III) at each complex by means of a first-principle DFT method. The Turbomole quantum package was selected for these calculations, in which a continuum solvent was also included. However, due to the calculations' expense, the system's size was reduced. We realized that in all the complex models, aluminum interacts with the 1–11 segment of the Aβ peptide, so to guarantee that most of the short-range effects were taken into account, the 1–16 segment of the Aβ peptides was extracted from each Al.Aβ structure, along with the Al(III) and the water molecules present (if any) in the cation's coordination shell. The interaction energy for each structure was evaluated according to the next equation:

$$[\mathrm{Al.(H_2O)_6}]^{+3} + [\mathrm{A}\beta_{1-16}]^{-1} \leftrightarrow [\mathrm{Al.A}\beta_{1-16}.(\mathrm{H_2O})_{6-\mathrm{m}}]^{+2} + \mathrm{m.\,H_2O} \qquad (11.9)$$

where $A\beta_{1-16}$ is the apoform of the 1–16 fragment of the $A\beta$ peptide, $[Al.A\beta_{1-16}.(H_2O)_{6-m}]$ the complex formed by the peptide fragment, Al(III) and the water molecules placed in its solvation sphere, and $m$ the number of water molecules displaced from $Al.(H_2O)_6$ by the $A\beta$ peptide. It must be remarked that the peptide's conformation in solution is the same as in the $[Al.A\beta_{1-16}.(H_2O)_{6-m}]$ complex, but certainly, its conformation would vary since the conditions are very different. Therefore, Reaction (11.9) does not compute the formation energy of the complex but rather the interaction energy between the peptide and the cation in the complex.

Once the interaction energies were computed for the 13 models of the Al.Aβ complex, they were sorted according to their interaction energy values, shown in Figure 11.9

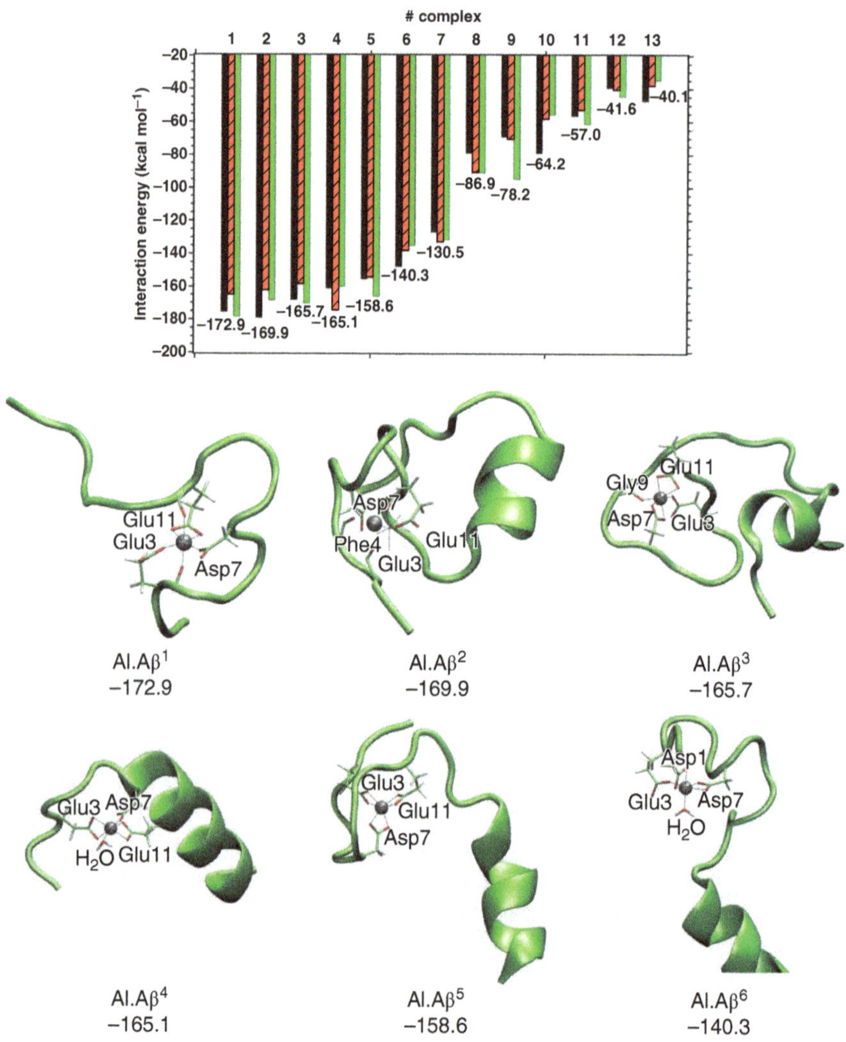


**Figure 11.9** Above, the interaction energies (in kcal $mol^{-1}$) between aluminum and the peptides computed for the 13 Al.Aβ complexes studied. Below, the optimized geometries of the complexes, along with their interaction energies in kcal $mol^{-1}$.

along with the optimized structures. A quick look at the plot is sufficient to observe that energies can split into two main groups: (i) the energies of the first seven complexes are in the −173/−130 kcal $mol^{-1}$ range, whereas (ii) the energies of the remaining six complexes are much higher, in the −87/−40 kcal $mol^{-1}$ range. The explanation for this splitting can be found by looking at the metal's first coordination shell: in the first set of complexes, the most stable ones, three carboxylic groups are interacting with aluminum; by contrast, in the less stable six complexes, only two carboxylic groups are interacting with the cation. It must be noted that this trend is in complete agreement with the results obtained from cluster model calculations in which the number of carboxylic groups was the critical factor determining the stability of a binding site.

A closer look at the most stable complexes renders another essential conclusion. In the five most stable structures (energies in the −173/−158 kcal $mol^{-1}$ range), the same three carboxylic groups are interacting with the cation, namely Glu3, Asp7, and Glu11. The precise coordination mode differs on each of the complexes, but in all of them, these three amino acids are coordinated by aluminum. Interestingly, in the less stable two complexes among the most stable seven ones (energy values of −140.3 and −130.5 kcal $mol^{-1}$), even though three carboxylic groups remain interacting with the metal, one of these three amino acids has been replaced by Asp1. Therefore, all these evidences confirm that the interaction energy is maximized when Glu3, Asp7, and Glu11 are coordinated by Al(III).

This is a very relevant piece of information since the study identifies the amino acids of the Aβ that are involved in the aluminum binding for the first time. It is worth noting that the proposed coordination mode of aluminum to Aβ is very different from the ones established for cations such as Zn(II) or Cu(II). This is not unexpected due to their different chemical character, which is reflected in their chemical preferences.

To finish this section, we want to stress the multi-level computational methodology used in this study. The combination of different methods has solved the technical problem that arose due to the lack of a trustable 3D experimental structure of the Al.Aβ complex. We remark that the methodology relies on identifying inside the conformational ensemble provided by experimental structures of the Aβ peptide those structures pre-organized to interact with aluminum. But this search can only be performed if some essential information is well defined in advance. In our strategy this information was extracted from QM cluster model calculations, but it could be obtained by other means. We hope that this protocol could inspire the study on other molecular systems for which 3D models are difficult to generate via standard methodologies.

## 11.5 *In Silico* Design of Novel Aluminum Chelators

As pointed out in the introduction of this chapter, aluminum has been associated with neurological disorders and other health disorders. The role of aluminum in human pathology has been clearly established in at least three diseases: dialysis dementia, osteomalacia, and microcytic anemia without iron deficiency. These diseases are produced by abnormally high concentrations of aluminum in the human body, so treating this condition requires a reduction of aluminum concentration.

For instance, in chronic hemodialysis patients, chelation therapy is indicated at serum Al(III) concentration higher than 80 $\mu g L^{-1}$.

Reducing free aluminum in the human body can be accomplished by administering chelator agents. Chelators are molecules specifically designed to interact very efficiently with the target molecule, such as aluminum. The formation of the Al. Chelator complex, therefore, reduces the concentration of free aluminum, along with its toxic effects. During the last decades, several chelator agents of aluminum have been used. One of the first ones to be commercialized was deferoxamine (DFO). Initially developed for the decorporation of iron in transfusional iron overload in thalassemia and sickle cell anemia patients, DFO has its drawbacks: the hydrophilic chelator DFO is poorly absorbed in the gastrointestinal tract and must be administered parenterally, that is, by intravenous or intramuscular injection, making difficult its self-administration. Moreover, DFO is expensive.

An extensive body of evidence from *in vitro* and experimental animal studies demonstrates the potential of hydroxypyridinone derivatives to reduce Al(III) toxicity and promote Al(III) decorporation. Among these compounds, 1,2-dimethyl-3-hydroxy-4-pyridone (deferiprone, DFP, see Figure 11.10) has become one of the most promising candidate. DFP has been used as an iron chelator, and in fact, it went into clinical trials initially for iron decorporation, and nowadays it is licensed in the United States and Europe for the treatment of iron overload in thalassemia major. Even if it coordinates aluminum less efficiently than DFO, DFP is also a good chelator. Its main advantages with respect to DFO are that it can be orally administrated, is somewhat less toxic, and is much cheaper.

Thus, it doesn't exist yet completely efficient, safe, and cheap aluminum chelators, so more studies are needed to find them. In this sense, mimosine(β-[N-(3-hydroxy-4-oxypyridyl)]-α-aminopropionic acid, see Figure 11.10), another hydroxypyridinone derivative, and more precisely peptides made of this non-proteic amino acid, could be a promising candidates. Mimosine belongs to the *mimosoideae* family, and this amino acid contains a side chain that it is not found in any of the twenty natural amino acids combined in proteins. Mimosine has been found to have various promising biological activities such as antibacterial, anti-inflammation, anti-fibrosis, anti-influenza, or anti-virus activities. In addition, over the past years of active research, mimosine evolved as promising agent for the treatment of cancer disease. Due to all these positive effects, mimosine has been extensively studied and tested in human organisms.

The side chain of mimosine is structurally very similar to DFP, so this residue is also an effective ligand for binding divalent, trivalent, and tetravalent transition metal ions. Moreover, joining three mimosine residues in the same peptide (as in the peptides shown in Figure 11.10) could significantly enhance aluminum complex stability, making these peptides very appealing as efficient chelators of aluminum. However, the size of these peptides might be critical for their successful absorption into the organism: a large Al.Peptide complex may hinder its crossing through cell membranes, but a too-short peptide could not allow the optimum interaction of the three mimosine residues with the cation. Therefore, the length of the mimosine-containing peptide may determine its final success as an effective chelator of aluminum.

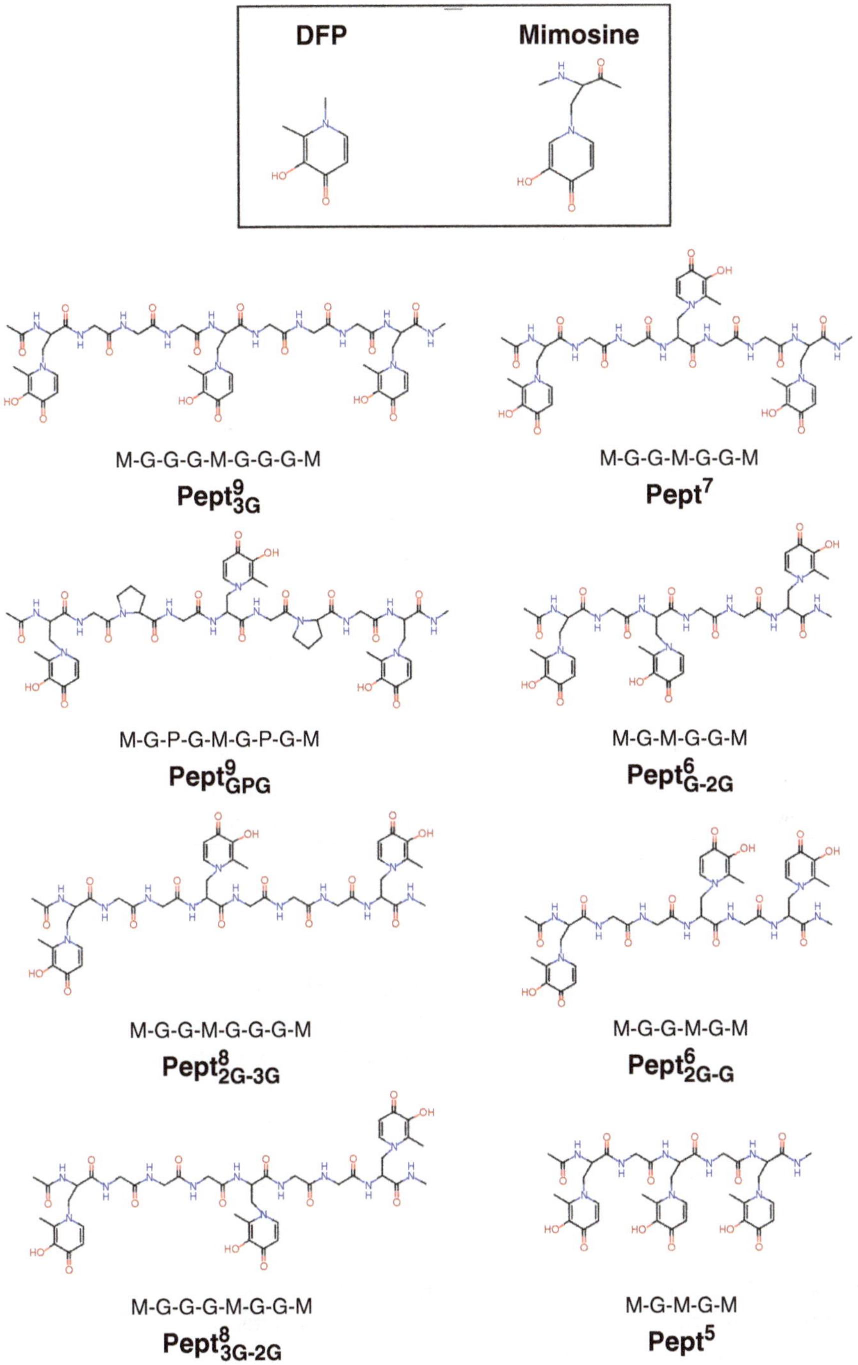


**Figure 11.10** Above, the molecular structure of DFP and mimosine. Below, sequences of the mimosine-containing peptides studied as potential efficient chelators of aluminum.

Hence, the goal of the computational study described in this section [6] was straightforward: to assess which must be the smallest length of the peptide to guarantee its optimum interaction with aluminum. With this aim, state-of-the-art computational methods were combined to systematically characterize the complexation of Al(III) to eight mimosine-containing peptides of different lengths (shown in Figure 11.10). The geometries and energetic stability of all the Al.Peptide complexes were analyzed, focusing on the interactions between the cation and the peptide. Moreover, all this data was compared with the complexes formed by DFP, so far, the most used chelator of aluminum. All this information allows us to propose for the first time, the shortest mimosine-based peptide with the highest coordination stability toward Al(III) ions. This will open the field to a new type of chelators, not only for Al(III), but for highly valent metals showing similar coordination features as Al(III), e.g. lanthanides with their peculiar magnetic properties used in PET imagining.

First, QM/MM MD simulations of the eight Al.Peptide complexes were carried out in order to equilibrate the structures and analyzes the geometries. Classical MD simulations are usually used for such simulations since they allow more extended ones. However, the QM/MM approach was chosen for different reasons: (i) as pointed out above, a QM treatment of the metal and its first shell ligands provides a much more realistic description of the interactions between them since, unlike a conventional MM force field (even that nowadays polarizable MM force fields exist), the electronic structure of this region is characterized. For that reason, the QM part includes aluminum and the side chains of the three mimosine residues; (ii) since mimosine is not a standard amino acid, the most popular force fields do not include specific parameters for this residue. Therefore, in order to perform MM MD simulations, these parameters should be obtained somehow, what would make the study more tedious and with a larger uncertainty. Due to all these reasons and due to the fact that the small size of the system does not require long MD simulations, QM/MM MD simulations seem the most convenient approach. The PM6 semiempirical method was chosen as QM method. As a semiempirical method, PM6 is a simplified version of a first-principle QM method, in which quite severe assumptions are applied and some parameters are added to calculate the energy of the system. The accuracy of a semiempirical method is lower than a first-principle QM method such as DFT or an *ab initio* method, but their accuracy has been extensively tested, with encouraging results. The main advantage is that semiempirical methods are computationally less demanding than a pure QM method, an important issue for QM/MM MD simulations, in which the QM energy must be calculated at each single simulation step.

The key point for an efficient chelator is its strong interaction with the target. In this case, the mimosine-containing peptide is sought to coordinate aluminum in an octahedral manner, with the three mimosine residues interacting through their two oxygen atoms. Therefore, the study analyzes the structures generated during the QM/MM MD simulations of the eight Al.Peptide complexes and determines which peptide sequence guarantees such a coordination mode. For such analysis, two geometric parameters were monitored: (i) the distances between aluminum and the six oxygen atoms (the two oxygen atoms of the three mimosine residues) should be around 2.0 Å, and (ii) The O—Al—O angle between two of the mimosine O atoms located at the aluminum first coordination shell and the metal. Note that in a perfect octahedral geometry, there are four equatorial ligands and two axial ligands; all equatorial-axial angles and the angles between adjacent equatorial ligands should be near 90°, whereas the angles between the three axial angles should be 180°. In this section, we focus on the distributions of all possible $O^{Mim}$—Al—$O^{Mim}$ angles (shown in Figure 11.11), and do not present the analysis of the Al—$O^{Mim}$ distances. However, the conclusions are the same with both geometric criteria.

The analysis of the $O^{Mim}$—Al—$O^{Mim}$ angles computed along the QM/MM MD simulations of the eight Al.Peptide complexes indicates that only the four largest peptides, made of eight or nine amino acids, can provide an octahedral coordination mode to Al(III). The other four peptides are too short for adequately placing the three mimosine residues in an octahedral arrangement. In the simulations of the complexes formed by these four peptides, one of the mimosine residues can interact only through one of its two oxygen atoms so that only five oxygen atoms interact with the cation. These different coordination modes can be observed clearly in Figure 11.11: the distributions of the $O^{Mim}$—Al—$O^{Mim}$ angles computed on the complexes formed by the four longer peptides match very well with an octahedral arrangement. There are two sets of values well defined, centered approximately at 95° and 160°, which correspond to typical values for the equatorial and axial angles of a distorted octahedral. On the other hand, with the four shortest peptides, a more scattered angle distribution is observed, indicating that these peptides cannot maintain an octahedral arrangement. A similar picture emerges when the distances are analyzed.

But this is a qualitative description of the different metal coordination modes. To quantify them, we counted the number of frames on a given QM/MM MD trajectory with the metal octahedral. The coordination mode of the metal in a frame was considered octahedral when at least the values of three angles are larger than 150°, i.e. they are considered axial angles. With this criterion, it was computed that in more than 90% of the frames, the four larger peptides could provide an

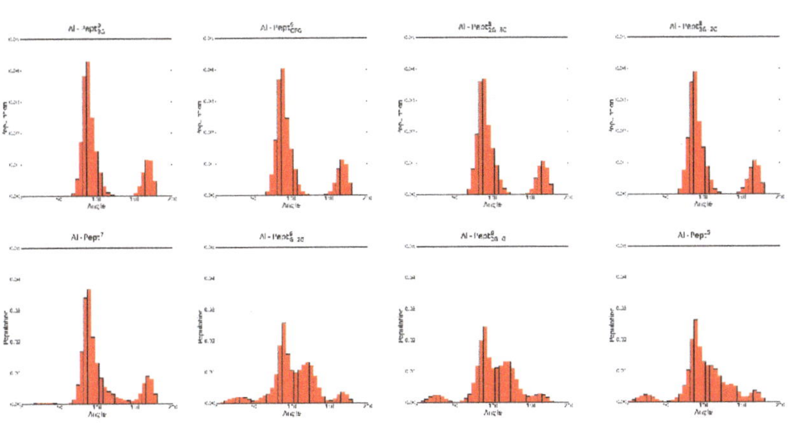

**Figure 11.11** Distribution of all $O^{Mim}$—Al—$O^{Mim}$ angles computed along the QM/MM MD simulations trajectories of the eight Al.Pept complexes. From left to right, upper line: $Pept^{9}_{3G}$, $Pept^{9}_{GPG}$, $Pept^{8}_{2G-3G}$, and $Pept^{8}_{3G-2G}$; bottom line: $Pept^{7}$, $Pept^{6}_{G-2G}$, $Pept^{6}_{2G-G}$, and $Pept^{5}$.

octahedral coordination mode to aluminum. With $Pept^7$, the percentage reduced to 70%, whereas with the two $Pept^6$ peptides and $Pept^5$, the percentage was negligible, less than 2%. Hence, these numbers confirm what a visual inspection of the $O^{Mim}$—Al—$O^{Mim}$ angle distribution suggested: only the four largest peptides guarantee an optimum interaction with aluminum.

The QM/MM MD simulations were not only used to perform the geometric analysis; they were also useful to estimate the stability of the complexes formed by each peptide. In order to estimate the relative stability of each Al.Peptide complex, the energy difference of the next reaction was computed:

$$\begin{aligned} &\text{Al.}\,[\text{Pept}_{x}.\,(\text{MIM})_3] + \text{Pept}_{y}.\,(\text{MIH})_3 \leftrightarrow \text{Al.}\,[\text{Pept}_{y}.\,(\text{MIM})_3] \\ &+ \text{Pept}_{x}.\,(\text{MIH})_3 \end{aligned} \tag{11.10}$$

where $Pept_x$ and $Pept_y$ refer to peptides of different sequences. The energy of each species was taken as the average value along the corresponding QM/MM MD simulation. This reaction, therefore, estimates the difference in enthalpy (ΔΔH) between the interaction of Al(III) with $Pept_x$ and $Pept_y$. In other words, it gives us relative stability between the two complexes. This reaction requires simulations of the peptides in solution (not interacting with aluminum). Note as well that the protonation state of the three mimosine residues changes: they are protonated in solution but deprotonated when they interact with the cation since the acidity of a molecule increases upon its interaction with a metal, that is, a substantial drop of its $pK_a$ is associated upon the binding of the peptide to aluminum.

The relative enthalpies computed for the eight Al.Peptide complexes are shown in Figure 11.12, taking the energy of the Al. $Pept^9{}_{3G}$ as reference. Interestingly, the energy trend confirms what the geometric analysis of the metal coordination shell already showed: the complexes formed by the four longest peptides, with aluminum octahedral, are significantly more stable than the complexes formed by the remaining shorter peptides. The energies of the complexes formed by the four longest peptides differ by

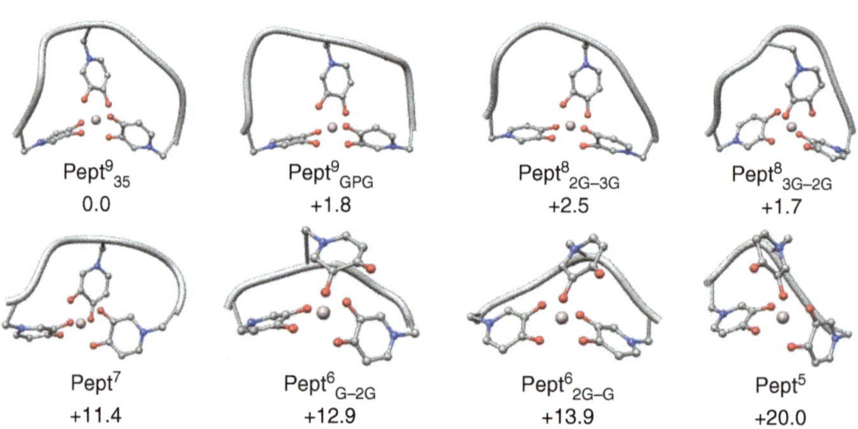


**Figure 11.12** Representative snapshots of the QM/MM MD simulations carried out for the eight Al.Pept complexes studied. The relative energies (ΔΔH in kcal $mol^{-1}$) computed according to Reaction (11.10) are also shown, taking the complex formed by $Pept^9{}_{3G}$ as reference. *Source*: Adapted from Ref. [10].

less than 2.5 kcal $mol^{-1}$. By contrast, the relative energies of the remaining four complexes span above 11 kcal $mol^{-1}$, the complex formed by the shortest peptide being the less stable one, with a $\Delta\Delta H$ value of 20.0 kcal $mol^{-1}$. Among them, the peptide with nine amino acids is the most stable one. In fact, the DFT calculations on representative structures of each complex indicate that the complex formed by this peptide is almost as stable as $Al.DFP_3$. Therefore, based on geometric and energetic analysis, the computational study proposes that this mimosine-containing peptide can be an efficient chelator for aluminum. However, more studies are still needed, including *in vitro* and *in vivo* experiments, to support this suggestion.

## 11.6 Concluding Remarks

In this chapter, we have described some computational studies conducted by our research group on the biophysics and biochemistry of aluminum. We hope these examples illustrate the capabilities of computational chemistry to provide significant insight into systems that are difficult to study by other means. With these examples, we want to highlight the importance of combining different computational techniques to conduct meaningful investigations. However, we also want to point out that in many cases, the use of computational chemistry alone can be insufficient to understand a research problem entirely. Therefore, the necessity of combining computational studies with experiments are increasingly evident. With this combined strategy, each field can provide solid evidence to induce and guide further complementary studies to be carried out by other fields and *vice versa*. This combination of different research fields allows a deeper exploration of the systems under study and guarantees an advance in the knowledge of unresolved problems.

## References


1. Tomljenovic, L. Aluminium and Alzheimer's disease: after a century of controversy, is there a plausible link? *J. Alzheimers Dis.* 2011;*23*:567–598.
2. Mujika, J.I., Ruipérez, F., Infante, I. et al. Pro-oxidant activity of aluminum: stabilization of the aluminum superoxide radical ion. *J. Phys. Chem. A.* 2011;*115*:6717–6723.
3. Ruipérez, F., Mujika, J.I., Ugalde, J.M. et al. Pro-oxidant activity of aluminum: promoting the Fenton reaction by reducing Fe(III) to Fe(II). *J. Inorg. Biochem.* 2012;*117*:118–123.
4. Mujika, J.I., Torre, G., and Lopez, X. Aluminum and Fenton reaction: how can the reaction be modulated by speciation? A computational study using citrate as a test case. *Phys. Chem. Chem. Phys.* 2018;*20*:16256–16265.
5. Mujika, J.I., Rodríguez-Guerra, J., Lopez, X. et al. Elucidating the 3D structures of Al(III)-Aβ complexes: a template free strategy based on the pre-organization hypothesis. *Chem. Sci.* 2017;5041–5049.

6 Mujika, J.I., Dalla Torre, G., Lachowicz, J.I. et al. In silico design of mimosine containing peptides as new efficient chelators of aluminum. *RSC Adv.* 2019;*9*:7688–7697.

7 Mujika, J.I., Rezabal, E., Mercero, J.M., et al. Aluminium in biological environments: a computational approach. *Comput. Struct. Biotechnol. J.* 2014;*9*:e201403002.

8 Grande-Aztatzi, R., Formoso, E., Mujika, J.I. et al. Phosphorylation promotes Al(III) binding to proteins: GEGEGSGG as a case study. *Phys. Chem. Chem. Phys.* 2016;7197–7207.

9 Luque, N.B., Mujika, J.I., Rezabal, E. et al. Mapping the affinity of aluminum(III) for biophosphates: interaction mode and binding affinity in 1: 1 complexes. *Phys. Chem. Chem. Phys.* 2014;*16*:20107–20119.

10 Mujika, J.I., Dalla Torre, G., Formoso, E. et al. Aluminum's preferential binding site in proteins: side chain of amino acids versus backbone interactions. *J. Inorg. Biochem.* 2018;111–116.

11 Mujika, J.I., Escribano, B., Akhmatskaya, E. et al. Molecular dynamics simulations of iron- and aluminum-loaded serum transferrin: protonation of Tyr188 is necessary to prompt metal release. *Biochemistry*. 2012;*51*:7017–27.

12 Mujika, J.I., Lopez, X., Rezabal, E. et al. A QM/MM study of the complexes formed by aluminum and iron with serum transferrin at neutral and acidic pH. *J. Inorg. Biochem.* 2011;*105*:1446–1456.

13 Mujika, J.I., Ugalde, J.M., and Lopez, X. Computational evaluation of pK a for oxygenated side chain containing amino acids interacting with aluminum. *Theor. Chem. Acc.* 2010;*128*:477–484.

14 Mujika, J.I., Ugalde, J.M., and Lopez, X. Aluminum speciation in biological environments. The deprotonation of free and aluminum bound citrate in aqueous solution. *Phys. Chem. Chem. Phys.* 2012;*14*:12465–12475.

15 Mujika, J.I., Ugalde, J.M., and Lopez, X. Aluminum interaction with glutamate and $\alpha$-ketoglutarate: a computational study. *J. Phys. Chem. B* 2014;*118*:6680–6686.

16 Torre, G.D., Mujika, J.I., Formoso, E. et al. Tuning the affinity of catechols and salicylic acids towards Al(III): characterization of Al–chelator interactions. *Dalton Trans.* 2018;*47*:9592–9607.

17 Torre, G.D., Mujika, J.I., Lachowicz, J.I., et al. The interaction of aluminum with catecholamine-based neurotransmitters: can the formation of these species be considered a potential risk factor for neurodegenerative diseases? *Dalton Trans.* 2019;*48*:6003–6018.

18 Exley, C. The pro-oxidant activity of aluminum. *Free Radical Biol. Med.* 2004;*36*:380–387.

19 Fukuzumi, S. and Ohkubo, K. Quantitative evaluation of Lewis acidity of metal ions derived from the g values of ESR spectra of superoxide: metal ion complexes in relation to the promoting effects in electron transfer reactions. *Chem. A Eur. J.* 2000;*6*:4532–4535.

# Index


## a

Ab Initio see Quantum Mechanics
Aluminum 23, 132, 170, 213–237
  pro-oxidant activity 217
Amavadin 62
AlphaFill 105
Alphafold 12, 105, 204
AMBER 124, 129, 132, 161, 169, 185–186
AMS 58
aMOx 108–117
Amyloid 9, 23, 131, 165–169, 183, 223
Antechamber 129
Atoms in Molecules 182–185

## b

*B3LYP see DFT*
*BP86 see DFT*
*BLYP see DFT*
*PBE0 see DFT*
*BioMetAll 15–16, 127, 197*
*Bonded model 106, 126, 134–135, 199*

## c

*Carboplatin 84, 180*
*Caboxypeptidase 18–19*
  *see also Vanadium*
*Catalase 71, 150*
*CASSCF 40, 47, 59, 70–72, 83, 219*
*C–H azidation 103*
*CHARMM 124, 167, 190, 229*
*Chelators 231–234*
*Chemical Bonding Analyses 38*
Chromium complexes 34
*Circular Dichroism (CD) 58, 81*
*Cisplatin 84, 127, 180–192*
*Clashes 17, 24, 228*
*Closed shell 4, 34, 58, 82, 160*
*Classical Molecular Dynamics – see Molecular Dynamics*
*Cluster models 22–23, 225–227*
  *See also truncated model*
*Cobalt-pyrazolylborate 72–75*
*Collective variables 142–145, 151–153*
*Configuration interaction (CI) 58*
  *CIS 59*
  *CID 59*
  *CISD 59*
*Conformational sampling 3, 123*
Copper
  with Amyloid 171
  with Human Serum Albumin 66
  with Prion Protein 67
  with polyhistidine peptides 126
  Metallohydratase 128–129
  Transporter 148
Continuum Solvation model 60, 219, 221
COSMO 41, 60, 86–87
COSMO-RS 60
Covalent model see bonded model
Covalentdock 17
$[Cu(CF_3)_4]^{1-}$ 39


*Computational Bioinorganics: From Description to Prediction*, First Edition. Edited by Jean-Didier Maréchal.

Published 2026 by John Wiley & Sons Ltd.

Cytochrome P450 99–102, 109–119
Cytochrome c carbene transferase 101

## d

*De novo* 7, 128, 159, 197, 204–206, 224
*Deep Learning* 143, 197
*Density Functional Theory (DFT) (selected)* 21, 37, 110, 128, 140, 164, 180
  *B3LYP* 47, 61, 74, 77, 87, 128, 134, 161, 181
  *BP86* 47, 63, 74–82,
  *BLYP* 79–80, 84, 143, 185
  *PBE0* 63–66, 71–80, 84
  *TD-DFT* 77
Desolvation 17, 159, 172
Dirac equation 60–62, 180
Directed Evolution 196–198
Docking see Protein–ligand dockings
Docktite 17
Drug design 6, 190
Dummy Atoms 18, 124, 131
  See also protein-ligand dockings
  See also non-bonded models

## e

*ECOSTBio* 38
*Electric Field*
  *In spectroscopy* 85–86
  *In Enzymatic reactions* 116–117
*Electron Absorption Spectroscopy* 77–80
*Electron Spin Echo Envelope Modulation (ESEEM)* 67–69
*Electron-Nuclear Double Resonance (ENDOR)* 67
*Electronic Paramagnetic Resonance (EPR)* 61
  *Zeeman Effect* 63
*Electron transfer* 34, 104, 107–108
*Electrostatic interactions* 2, 10, 17, 61, 104, 187, 197
*Enhanced sampling* 102, 140, 143, 147, 170, 186, 203
  *See also Gaussian accelerated Molecular Dynamics*
  *See also Umbrella Sampling*
*Enzyme design* 101, 117, 196–205, 224

## f

Feature 12
First coordination sphere 2–3, 6, 128, 162, 222, 227
*Force fields*
  *Classical MD* 102–106, 123–134
  *Generalities* 2
  *Machine learning based* 135
  *Bonded models for metals – see bonded models*
  *Non-bonded models and dummy atoms* 106, 131
  *Simplified for Protein-ligand dockings* 17
*Fuerza* 125

## g

G factor 61–65, 69–70, 84
Gabedit 58
Gamess 58
GaudiMM 19, 21, 23–24
Gaussian
  Software 60, 128, 134
  Type orbital function 58, 145–146
Gaussian accelerated Molecular Dynamics 105, 170
Gaussview 58, 83
Generalized Amber Force Field (GAFF) 129
Genetic Algorithms 6, 17, 228
Gold 17–19, 127, 191
GoldScore 18–19, 191
Grimme dispersion 65, 184

## h

Hartree-Fock 58, 179–180
Heme 11, 100, 106–108, 150
  In Artificial enzyme 203–205
  In Heme-oxygenase 14
  Oxidation state 36
  In P450 see Cytochromes P450
HOMO 185
homology modeling 166

Human Serum Albumin (HSA) 15, 65, 165
Hydrogen bond
In Protein-ligand dockings 18, 190
In enzyme design 129, 185–187, 201–202
Hyperfine coupling constant 67

*i*
Iboview 39
IonCom 12
Iron
*pincer 47*
*Mössbauer 86–87*
*carbene intermediates 101*
Oxidation states 34
$[Fe(O)]^q$ 41
Fe-oxo 45, 103
*see also HEME, Cytochrome P450*
Insulin 125–128

*j*
Jaguar 58

*k*
Kohn-Sham 62, 67, 180, 184

*l*
Linear Combination of Atomic Orbitals (LCAO) 33, 58
Lennard-Jones Potentials 124, 132–133,
3DLigandsite 12
Local Electric Fields 118–119
Long-chain fatty acids 15
LUMO 185

*m*
Magnetic circular dichroism (MCD) 58, 81
MBS 15–16
MCSCF 59
MCPB.py 125, 199
Mean percentage deviation (MPD) 60
Mean absolute percentage deviation (MAPD) 60
Metal3D 12
Metal binding site 9
Experimental approaches 10
Computational Approaches 10–24
Metalloporphyin 36
see also Heme
MetalPredator 11
MetSite 12
MIB(2) 11–12
Mn(III)-complex (red ox) 43
Molecular Mechanics Parameters – see force fields
Molecular dynamics (MD) 22–24, 102, 112, 123, 140, 161, 166
Molecular Orbitals 33, 38, 42, 58
Monte Carlo 3, 105, 123, 140, 161
Mössbauer effect 84
Mössbauer spectroscopy 84
Møller–Plesset 59, 179
Mulliken analysis 38, 41
Multi-objectives (as applied to Genetic Algorithms) 19, 228
Multi-scale modeling 4, 7, 22–24, 100, 107, 118, 160, 199, 203,

*n*
NAMI 74–76
[Ni–Fe] hydrogenase 87–88
Nickel complexes oxidation states 38,46
Non-bonded models see force fields
Noninnocent ligands 33–35, 47, 48
Normal mode 5, 111
NTS 15
Nuclear Magnetic Resonance (NMR) 57, 69, 123
NWCHEM 41

*o*
Open-Shell 10, 34, 67, 82, 86, 162,
Open-close transition 102–103
OPLS 124
ORCA 58, 62, 81, 229
Oxaliplatin 125–128, 185–188
Oxidation state 33–49

**p**

Polarizable continuum model (PCM) 60, 182
Peptides
  βAmyloid see Amyloid
  Polyglycine 160–162
  Polyhistidine 160–162
  Preorganization
  Metal binding site 16
  Catalysis 104–106, 115–117, 197
  Folding 159
Peroxidase 71
Principal component analysis (PCA) 187
Prion Protein 65
Protein Data Bank 127, 185
Protein-ligand dockings 6, 17, 128–129, 108, 197–202

**q**

QSAR 179, 190–192
Quantum Chemistry see Quantum Mechanics
Quantum Mechanics
  Ab initio 12, 47, 58, 161, 179
  Ab initio MD 22, 129, 143, 147, 150, 199
  Density functional theory see Density functional theory
  Semi-empirical 58, 153, 189, 234
  Wave function 37, 58–59
Quantum mechanics/molecular mechanics (QM/MM) 3, 103–117, 128, 140
  for metal binding site prediction 22
  partitioning schemes, 106
  link atom approaches, 223–226
  QM/MM MD 151
  For metallodrugs 179, 184–185
  For Artificial Metalloenzyme design 199–200
  For Amyloid studies 228–229, 234–236

**r**

Radical 103, 108–116
RASMOT-3D PRO 13
Rational Design 100, 118, 195–197
Restrained Electrostatic Potential (RESP) 128
RoseTTAFold 105, 204
RFDiffusion 204–205
Rotamers 15, 227

**s**

Scandium complexes 34,46
Self-consistent field (SCF) 58
Semi-empirical 58, 189
SeqCHED 11
Sit Analysis 41
Slater-type orbitals 58
Slater determinant 59
Speciation
  Aluminum 215–217
Spin state 21, 36–43, 59–61, 69–71, 159
Spartan 58
STAMPS 13
Starting structure 104, 133
Substrate
  Specificity 100–118, 129–130, 200–203
  Enantioselectivity 6, 100, 113–114, 128–130, 203
  Orientation 104
  Near-Attack conformation 106, 114, 198
Superoxide reductase 133

**t**

TD-DFT see DFT
TEMSP 12
Theozyme 103, 196–202
Transition state 6, 48, 84, 100, 141
Turbomole 58, 229
Transferrin
  Interaction with vanadium see 68
  Interaction with Aluminium see 215
Truncated models *109–116*

**u**

Umbrella sampling 145–148

## v

Vanadium
  Interaction with caboxypeptidase 18
  EPR analysis 63
  NMR analysis 68–69
  EAS analysis 77
  Interaction with transferrin 67
  Interaction with spinach chloroplast CF1–APTase 68
Van der Waals 17, 106, 124–125
Vibrational spectroscopy 83
VFFDT 125

## w

Wavefunctions 34, 38, 41, 58, 182

## x

X-ray
  Crystallography 10, 101, 187–188
  Absorption Spectroscopy XAS 35
  Absorption near edge structure XANES 35
  Emission spectroscopy XES 36

## y

Yield 114

## z

ZincFinder 11
ZORA 41, 62

Printed in the USA/Agawam, MA
July 20, 2026

907143.104